Foundations of Engineering Mechanics

V. I. Babitsky, Theory of Vibro-Impact Systems and Applications

Springer-Verlag
Berlin Heidelberg GmbH

V. I. Babitsky

Theory of Vibro-Impact Systems and Applications

Translated by N. Birkett

With 146 Figures

Springer

Series Editors:
Prof. Dr V. I. Babitsky
Loughborough University
Department of Mechanical Engineering
SE11 3TU Loughborough, Leicestershire
UK

Prof. Dr J. Wittenburg
Universität Karlsruhe (TH)
Institut für Mechanik
Kaiserstraße 12
D-76128 Karlsruhe/Germany

Author:
Prof. Dr V. I. Babitsky

Translator:
Nicolas Birkett
Grangefield Drive 28
LE7 7NB Rothley, Leicester
UK

ISBN 978-3-662-22534-9 ISBN 978-3-540-69635-3 (eBook)
DOI 10.1007/978-3-540-69635-3

Cataloging-in-Publication Data applied for
Die Deutsche Bibliothek – CIP Einheitsaufnahme
Babickij, Vladimir I.:
Theory of vibro-impact systems with applications / V. I. Babitsky.
Transl. by N. Birkett. –Berlin; Heidelberg; New York; Barcelona; Budapest; Hong Kong; London; Milan; Paris; Singapore; Tokyo:
Springer, 1998
(Foundations of engineering mechanics)

Originally published by Springer-Verlag Berlin Heidelberg New York in 1998.
Softcover reprint of the hardcover 1st edition 1998

Cover design: de'blik, Berlin
Typesetting: Camera ready by author

SPIN: 10630548 62/3020 - 5 4 3 2 1 0 - Printed on acid-free paper

To Eleonora and Ilya

Preface to the English edition

The drop of rain maketh a hole in the stone, not by violence, but by oft falling.

Bishop Hugh Latimer
(Born 1485 in Leicestershire, was burned in 1555 for heresy at Oxford)

The first edition of this book, published in Russian, generated wide interest in the problem within Russia but was not widely known outside. The aim of the author in writing this book on the synergistic effects of multiple impacts was to achieve the following:

- to understand the unusual resonant behaviour of vibro-impact systems,
- to accommodate vibro-impact systems within the general theory of nonlinear vibrations for all forms of mechanical model and excitations,
- to compare different mathematical methods for the analysis of vibro-impact systems and to develop effective approximate methods based on frequency and structural approaches,
- to apply frequency methods for analysis of vibro-impact processes in complex mechanical structures interacting with driving and control units,
- to develop an approach for the structural synthesis of vibro-impact systems,
- to explain the dynamics of some important engineering systems.

In order to solve these problems, the author extended the mechanical description of vibro-impact systems by applying contact compliances, distributed flexible mechanical structures and complex excitations.

The theory of vibro-impact systems that was developed, revealed new nonlinear phenomena. The exploitation of these phenomena gave an impetus for the promotion of different engineering applications. To achieve this the author founded the Laboratory at the Blagonravov Institute of Machine Studies of the Russian Academy of Sciences which carried out research and development on a wide range of engineering systems that resulted in the design of effective rigs. These activities were generally described in the article included as Appendix II of this book, which we recommend to the reader for initial acquaintance with the problem. Regular Symposiums and Seminars on "Dynamics of Vibro-Impact Systems" were conducted within the mentioned activity, which stimulated intensive development of this area and mutual applications.

The following lectures and demonstrations in different countries together with interest in this topic from industry, inspired the author to propose the revised English translation of this book. The development in recent years of powerful software such as Matlab-Simulink gave new opportunities to simulate vibro-impact structures using frequency and structural methods. The book can also be used as a foundation for these new applications. In order to demonstrate these possibilities, we have added to the book Appendix I, written together with Dr A Veprik who also helped the author with the description of recently discovered vibro-impact phenomena in §14 section 5 and §16 section 4. The book is further upgraded with an additional list of new references.

We hope that the book will introduce these general approaches for important areas of mechanical engineering to English speaking readers and help the generation of new research and development specialists in the analysis of such complex mechanical structures. From the book, students and their teachers can acquire knowledge of the mechanical and engineering aspects of dynamics of strongly nonlinear systems.

The translator of the book, Mr Birkett, and the staff of Springer-Verlag were extremely co-operative during all stages in the preparation of the book. The author is very grateful to all the individuals who participated in accomplishing this project.

Vladimir Babitsky Loughborough, Leicestershire

Preface to the Russian edition

It is difficult to think of a physical effect other than impact that generates more significant results for so small a resource requirement. It is constantly attracts interest to study it and its varied applications. The use of impact principles in household and industrial practices has evolved on its own over the years. For thousands of years, impact tools have developed in a complex manner, remaining one of the most productive and effective tools throughout. From the first primitive hammers, to contemporary forging and riveting machines, from ancient pick-axes to novel types of hammers and pneumatic drills, from battering rams to perfectly designed concrete breakers and perforators, these represent the extremes of this long lasting process.

From the wide circle of engineering problems linked to the study of impact interactions, it is reasonable to separate the class of problems where *systematic* collisions occur. Systems in which such a form of interaction occurs are known as *vibro-impact*. This type of interaction becomes significant compared to a single impact only when the collision frequency matches the natural frequency of the interacting elastic systems. In the state of affairs specified, collisions occurring in the excited system, which has some energy resource that can be built up from impact to impact, cause various forms of vibration effects. What has been said defines the subject of the *theory of vibro-impact systems* as a self-contained section within the theory of mechanical vibrations.

The book systematically presents the theory of vibration for mechanical systems whose motion is accompanied by collisions between elements. Loss of kinematic links within machines and collisions between the elements of kinematic pairs and structures cause the establishment of intensive vibro-impact processes, as the result of which the tuning of elastic systems changes significantly, as do their natural frequency distributions and the character of transient processes, and a variety of nonlinear effects can arise. Thus the study of problems in dynamics in terms of accuracy, stability and vibration activity, requires vibro-impact processes to be considered in systems with various structural forms, having both distributed and lumped masses, nonlinearities and being exposed to the effects of a variety of regular and random forces.

In addition, vibro-impact regimes form the basic working process for a broad class of machines, drives and installations with the widest functional purposes. To this class belong machines for driving and extracting piles, tamping and cultivating ground, crushing and grinding materials, compacting concrete and foundry mixtures, vibro-impact test devices, various examples of drill, percussion devices, riveters, nut-tightening machines, etc. For this class of system, the characteristic periodic motion arises either due to forced vibration or self-sustaining vibration.

The range of constructional variations for vibro-impact mechanisms and systems requires the development of methods not only to solve the problems of analysis and tuning, but also the development of methods to synthesise dynamic structures which best fulfil the optimum form of motion.

Nowadays, the basic techniques for the theoretical study of vibro-impact processes are the exact methods of nonlinear mechanics, based on stitching solutions describing adjacent intervals of motion, which are separated by the instants of impact. These methods allowed a detailed study of the complicated dynamics picture for the motion of a series of systems and revealed a number of fundamental properties inherent to them. Results obtained in this way are reflected to a considerable extent in the monograph by A E and A A Kobrinsky [118]*.

Searching for solutions by exact methods is a very tedious procedure, which limits the scope of their application. It is particularly difficult to use them as the system dimensions increase in number, when it becomes necessary to take into account additional nonlinear factors and the complicated character of excitation from the effects of non-periodic and random forces.

In addition, to solve a large circle of problems in the theory or vibro-impact systems, and generally in the theory of nonlinear vibrations, it is unnecessary to define motion in a detailed manner. Foremost in this context, are problems connected with the *tuning* of such systems: finding their resonance conditions, the number and type of resonance bands, studying questions of stability and self-excitation etc. All these effects already appear when the basic harmonic components of vibration are analysed, since one of the effective methods of studying nonlinear systems is to separate from the general mathematical description the simpler relationships which characterise most of the corresponding patterns of vibration. These ideas, which come from the general procedure for the asymptotic presentation of solutions as power series of small parameters, have shown themselves to be particularly effective when transformed to spectral representation and the ideas of equivalent linearisation.

The first reported connection was established by N M Krylov and N N Bogolyubov, who provided the first approximation for the asymptotic method they developed in the form of *equivalent (harmonic)* linearisation. Later the harmonic linearisation method was widely developed and applied in the works of V A Kotelnikov, L S Goldfarb, W Oppelt, R Kochenburger, J D Loeb, E P Popov, K Magnus, L A Zadeh, and many other investigators, principally being applied to problems in the theory of automatic control. The contemporary state of this question and a bibliography appear in a series of collective monographs [162, 163, 173] published under the general editorship of E P Popov.

In the 1950s, P Booton [66] and I E Kazakov [112] independently presented an effective form of equivalent linearisation for the study of nonlinear *stochastic* processes, which is known as the *statistical* linearisation method. Techniques for applying the method were developed in the books by I E Kazakov and B G Dostupov [113] and A A Pervozvansky [177]. Finally we must mention the suggestions of M Z Kolovsky and A A Pervozvansky for the original form of

* See also Chapters 7 to 9 in [117] written by R E Brunstein and A E Kobrinsky,

equivalent linearisation by *distribution function* [128, 130] which made it possible to study multiharmonic processes in nonlinear systems. This method and its application to problems in the theory of nonlinear vibration protection is described in the monograph by M Z Kolovsky [111].

Systematic application of the range of methods described for equivalent linearisation and the solution of problems in the theory of vibro-impact systems forms the basis of the material in this book. In this way it was possible to consider the basic forms of vibration motion for vibro-impact systems of arbitrary dimensions in a unified manner using simple methods and to develop a sufficiently general theory for tuning a wide range of systems, such as impact vibration absorbers, pneumatic impact machines, devices for ultrasonic machining of materials, etc., and to reveal new nonlinear effects.

Sometimes in the analysis of a range of problems in the theory of vibro-impact systems, information yielded by equivalent linearisation methods is insufficient. This happens when it is necessary to obtain an idea of the law of motion or the trajectory of the system elements, for example, when estimating collision velocities, studying the spectral characteristics of motion, defining existence domains of dynamic models in configurational space, etc. For all of these, methods are suggested, based on full approximations of the laws of motion for vibro-impact systems with the help of so-called *impulse-frequency characteristics*, widely studied in the works of E N Rosenvasser [191, 192,]. Following the traditions of mechanics these characteristics will be known as *periodic Green's* functions in this book

Together with these, procedures for exact methods are presented and applied, particularly in those cases when they can help the understanding of the nature and accuracy of approximate methods, but also for the better understanding of the effects studied. All the methods described and applied in this book do not assume that the reader has any previous knowledge of them.

The book consists of five chapters. The first chapter presents a short summary of contemporary ideas on the theory of collisions between solids and gives examples of simple vibration systems showing exact methods of analysis for the basic nonlinear behaviour of these systems in the presence of collisions. The character of resonance phenomena in vibro-impact systems and dynamics of the basic structure of vibro-impact machines are considered.

The second chapter introduces equations describing the process of vibro-impact interaction in general types of system undergoing arbitrary excitation and deals with the analysis of frequency characteristics for vibro-impact systems and the processes they cause. The concept of equivalent linearisation of vibro-impact systems is explained and harmonic and statistical linearisation coefficients for basic types of link between colliding elements are calculated.

The third chapter gives a systematic presentation of vibro-impact processes in systems with one degree of freedom based on approximate methods. Dynamics, stability, transient processes of periodic force vibration, parametric vibration and self-sustaining vibration are considered. Systems with random excitation are also studied. Special attention is given to sub-periodic regimes, and effective techniques for calculating these are developed. Throughout the explanations, descriptions of corresponding calculation methods are given and amplitude-frequency characteristics are revealed and analysed, and a comparison of approximate and exact solutions is developed. The influence of the characteristics of the source of excitation on vibro-impact processes is estimated.

The fourth chapter develops analytical methods extended to systems with arbitrary degrees of freedom and is applied to the study of vibro-impact processes occurring during axial vibration of rods and bending vibration of beams for the analysis of systems with impact dynamic absorbers.

Finally, the fifth chapter presents problems in the structural synthesis of self-sustaining vibro-impact systems. Exact solutions are presented as a series of problems based on optimal control methods, and approximate methods to find the structure of controllers for general types of system are given using harmonic linearisation. These methods are applied to the solution of problems of excitation and stabilisation of autoresonant regimes in vibro-impact systems.

Trying to reflect the qualitative features of vibro-impact systems as objectively as possible, we have tried everywhere to develop the results up to the final analytical expressions, giving clear physical interpretations. In addition, experimental data is presented to confirm the basic effects.

The book is designed for engineers and researchers immersed in the study and design of systems involving mechanical vibrations, and also for lecturers and students interested in nonlinear vibrations. To study the book a reasonable knowledge of engineering based general mathematics will be adequate, covering elements of Fourier and Laplace transform theory and probability theory, together with a knowledge of basic theory of mechanical vibrations and stability of motion, as presented in the book by I M Babakov [24]*. All additional mathematical information is presented in the text during the explanation with references to the source where a detailed explanation is given.

The concluding bibliography lists only works either mentioned or used in the text, together with a series of related works of a general character.

Those wishing to use approximate methods immediately can begin their study with Chapter 2. To a considerable extent, Chapter 5 can be read as stand-alone. A self-contained concentrated explanation is also give for the statistical aspects of vibro-impact theory, to be found in §8, §13 and partially in §14 section 6 and §17 section 4. Before beginning to study these topics it is better to read through §6 first.

A E Kobrinsky first raised the author's interest in vibro-impact systems. The original ideas for the approximate methods developed in this book started as a series of papers completed in conjunction with M Z Kolovsky; whose advice and constant support was given freely to the author both during the planning of this book and during the writing process. V O Kononenko and Y G Panovko endorsed the plan of the book, and P F Nagaev made kind comments when the book was in manuscript form. V K Astashev helped greatly and §1 section 7, §2, §4, §15 were written with his help. The deep interest in the methods developed, shown by my colleagues in the Institute of Machine Studies such as T S Akinfiev, B A

*Translator's Note. For texts in English covering similar material the reader should consult W T Thomson's "Theory of vibrations with applications", 4th edition, (Chapman and Hall) or S S Rao's "Mechanical vibrations", 3rd edition (Addison-Wesley).

Borovkov, M E Hertz. M Y Israilovich, V L Krupenin and A N Tresviatsky, who were concerned in the development of a series of problems and in discussion of the results, was very stimulating for the author. The author sincerely thanks all the people named.

V Babitsky Moscow

Contents

Preface to English edition vii
Preface to Russian edition ix
Contents xv

Chapter 1 Vibro-impact systems

§1. Collision of solids 1
§2. Oscillator with fixed stop 20
§3. Oscillator with symmetrical stops 37
§4. Principles of percussion machine dynamics 53
§5. Stitching method 69

Chapter 2 Equivalent linearisation of vibro-impact systems

§6. Equations of vibro-impact systems 75
§7. Harmonic linearisation of vibro-impact systems 94
§8. Statistical linearisation of vibro-impact systems 108

Chapter 3 Vibro-impact processes in systems with single degree of freedom

§9. Forced vibration 125
§10. Stability and transients of forced vibration 140
§11. Parametric vibration 148
§12. Vibration of self-sustaining systems 165
§13. Random vibration 185

Chapter 4 Vibro-impact processes in systems with multiple degrees of freedom

§14. Vibro-impact interactions of elastic systems 193
§15. Vibro-impact interactions of visco-elastic bars 212
§16. Vibro-impact regimes in vibrating beams 229
§17. Theory of impact dynamic absorber 243

Chapter 5 Structural synthesis of self-sustaining vibro-impact systems

§18. Optimal periodic motions of vibro-impact systems 251
§19 Synthesis of self-sustaining systems 260
§20. Synthesis of autoresonant systems 271

Bibliography 279
Additional Bibliography 287

Index 292

Appendix I AI-1
Appendix II AII-1

Chapter 1 Vibro-impact systems

Awaiting the sensation of a short, sharp knock,
From a cheap and chippy chopper on a big black block.

W S Gilbert: Bab Ballads

§1. Collision of solids

1. We shall examine some aspects of impact theory for solids [60, 80, 103, 115], limiting the analysis to those points necessary to understand the processes occurring in vibro-impact systems, and generate mathematical models for them. We shall mainly be interested in the force changes and the kinematic characteristics of the motion of solids and their elements: particles, plane sections, etc.

Impact is characterised by the appearance of strain in the contact zone due to the transfer of kinetic energy arising from the relative motion of the contacting elements into deformation energy, and the spread of waves of strain throughout the reacting bodies from these zones. As a result, the energy spreads through the volume of the bodies, causing vibration of the elements and gradually dissipating throughout the material.

Consider initially the longitudinal impact of a short rod (Fig. 1.1, *a*) of length l_0, moving with velocity v_0, against a long rod of length l of identical cross-section *S*, fixed at the right hand end and initially at rest. At the moment of impact between the rods, interaction forces of equal magnitude, *F*, are developed, acting internally in the body of the rods. These forces deform a portion of the rod of length *dl* for time *dt*. The value

$$\gamma = dl / dt \tag{1.1}$$

defines the velocity at which the disturbance in the rods spreads.

Let us now examine in greater detail how these waves of strain travel through one of the rods. In time *dt* its deformed element is compressed by an amount $du = \varepsilon dl$, where ε is the relative strain. As a result, as the wave reaches sections of the rod, they move with velocity

$$v = du / dt = \varepsilon dl / dt = \varepsilon\gamma, \tag{1.2}$$

or, assuming that the deformation is elastic,

$$v = \sigma\gamma / E, \tag{1.3}$$

where $\sigma = F / S$ is the stress in the disturbed rod and *E* is the elastic modulus.

Denoting the mass of the deformed element as *dm*, we can write an expression, using the principle of impulse and momentum, for the vibrating motion of this element as $v dm = v\rho S dl = F dt$ or

$$v = \sigma/(\rho\gamma) \tag{1.4}$$

where ρ is the density of the rod material. Eliminating v from equations (1.3), (1.4) we obtain

$$\gamma = \sqrt{E/\rho} \tag{1.5}$$

For steel, $E \approx 1.96.10^{11} N/m^2; \rho \approx 7.85.10^3 kg/m^3$, and as a result, $\gamma \approx 5.10^3 m/s$. Using these values in equation (1.2), we can now calculate the velocity of displacement, v^*, at which plastic deformation will occur for an element of the material, Assuming that at the elastic limit for steel $\varepsilon^* \approx 0.001$, we find $v^* \approx 5 m/s$.

The working velocity range of tools should not exceed this critical value. By analogy, this imposes a limit on the collision velocity that can be allowed to occur between machine parts with clearances or in linkages. However, it must be mentioned that in systems where vibro-impact occurs, plastic deformation can occur even at low impact velocities as a result of an accumulation of deformation under regular collisions. Some special tools permit impact velocities of the order of 10 *m/s*.

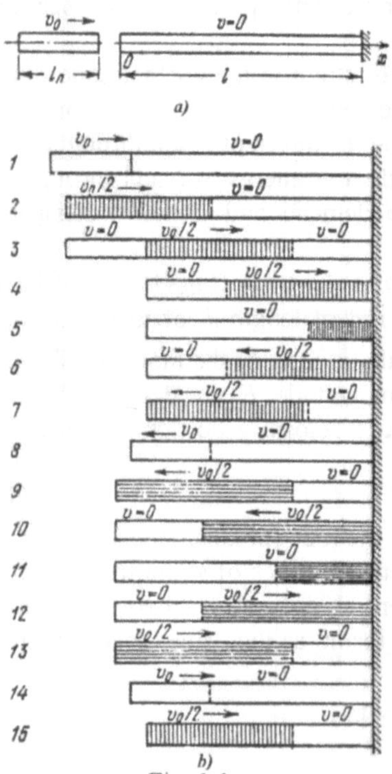

Fig. 1.1

In the example studied, the motion of the material elements occurred in the same direction as the motion of the wave, hence such waves are called *longitudinal* or *expansion waves*, since they occur as a result of the resistance of the material to change in volume. In cases when the transverse dimensions of the waveguides are equal to or greater than the length of the longitudinal waves, their velocity increases because of the complementary material resistance, expanding into the transverse direction and becomes equal to $E[(1-\nu)/(1+\nu)(1-2\nu)\rho]^{1/2}$, where ν is Poisson's ratio.

Another kind of waves, in which portions of the material move perpendicularly to the direction of the spreading waves, occurs due to the resistance of solids to changes of shape. These *transverse* or *shear waves* have a velocity of $(G/\rho)^{1/2}$, where G is the shear modulus. The values of E, G and ν are related through the expression $G = E/[2(1+\nu)]$.

In addition, on the surfaces of solids, waves of a different type can occur, whose importance was first recognised by Rayleigh. These produce trajectories in the element that are almost circular, and lie on planes normal to the surface and coinciding with the direction of the longitudinal waves. The amplitude of the *surface* waves decreases exponentially with depth, while their velocity is about 90% of that of the shear waves [131]. The types of waves mentioned determine the general picture of the propagation of disturbances in solids, which can be obtained from the theory of elasticity [201].

2. Let us continue to examine the one-dimensional case (Fig. 1.1 *a*). We shall take the x-axis along the central axis of the longer rod, with the origin at the interaction section at the instant of the start of impact, when $t = 0$. We shall write the absolute displacement of an arbitrary section as a function $u(x,t)$, and the axial force occurring in it will be called $F(x,t)$.

According to Hooke's law

$$F(x,t) = ES\varepsilon = ES\frac{\partial u(x,t)}{\partial x} \tag{1.6}$$

The change in axial force over an element dx is as follows:

$$F(x+dx,t) - F(x,t) = ES\frac{\partial^2 u(x,t)}{\partial x^2}dx \tag{1.7}$$

This force difference produces an acceleration $\partial^2 u(x,t)/\partial t^2$ transferred to the element of mass $\rho S dx$. From Newton's second law, we can write

$$\rho S dx\frac{\partial^2 u(x,t)}{\partial t^2} = ES\frac{\partial^2 u(x,t)}{\partial x^2}dx \tag{1.8}$$

Dividing both sides of (1.8) by Sdx and using (1.5), we obtain the differential equation for the longitudinal vibration of a uniform rod as

$$\frac{\partial^2 u(x,t)}{\partial t^2} = \gamma^2\frac{\partial^2 u(x,t)}{\partial x^2}. \tag{1.9}$$

Direct substitution shows that equation (1.9) is satisfied by a general solution of the form

$$u(x,t) = \varphi(\gamma t - x) + \psi(\gamma t + x) \tag{1.10}$$

where φ and ψ are arbitrary double differentiable functions. These functions represent waves of deformation moving to right and to left within the rods at velocity γ.

Immediately after impact the compression wave in the short rod moves to the left while in the longer rod it moves to the right away from the contact zone. The waveform is defined by the character of the force of interaction between the contacting surfaces of the rods. As long as these forces are equal in magnitude they produce equal waves, hence the wave propagating along the longer rod takes the form $\varphi(\gamma t - x)$, while that in the shorter rod is represented by $-\varphi(\gamma t - x)$. The first wave involves motion into the elements of the rod under investigation

$$u_1^{\mathrm{I}}(x,t) = \varphi(\gamma t - x) \qquad (0 \le x < l) \tag{1.11}$$

while conversely, the second causes deceleration in the corresponding section of the short rod given by the relationship

$$u_1^{\mathrm{II}}(x,t) = v_0 t - \varphi(\gamma t + x) \qquad (-l_0 < x < 0) \tag{1.12}$$

Here the superscript shows the rod number while the suffix shows the time interval examined.

The relationship between the shape of the generated wave $\varphi(\gamma t - x)$ and the impact force *F(t)* can be found from the following relationship for the longitudinal force at the zero section of the longer rod:

$$-F(t) = ES\, \partial u^{\mathrm{I}}(x,t)/\partial x\,|_{x=0} \tag{1.13}$$

where the negative sign indicates a compressive force. Substituting (1.11) into (1.13), we obtain

$$\varphi'(\gamma t) = F(t)/(ES) \tag{1.14}$$

where the dash indicates differentiation with respect to the argument. Writing the argument as α and integrating (1.14) using the value $\varphi(0) = 0$, since the section at $x = 0$ was static at time $t = 0$, we obtain

$$\varphi(\alpha) = \frac{1}{ES}\int_0^{\alpha} F\left(\frac{\alpha}{\gamma}\right) d\alpha \tag{1.15}$$

For the example we have been examining (Fig. 1.1, *a*) let us define the velocity, v, of the displaced section taking part in the wave process. From the law of conservation of momentum, we can put

$$v_0 dm = 2v dm\,, \tag{1.16}$$

from which

$$v = v_0/2 \tag{1.17}$$

Comparing (1.17) with (1.3) we find

$$F = ESv_0/(2\gamma) \tag{1.18}$$

Substituting (1.18) into (1.5), we obtain after integration for $\alpha = \gamma t - x$:

$$\varphi(\gamma t - x) = \frac{v_0}{2}\left(t - \frac{x}{\gamma}\right). \tag{1.19}$$

Using (1.19), we find from (1.11) and (1.12)

$$u_1^{\mathrm{I}}(x,t) = \frac{v_0}{2}\left(t - \frac{x}{\gamma}\right) \qquad (0 \le x < l), \tag{1.20}$$

$$u_1^{\mathrm{II}}(x,t) = v_0 t - \frac{v_0}{2}\left(t + \frac{x}{\gamma}\right) \qquad (-l_0 < x < 0) \tag{1.21}$$

Suppose we have impact of a rigid body of mass M_0 at velocity v_0, onto a long rod. Assuming that after impact the rigid body moves together with the section of the rod $x = 0$, we have, according to (1.13)

$$-F(t) = -M_0 \left.\frac{\partial^2 u^{l}(x,t)}{\partial t^2}\right|_{x=0} = ES \left.\frac{\partial^2 u^{l}(x,t)}{\partial x}\right|_{x=0}$$

or, introducing a term covering the relation of the mass of the rod and a striker $\mu = \rho Sl / M_0$ and using (1.5), we obtain

$$\left[\frac{\partial^2 u^I(x,t)}{\partial t^2} - \mu \frac{\gamma^2}{l} \frac{\partial u^I(x,t)}{\partial x}\right]_{x=0} = 0 . \tag{1.22}$$

Substituting (1.11) into (1.22) we obtain the following equation for the shape of the wave after impact

$$\varphi''(\alpha) + \frac{\mu}{l}\varphi'(\alpha) = 0 .$$

Integrating this, we find

$$\varphi'(\alpha) = C\exp(-\mu\alpha/l) ,$$

where C is an arbitrary constant. Making use of the initial condition $\partial u^{\mathrm{I}}(0,t)/\partial t\big|_{t=0} = v_0$, using (1.11) we obtain

$$\varphi'(0) = v_0/\gamma = C$$

Hence,

$$\varphi'(\alpha) = \frac{v_0}{\gamma}\exp\left(-\frac{\mu}{l}\alpha\right). \tag{1.23}$$

Integrating (1.23) with the initial condition $\varphi(0) = 0$, we find

$$\varphi(\alpha) = \frac{v_0 l}{\gamma\mu}\left[1 - \exp\left(-\frac{\mu}{l}\alpha\right)\right].$$

When $\alpha = \gamma t - x$ we finally obtain

$$u_1^{\mathrm{I}}(x,t) = \frac{v_0 l}{\gamma\mu}\left\{1 - \exp\left[-\frac{\mu\gamma}{l}\left(t - \frac{x}{\gamma}\right)\right]\right\} \qquad (0 \le x < l) \tag{1.24}$$

At the ends of the rods the waves are reflected. For the fixed end $(x = l)$ according to (1.10) we find

$$u(l,t) = \varphi(\gamma t - l) + \psi(\gamma t + l) \equiv 0 ,$$

from which

$$\psi(\gamma t + l) = -\varphi(\gamma t - l) . \tag{1.25}$$

The result is that the reflection at the fixed end causes the appearance of an additional wave, similar to the incident wave and spreading to meet them with opposite phase. Using (1.25), equation (1.10) takes the form

$$u_2^{\mathrm{I}}(x,t) = \varphi(\gamma t - x) - \varphi(\gamma t + x - 2l) . \tag{1.26}$$

(The suffix 2 shows the time interval after reflection of the waves.) In the same way, it is possible to show that reflection from the free end is characterised by absence of axial force: $ES\partial u / \partial x = 0$, acting in phase, since for the short rod the following will apply after reflection:

$$u_2^{\mathrm{II}}(x,t) = v_0 t - \varphi(\gamma t + x) - \varphi(\gamma t - x - 2l_0) . \tag{1.27}$$

Bearing in mind (1.19), from (1.27) we find

$$u_2^{\mathrm{II}}(x,t)=v_0t-\frac{v_0}{2}\left(t+\frac{x}{\gamma}\right)-\frac{v_0}{2}\left(t-\frac{x-2l_0}{\gamma}\right)=\frac{v_0l_0}{\gamma}=\text{const}$$

Hence, as the initial compression wave reaches the left end of the short rod, it is reflected as an expansion wave moving towards the right which relieves the rod sections from their deformed conditions.

The wave behaviour we have been examining can be visualised graphically. Fig 1.1, *b* shows the spread of waves through the rod at successive moments in time. Transverse shading shows compressed sections, axial shading – expansion. Areas without shading show that elements are relatively undeformed.

Once the compression waves have passed through the contact zone (position 3, Fig. 1.1, *b*), the striker has been completely relieved of stress and all its energy has been transferred to wave energy moving along the longer rod. However, the rods remain in contact until the wave from the fixed end returns as a reflection.

The duration of impact, τ, is taken to represent the time during which the strained regime in the contact zone is maintained. In this case, $\tau=2l_0/\gamma$. If the short rod is removed at the end of impact, then wave (1.20) or (1.24) is reflected as it reaches the fixed end and according to (1.26) will circulate along the long rod, causing vibration of its elements with period $4l/\gamma$ (Fig. 1.1, *b*).

The periodicity that appears in our model fails to account for the real imperfections of elastic conditions, which become apparent when energy is absorbed by *internal damping* of the material. This idea of internal damping is common to a range of mechanisms that occur in the deformation of materials that are cyclically strained; in particular, internal damping is the cause of hysterisis loops that occur in the stress/strain curve. Creep and relaxation influence energy dissipation and make it impossible to obtain equations for force distribution as functions of current deformations. Using integral-differential equations to describe the evolution of material properties makes the study of wave processes more difficult.

At the same time, numerous tests have shown that under a periodic deformation the area of hysterisis loops for most materials depend on the amplitude of the deformation for a wide range of change and not, in practice, on the frequency. For this reason, in the study of periodic processes, which form the basis of this book, we will usually introduce an equivalent force proportional to the velocity of deformation, selecting a suitable factor so that the general dissipative effect corresponds to the experimental data [176]. As a result, equation (1.9) for the study of *periodic* vibration of the rod will be written in the form

$$\frac{\partial^2u(x,t)}{\partial t^2}-\gamma^2\frac{\partial^2u(x,t)}{\partial x^2}-b\frac{\partial^3u(x,t)}{\partial x^2\partial t}=0\,, \tag{1.28}$$

where $b=\gamma^2\varkappa/(2\pi\omega)$ is the linearised coefficient of the force of internal damping; ω the frequency; $\varkappa$ the absorbtion coefficient found from tests as the ratio of absorbtion of the energy during the cycle (proportional to the area of the hysterisis loop) to the basic deformation energy. For steel, $\varkappa\approx0.015$.

Because of internal damping, the periodic process can only be established when there is a regular addition of energy into the system. In the example considered, this depends on periodic motion of the short rod caused by an external source to induce systematic impacts.

Under periodic excitation due to the result of superposing direct and reflected waves to the system, stationary fields of waves can develop (*standing waves*), whose intensity can greatly exceed the amplitude of the generating waves. As an example, for a contact force $F(t) = a_F \sin \omega t$ we find from (1.15)

$$\varphi(\alpha) = \frac{a_F}{ES} \int_0^{\alpha} \sin \frac{\omega\alpha}{\gamma} d\alpha = -\frac{a_F \gamma}{\omega ES} \cos \frac{\omega\alpha}{\gamma},$$

Hence a direct wave along the long rod is generated, given by

$$\varphi(\gamma t - x) = -\frac{\alpha_F \gamma}{\omega ES} \cos\omega\left(t - \frac{x}{\gamma}\right) = -\frac{\alpha_F \gamma}{\omega ES} \cos\left(\omega t - \frac{x\omega}{\gamma}\right)$$

We define the amplitude of the standing waves, formed as the result of repeated reflection of the generating waves from the fixed and free ends, as a. According to (1.26), by superposing the direct and reflected waves, to satisfy the conditions for the fixed right hand end

$$u^1(x,t) = a\cos(\omega t - x\omega/\gamma) - a\cos[\omega t + (x - 2l)\omega/\gamma]$$

or, using the well-known trigonometric formula

$$\cos\alpha - \cos\beta = 2\sin\frac{\alpha+\beta}{2}\sin\frac{\beta-\alpha}{2},$$

we obtain the equation of the standing wave

$$u^1(x,t) = 2a\sin\frac{\omega}{\gamma}(l - x)\sin\omega\left(t - \frac{l}{\gamma}\right). \tag{1.29}$$

Introducing phase ψ into the exciting force $F(t)$,

$$F(t) = a_F \sin(\omega t - \psi), \tag{1.30}$$

and imposing the free end conditions (1.13) onto (1.29) and (1.30), we obtain

$$2a\frac{\omega}{l} ES\cos\frac{\omega l}{\gamma}\sin\left(\omega t - \frac{\omega l}{\gamma}\right) = a_F \sin(\omega t - \psi),$$

whence

$$a = \frac{a_F \gamma}{2\omega ES\cos(\omega l/\gamma)}.$$

As a result, at values of

$$\omega = \frac{\pi\gamma}{2l}(2m - 1) \qquad (m = 1,2,\ldots)$$

the amplitude of the standing waves increases infinitely. The values of ω correspond to the natural frequencies of vibration of the rod. Stability of the standing waves during resonance is due to the forces of internal damping.

This standing wave phenomenon that occurs during vibro-impact interaction of rods will be examined in §15.

3. So far, we have assumed that during impact the contacting elements start to move instantaneously at a velocity determined by the law of conservation of momentum and move together before the impact is finished. Such an interpretation

for the theory of axial impact of rods was developed by Navier, Poisson, Boussinesque, St Venant and Flamant [103]. This theory gives a convenient approach to study the spread of disturbances produced by impact.

Coupled with this disturbance, as we have shown above, contact forces occur, and their idealised representation can lead to the essential distinction in the explanation of the behaviour of the strained elements involved in the wave process. In particular, in the examples considered, the effects we neglected in the contact zone cause the appearance of discontinuities in deformation and strain in front of the wave due to a sudden increase in the velocity at $t = 0$. For example, according to (1.24)

$$\varepsilon\left(x, t = \frac{x}{\gamma}\right) = \left[\frac{\partial u_1^{\mathrm{I}}(x,t)}{\partial x}\right]_{t=x/\gamma} = -\frac{v_0}{\gamma}$$

These results stimulated research on the development of models of contact interaction of bodies during impact and the matching of them with the equation of wave propagation.

Contact interaction is defined as a combination of effects that occur in, and near, the contact zone and are related in a complicated manner to the spread of the disturbance through the actual geometry of contact elements in the colliding bodies and their mechanical characteristics. The curvature of the contacting surfaces, their micro-geometry, the presence of strengthening layers, any special processing called for, etc., all have important influences. When the interacting surfaces are small in size and complicated in configuration changes of the strained condition in the neighbourhood of the contact zone occur more rapidly than in the main part of the colliding bodies. These circumstances led Sears to advance the hypothesis of quasi-static stresses in this region at impact and to describe contact effects by using a static solution [197]. Sears examined the problem of axial impact of rods with circular ends, using the Hertz equations to determine the contact deformations. Earlier, Hertz himself used the same technique to study impact of short bodies, in which the natural periods are considerably shorter than the contact time, and for this study ignored basic wave processes (see section 5).

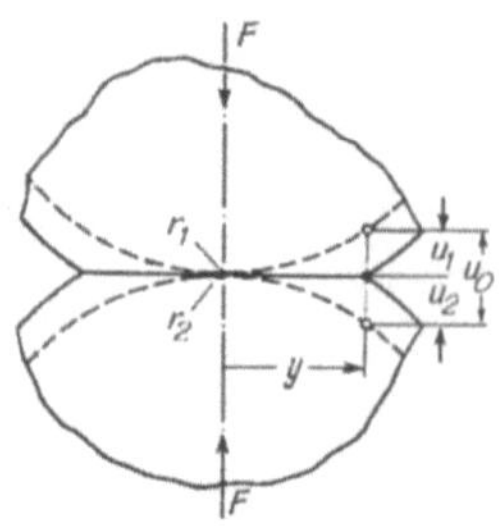

Fig. 1.2

We shall develop the simple form of the Hertzian formula, as produced by I. Y. Staerman [202]. Let us assume that under compression by a force F (Fig. 1.2), two elastic bodies deform so that the surface points which lie on a common normal to the plane constructed through the point of initial contact, are displaced along this plane. The relationship between the increase of force dF and the change of axial displacement u_0 due to deformation of value du_0 takes the form

$$dF = c\frac{E}{1-\nu^2}\Theta du_0, \tag{1.31}$$

where c is a constant of proportionality, Θ is the length of the perimeter of the compressed surface, and $(1-\nu^2)^{-1}$ is a factor accounting for the conditions of three-dimensional compression. Assuming that the cross-section of the contacting bodies can be represented by a second degree curve, and putting (Fig. 1.2) $u_1 = y^2/(2r_1)$, $u_2 = y^2/(2r_2)$, it follows that

$$u_0 = u_1 + u_2 = \left(\frac{1}{r_1} + \frac{1}{r_2}\right)\frac{y^2}{2}. \tag{1.32}$$

For a sphere $\Theta = 2\pi y$, hence integrating (1.31) using (1.32) we find

$$F = \frac{2\pi}{3} c \frac{E}{1-\nu^2}\left(\frac{1}{r_1} + \frac{1}{r_2}\right) y^3. \tag{1.33}$$

Substituting in (1.33) for y in terms of u using (1.32), we finally obtain

$$F = \frac{4\sqrt{2}\pi}{3} c \frac{E}{1-\nu^2}\left(\frac{1}{r_1} + \frac{1}{r_2}\right)^{-1/2} u_0^{3/2}. \tag{1.34}$$

The coefficient c is determined experimentally. We note that when $c = (\pi\sqrt{8})^{-1}$ equation (1.34) takes the Hertzian form. Writing

$$c_1 = \frac{4\sqrt{2}\pi}{3} c \frac{E}{1-\nu^2}\left(\frac{1}{r_1} + \frac{1}{r_2}\right)^{-1/2} \tag{1.35}$$

we obtain

$$F = c_1 u_0^{3/2} \tag{1.36}$$

Hence, using this assumption, the contact interaction of colliding solids can be represented by introducing an elastic "buffer" zone between them, as represented by (1.36) - an expression that shows the smoothing effect of the contact force increases during impact. According to (1.35) the slope of the impact wave front depends significantly on the curvature of the contacting surfaces and increases as the surfaces become less curved, reaching the limiting situation examined earlier for impact of rods with flat ends. These qualitative ideas agree well with experimental data [5].

According to I Y Staerman [202] as the bodies approach more closely to the limiting case of flat surfaces, equation (1.36) can be transformed into the following:

$$F = c_2 u_0^{\frac{2m+1}{2m}} \tag{1.37}$$

(m = 1, 2, ...), which increases due to increase in the adjoining value of m. Since $1 \le (2m+1)/2m \le 3/2$, then as the contact surface becomes flatter, equation (1.37) becomes more linear. In vibro-impact systems, this often becomes the situation because of additional wear of the surfaces under repeated impact. Coefficient c_2 has the order of the value of c_1.

As a result, the buffer zone mentioned earlier can be considered as a thin boundary layer that may be neglected when studying wave effects during impact between bodies. The static elastic behaviour of the layer, represented by a monotonically increasing function, can be taken as linear over its working range. As mentioned in [115], the strengthening effect of the surface layer influences its elastic properties to a great extent. This can cause additional effects, for example generating pure transverse waves on surfaces (Love's waves), vibration of shell surfaces on elastic foundations etc. For our purposes, however, it is important that these effects occur over a faster time scale, both because of the small dimensions of the contact zone and because of the increase in elastic properties, and the effect that they have on the spread of excitations can be calculated using a generalised model of static elastic behaviour of the form

$$F = \Psi(u_0), \tag{1.38}$$

which we will call the *static force characteristic of contact interaction*. Because of the strength of the filtering effects of vibro-impact systems (see §6) the nature of the vibration process shows little sensitivity to the basic static character of the contact interaction, so that various approximations, whose accuracy depends on the system frequency, can be used when creating dynamic models of vibro-impact systems. Of course, contact phenomena require more careful consideration in the analysis of strength due to impact [219, 108, 115, 187].

Note that force characteristics can be used to describe contact interaction of bodies during impact and in cases when plastic deformation begins. T Pöschl proposed the following relationship [182]:

$$F = c_1 u_0^{3/2} \pm b_1 (du_0/dt)^2,$$

where the positive sign corresponds to compression up to the maximum and the negative sign to unloading. Other relationships derived as approximations of experimental data, are given in [49, 217]. Obviously, when calculating the plastic deformation, material behaviour must take account of hysterisis.

Accordingly, a more general form of the force characteristic of contact interaction can be written as

$$F = \Psi(u_0, du_0/dt). \tag{1.39}$$

In contrast to (1.38), equation (1.39) will be called the *dynamic characteristic of contact interaction.*

4. A convenient way to describe impact taking into account contact and wave behaviour using integral equations, was suggested by S. P. Timoshenko when studying the bending impact of a sphere on a beam [208]. Such an approach has a universal character and can be used to study similar processes in systems of differing structure. We shall show the application of this method to the example problem of axial collision of rods with rounded ends (Fig. 1.3), one of which is stationary while the other moves with an initial velocity v_0. As before, we will assume that the impact interaction starts at section $x = 0$ at $t = 0$.

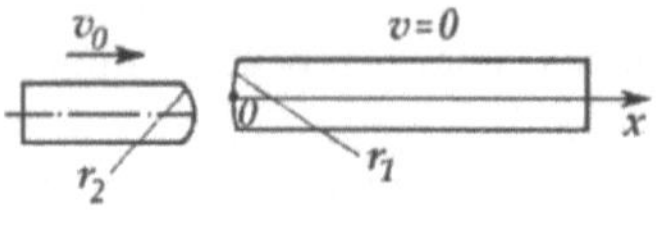

Fig. 1.3

We define the absolute displacement of the end elements of the relative rods, neglecting any compression on contact, as $u_{10}(t)$ and $u_{20}(t)$ and the value of the contact deformation on impact between the bodies as

u_0.

$$u_{10}(t)+u_{20}(t)+u_0(t)=v_0 t\,. \qquad (1.40)$$

We shall write the displacement of the end elements under the influence of the contact force *F(t)* through their reactions to a single impulse

$$u_{10}(t)=\int_0^t F(\tau)h_1(t-\tau)d\tau, \qquad u_{20}(t)=\int_0^t F(\tau)h_2(t-\tau)d\tau\,, \qquad (1.41)$$

where $h_1(t)$ and $h_2(t)$ are the *impulse response functions* of the rods, occurring as reactions at the end elements on the Dirac delta-function excitation. Impulse response functions are non-zero only for $t > 0$. Substituting (1.41) into (1.40) and taking into account that

$$u_0 = \mathrm{X}(F), \qquad (1.42)$$

obtained using the static characteristic of contact interaction (1.38), we obtain eventually

$$\int_0^t h(t-\tau)F(\tau)d\tau + \mathrm{X}[F(t)] = v_0 t\,. \qquad (1.43)$$

Here $h(t)=h_1(t)+h_2(t)$. Equation (1.43) defines the law of changing contact force. Once it has been solved, the wave process in each of the rods can be investigated separately by solving the boundary problem with the given force *F*(*t*) acting on the ends.

Note that we can obtain integral equations in the form of (1.43) for axial collision in systems of arbitrary configuration by using the proper impulse response function in (1.41). Hence, the framework of the present example gives us a general method of describing impact processes using nonlinear non-homogeneous integral equations. The lack of standard methods of finding solutions led to the development of numerical methods to find particular solutions, by substituting discrete summation in place of integration and the solving the resulting difference equations iteratively [178].

The right hand side of equation (1.43) defines the influence of external effects on the system. Denoting this in the general case as $\vartheta(t)$ and using (1.38) and (1.42), we obtain the equation of contact deformation from (1.43) as

$$\int_0^t h(t-\tau)\Psi[u_0(\tau)]d\tau + u_0(t)=\vartheta(t)\,. \qquad (1.44)$$

We will use integral equations like (1.44) later to investigate vibro-impact systems. We shall apply approximate analytical methods to find solutions.

5. Examining effects in the contact zone together with the way the waves spread throughout the structure of interacting bodies gives an adequate physical picture of impact processes. It reveals the mechanism of such processes, including the way that forces change together with the kinematic factors, elastic and dissipative effects of impact. To develop these objectives for the theory of impact requires a study of waves produced in impact of bodies of nonlinear and

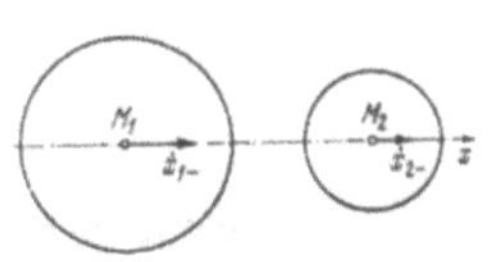

Fig. 1.4

elastic-plastic materials, or materials with post-effects and relaxation etc. [187].

Studying such examples is rather time-consuming and can be made more straightforward only for simple configurations of colliding bodies. For this reason, there has been a wide spread examination of various methods to simplify the study of impact [60,115]. For axial impact (*direct impact*) of short bodies with reasonably compliant contact zones, for example, small diameter elastic balls (Fig. 1.4), impact acts for a considerably longer time than that required for the repeated passage of the waves through the interacting bodies. In such cases, it is usual to neglect the finite time for the spread of the excitation and to apply the Hertzian theory outlined earlier, assuming that loading follows the static loading law.

Defining the masses of the impacting bodies as M_1 and M_2, the axial co-ordinate of the centre of mass as x_1 and x_2 and neglecting changes in the position of the centre of mass due to contact deformation, for *central* impact (the line of action of the principal vector of the contact force passes through the centres of masses)

$$M_1 \ddot{x}_1 = \Psi(x), \quad M_2 \ddot{x}_2 = -\Psi(x), \tag{1.45}$$

where $x = x_2 - x_1$; $\Psi(x)$ is the static characteristic of contact interaction. The dot above the function indicates differentiation with respect to time. Subtracting the second of these equations from the first, we obtain

$$M\ddot{x} + \Psi(x) = 0. \tag{1.46}$$

Here $M = M_1 M_2 / (M_1 + M_2)$ represents the reduced mass of the colliding bodies. Taking into account that at the moment of contact, $t = 0$, the relative co-ordinate $x = 0$, while velocity is given by $\dot{x} = v_0$, we obtain the first integral of equation (1.46) in the form

$$\dot{x}^2 - v_0^2 = -\frac{2}{M}\int_0^x \Psi(x)dx. \tag{1.47}$$

The maximum deformation x^* corresponds to $\dot{x} = 0$, i.e.

$$v_0^2 = \frac{2}{M}\int_0^{x^*} \Psi(x)dx. \tag{1.48}$$

Defining the impact duration from (1.47) as τ, we find

$$\tau = 2\int_0^{x^*} \left[v_0^2 - \frac{2}{m}\int_0^{x^*} \Psi(x)dx \right]^{-1/2} dx. \tag{1.49}$$

Assuming that deformation occurs according to the Hertz equation (1.36), we find from (1.48)

$$x^* = \left(\frac{5}{4} \frac{M v_0^2}{c_1} \right)^{2/5}, \tag{1.50}$$

and as a consequence, the maximum force

$$F^* = c_1^{2/5} \left(\frac{5}{4} M v_0^2 \right)^{3/5}. \tag{1.51}$$

Integrating (1.49) gives [115]

$$\tau = 2.9432\frac{x^*}{v_0} = 2.9432\left(\frac{5M}{4c_1}\right)^{2/5} v_0^{-1/5}; \tag{1.52}$$

for example, for collision of two identical spheres with radii $r_1 = r_2 = 0.01\ m$, moving with relative velocity $v_0 = 0.1 m/s$, putting $c = (\pi\sqrt{8})^{-1}$ in (1.35), then from equations (1.50) to (1.52), when $\nu = 0.28$, we find $x^* = 3.3\times10^{-6}\ m$; $F^* = 61.4\,N$; $\tau = 9.8\times10^{-5}\ s$.

According to (1.52), the duration of contact depends weakly on the velocity of impact. For a linear static characteristic, it will of course be constant. Estimating the time of impact for real short bodies to be 10^{-3} to 10^{-5} *s*, we find that for Hertzian theory the characteristic natural period of bodies must be shorter than 10^{-5} to 10^{-7} *s*.

6. The relatively brief duration of impact for short bodies allows us to exclude the impact process from consideration when defining the resulting kinematic characteristics of motion and we obtain the required relationship from basic theory of mechanics for rigid bodies. The impact can be evaluated from its impulse

$$J = \int_0^\tau F(t)dt$$

a function that is assumed instantaneous. All additional slowly changing forces have a vanishingly small impulse during impact, whose value on the effects of impact can be ignored. The resulting relationships for pre- and post-impact velocities of motion of the impacting bodies can be obtained by examining autonomous systems.

The theory of impact, based on assumptions by Huygens and Newton, has been called *stereomechanics* [103]. In stereomechanics, using integral estimates of the force factors occurring during impact does not allow the study of the strength aspects of impact, hence the field of application is limited to the study of the motion of bodies under the action of impulses. Such motion must occur very slowly by comparison to impact processes. The natural period of such motion should not exceed 10^{-2} to 10^{-3} *s*, according to previous estimates for the duration of impact. Under these conditions motion can be separated into pre- and post-impact, coupled with kinematic conditions of impact. The substance of the last approach is that the co-ordinates of the centres of mass remain unchanged but their velocity undergoes sudden instantaneous change, as defined by the laws of conservation of impulse and energy.

We shall examine several typical examples. Under straight central impact of two bodies moving with velocities $\dot{x}_1$ and $\dot{x}_2$ (Fig. 1.4), indicating pre- and post-impact velocities of each body with minus and plus signs and assuming $\dot{x}_{1-} > \dot{x}_{2-}$, we can write the law of conservation of momentum as

$$M_1\dot{x}_{1-} + M_2\dot{x}_{2-} = M_1\dot{x}_{1+} + M_2\dot{x}_{2+}\,. \tag{1.53}$$

To determine the unknown post-impact velocities, an additional equation is required. This can be obtained from the law of conservation of energy, defining the proportion of kinetic energy remaining in the post-impact motion to the energy dissipated during impact. Newton proposed a convenient way of showing this proportion as a *coefficient of restitution* for the relative velocities of the colliding bodies. Calling this R, we find

$$R = \frac{\dot{x}_{2+} - \dot{x}_{1+}}{\dot{x}_{1-} - \dot{x}_{2-}}, \tag{1.54}$$

in which $0 \le R \le 1$.

Stereomechanic theory describes all forms of direct impact between bodies uniformly in the range between absolutely elastic ($R = 1$) and completely plastic ($R = 0$). A reduction in the coefficient of restitution is related to an increase in the proportion of energy lost irreversibly during impact. The energy of the waves remaining in the bodies after the impact has ended come in this latter proportion so that, even during collision of completely elastic bodies, the value of R can differ from unity.

In agreement with the basic ideas presented in this section, the true values of the coefficient of restitution must be defined by the entire combination of parameters reflecting the conditions of collision. For this reason, it is hard to calculate true values because it often involves equations that are difficult to solve. The problem can be considerably simplified by finding the data from experiments. The experimental conditions must represent as closely as possible those occurring during the intended collision. Taking into account the relatively weak relationship between the coefficient of restitution and the velocity of collision can reduce the volume of experiments [206].

From the system of equations (1.53), (1.54), we find post-impact velocities given by:

$$\dot{x}_{1+} = \frac{(M_1 - RM_2)\dot{x}_{1-} + M_2(1+R)\dot{x}_{2-}}{M_1 + M_2}, \tag{1.55}$$

$$\dot{x}_{2+} = \frac{M_2(1+R)\dot{x}_{1-} + (M_2 - RM_1)\dot{x}_{2-}}{M_1 + M_2}. \tag{1.56}$$

The conditions for impact of a body onto a solid stop moving with velocity v_0 are obtained from (1.55), by putting $M_2 \to \infty$, $\dot{x}_{2-} = v_0$:

$$\dot{x}_{1+} = v_0 + R(v_0 - \dot{x}_{1-}). \tag{1.57}$$

If the stop does not move $(v_0 = 0)$, then

$$\dot{x}_{1+} = -R\dot{x}_{1-}. \tag{1.58}$$

Turning now to the dynamic treatment of the coefficient of restitution [149], we will make use of the concept of centre of inertia of colliding bodies. In view of the autonomous nature of such systems, the velocity of their centres of inertia does not change. For the case considered (Fig. 1.4), calling this velocity v_0, we find

$$v_0 = \frac{M_1\dot{x}_{1-} + M_2\dot{x}_{2-}}{M_1 + M_2} = \frac{M_1\dot{x}_{1+} + M_2\dot{x}_{2+}}{M_1 + M_2}. \tag{1.59}$$

Using (1.59), equations (1.55) and (1.56) take the form

$$\dot{x}_{1+} = v_0 + R(v_0 - \dot{x}_{1-}), \qquad \dot{x}_{2+} = v_0 + R(v_0 - \dot{x}_{2-}). \tag{1.60}$$

Comparing (1.60) and (1.57), we see that direct central impact of two bodies can be interpreted as collision of each of them with a stop, moving at the velocity of the centre of mass.

We can distinguish two phases of impact. Firstly - a *loading phase*, from the start of contact up to the maximum deformation, when the relative velocity of the bodies is zero, but their general absolute velocity is v_0 and secondly - an *unloading phase*, from the moment of maximum deformation until the bodies separate. Writing the impulse of the contact force in the first and second phases as J_1 and J_2, using the impulse-momentum equations, we can write for the first body

$$J_1 = -M_1(\dot{x}_{1-} - v_0),\ J_2 = -M_1(v_0 - \dot{x}_{1+}).$$

Examining the ratio between them, we obtain, using (1.59)

$$\frac{J_2}{J_1} = \frac{v_0 - \dot{x}_{1+}}{\dot{x}_{1-} - v_0} = \frac{\dot{x}_{2+} - \dot{x}_{1+}}{\dot{x}_{1-} - \dot{x}_{2-}} = R \tag{1.61}$$

Since the value of the impulse acting on each body is equal, then we find an identical result when considering the second body. Hence, the coefficient of restitution can be treated as the relation of the normal impulse of the contact force during the first and second phases of impact. Such a representation is known as Poisson's hypothesis.

7. We will use this latter treatment to examine oblique impact, retaining x as the normal displacement and defining the tangential displacement as y. We shall begin with the simplest case of impact of a solid point of mass M onto a non-moving stop (Fig. 1.5).

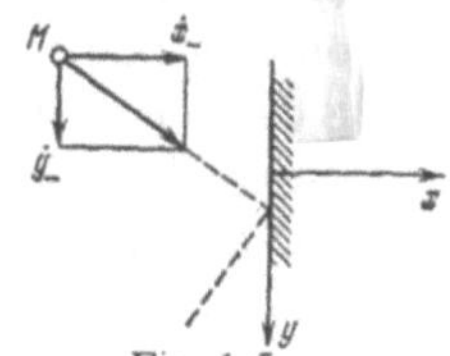

Fig. 1.5

When considering oblique impact we assume that the change in the normal component of velocity occurs independently from the tangential one, and follows the law of direct impact. To determine the tangential component, we use the hypothesis of dry friction, establishing a relationship between tangential and normal impulse according to Coulomb's law.

Applying the impulse-momentum equation to the two load phases, we find

$$J_{1x} = M\dot{x}_-, \tag{1.62}$$

$$J_{1y} = M(\dot{y}_0 - \dot{y}_-). \tag{1.63}$$

Here y_0 is the sliding velocity at the instant when the first phase ends; the sense of the remaining quantities is obvious. In agreement with the accepted hypothesis

$$J_{1y} = \theta J_{1x} \tag{1.64}$$

where θ is the coefficient of friction. Substituting (1.62) and (1.63) into (1.64), we obtain

$$\dot{y}_0 = \dot{y}_- - \theta\dot{x}_-. \tag{1.65}$$

Since the tangential velocity at the moment of impact can not change direction, then $\dot{y}_0 \geq 0$, and from (1.65), we find

$$\dot{y}_- / \dot{x}_- \geq 0 . \tag{1.66}$$

This is the condition for the absence of jamming. When it is not satisfied or is satisfied at the boundary, the tangential displacement stops even during the first phase of impact and the point returns along the normal to the obstacle. If this does not occur, then after the end of the second phase, we find

$$J_{2x} = M\dot{x}_+ \ , \quad J_{2y} = (\dot{y}_+ - \dot{y}_0) . \tag{1.67}$$

Attached to equations (1.67) are conditions equivalent to (1.64), and when we take into account that the normal impulse in the first and second phases becomes subordinate to equation (1.61), we eventually obtain, taking account of (1.62) and (1.65),

$$\dot{y}_+ = \dot{y}_- - \theta \dot{x}_- (1 + R) . \tag{1.68}$$

Putting $\dot{y}_+ > 0$ in (1.68), we find the condition for the absence of jamming during impact as

$$\dot{y}_- / \dot{x}_- > \theta(1 + R) . \tag{1.69}$$

We will call oblique impact when the tangential component does not become zero, *sliding impact.*

A more complicated picture emerges during *constrained oblique impact*, when supplementary links are present on the colliding bodies. A series of typical situations of this kind is considered below.

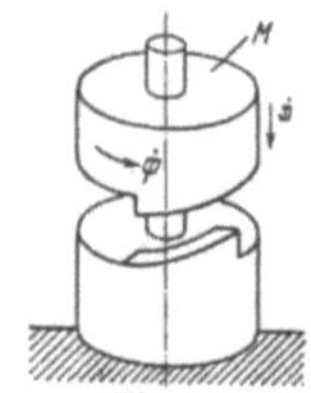

Fig. 1.6

First of all let us consider a mass M (Fig. 1.6) sliding freely on a guiding bar and colliding with an inclined cam on a fixed anvil.

The relative approach velocity of the colliding elements along the normal to the cam profile will be

$$v_N = -\dot{x} \cos\psi - \dot{\varphi} r \sin\psi \tag{1.70}$$

where $\dot{x}$ is the linear velocity of the striker along the bar; $\dot{\varphi}$ is the angular velocity of the striker; ψ is the angle of the oblique profile, r is the average radius of the cam. The impulse-momentum equation for the first phase of impact takes the form

$$\begin{aligned} M(\dot{x}_0 - \dot{x}_-) &= -J_1 \cos\psi \\ I(\dot{\varphi}_0 - \dot{\varphi}_-) &= -J_1 \sin\psi \end{aligned} \tag{1.71}$$

where I is the moment of inertia of the body; J_1 is the normal impulse during the first phase of impact.

At the end of the first phase of impact $v_N = 0$, hence from equation (1.70) we find at that instant

$$\dot{\varphi}_0 = -\frac{\dot{x}_0}{r} \cot\psi \tag{1.72}$$

From equations (1.71),(1.72) we find the value of the normal impulse of the first phase of impact as

$$J_1 = \frac{M(\dot{x}_- \cos\psi + \dot{\varphi}_- r \sin\psi)}{\mu \sin^2\psi + \cos^2\psi} , \tag{1.73}$$

where $\mu = Mr^2 / I$.

The post-impact velocity of the bodies can be found by writing the impulse-momentum equation for impact of the whole as

$$M(x_+ - x_-) = -J\cos\psi, \quad I(\varphi_+ - \dot\varphi_-) = -Jr\sin\psi \tag{1.74}$$

and noting that the complete normal impulse is, according to (1.61)

$$J = J_1 + J_2 = (1+R)J_1. \tag{1.75}$$

Eventually, using (1.73) to (1.75) we obtain

$$\dot x_+ = \frac{\dot x_-(\mu\sin^2\psi - R\cos^2\psi) - 0.5(1+R)\dot\varphi_- r\sin 2\psi}{\mu\sin^2\psi + \cos^2\psi}$$

$$\dot\varphi_+ = \frac{\dot\varphi_- r(\mu\cos^2\psi - R\mu\sin^2\psi) - 0.5(1+R)\mu\dot x_- r\sin 2\psi}{r(\mu\sin^2\psi + \cos^2\psi)}$$

We note that similar relationships can be written for the impact of a free body, without a guiding bar, when the cams are symmetrical by assuming that contact of the cams occurs simultaneously.

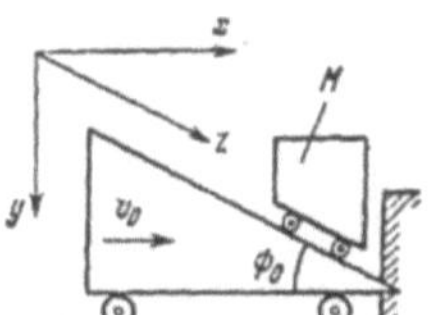

Fig. 1.7

When calculating constrained impact with friction, the problem becomes complicated. In Fig. 1.7 a body of mass M, impacting against a stop, is coupled kinematically to an inclined plane having an unlimited large mass and moving with velocity v_0. The constraint is assumed to be ideal.

Denoting by $\dot x$ and $\dot y$ the normal and tangential components of the absolute velocity of the impacting body and introducing the value $\dot z$ for its relative velocity along the oblique plane, we find from Fig. 1.7

$$\dot x = v_0 + \dot z\cos\psi_0 \tag{1.76}$$

$$\dot y = \dot z\sin\psi_0 \tag{1.77}$$

where ψ_0 is the slope of the oblique plane. At the end of the first phase of impact $\dot x = 0$, hence from (1.76), we find at that instant

$$\dot z_0 = -v_0 / \cos\psi_0 \tag{1.78}$$

Comparing (1.78) and (1.77), we see that in the first phase of impact $\dot z$, and consequently $\dot y$, change sign, hence we can divide the phase into two intervals having different signs for these velocities. To agree with (1.77) on the interval boundaries $\dot y = \dot z = 0$.

The impulse-momentum equation for the first interval of the first phase takes the form

$$-M\dot z_- = J_{1x}^{\mathrm{I}}\cos\psi_0 + \theta J_{1x}^{\mathrm{I}}\sin\psi_0 \tag{1.79}$$

where J_{1x}^{I} is the normal impulse during the first interval. From (1.79), we find

$$J_{1x}^{\mathrm{I}} = -\frac{M\dot z_-}{\cos\psi_0 + \theta\sin\psi_0}. \tag{1.80}$$

The impulse-momentum equation for the second interval is written as

$$M\dot{z}_0 = J_{1x}^{II}\cos\psi_0 - \theta J_{1x}^{II}\sin\psi_0 .$$

From this, and using (1.78) we obtain

$$J_{1x}^{II} = -\frac{Mv_0}{\cos\psi_0(\cos\psi_0 - \theta\sin\psi_0)} .$$

Since the normal impulse can not change sign during impact, we obtain the condition

$$\theta < \cot\psi_0 . \qquad (1.81)$$

This condition ensures that jamming does not occur.

The full normal impulse in the first phase of impact is as follows

$$J_{1x} = J_{1x}^{I} + J_{1x}^{II} = -M\left[\frac{\dot{z}_-}{\cos\psi_0 + \theta\sin\psi_0} + \frac{v_0}{\cos\psi_0(\cos\psi_0 - \theta\sin\psi_0)}\right] \qquad (1.82)$$

The impulse-momentum equation for the second phase of impact gives

$$M(\dot{z}_+ - \dot{z}_0) = J_{2x}\cos\psi_0 - \theta J_{2x}\sin\psi_0$$

From which, using (1.78), we find

$$J_{2x} = \frac{M(\dot{z}_+ + v_0/\cos\psi_0)}{\cos\psi_0 - \theta\sin\psi_0} .$$

Writing the expressions for the impulses during the first and second phases in terms of the coefficient of restitution (1.61) we eventually obtain

$$\dot{z}_+ = -R\dot{z}_- \frac{\cos\psi_0 - \theta\sin\psi_0}{\cos\psi_0 + \theta\sin\psi_0} - (1+R)\frac{v_0}{\cos\psi_0} . \qquad (1.83)$$

In reference [17], it is shown that, for the impact arrangements of Fig. 1.8, *a* the expressions for the linear velocity $\dot{z}$ of mass M before and after impact are similar to (1.83), where ψ_0 is the value of the angle of crank rotation ψ, at the moment of impact, measured from the vertical axis counter-clockwise. Bearing in mind that in this case $\dot{z} = \dot{\psi} l$, where l is the length of the crank, we obtain the angular velocity as

$$\dot{\psi}_+ = -R\dot{\psi}_- \frac{\cos\psi_0 - \theta\sin\psi_0}{\cos\psi_0 + \theta\sin\psi_0} - (1+R)\frac{v_0}{l\cos\psi_0} \qquad (1.84)$$

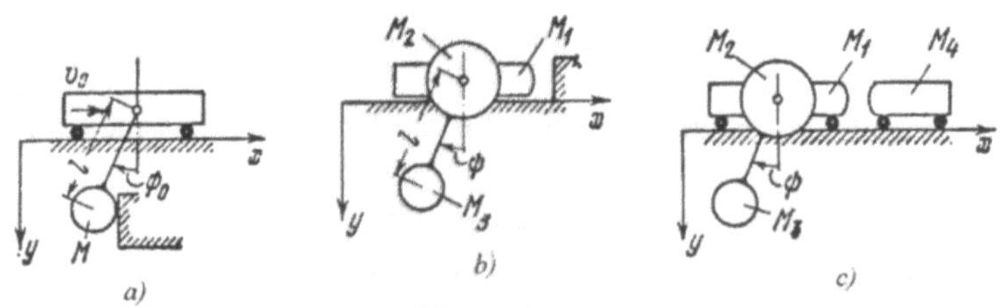

Fig. 1.8

Let us examine a type of machine arrangement frequently found in vibro-impact equipment (Fig. 1.8, *b*). Carriage M_1 carries flywheel M_2 with an unbalanced mass M_3. We shall assume that a dry friction force acts between the carriage and the guiding surface. The impulse-momentum equations, describing an increase in the velocity $\delta\dot{x}_1$ of the carriage and the angular velocity $\delta\dot{\psi}$ of the flywheel, can be written as

$$J_x = \delta\dot{x}_1(M_1 + M_2 + M_3) - \delta\dot{\psi}M_3 l\cos\psi - \theta|J_y|\text{sign}\,\dot{x}_1\ ,$$
$$J_y = \delta\dot{\psi}M_3 l\sin\psi \qquad (1.85)$$
$$\delta\dot{x}_1 M_3 l\cos\psi - \delta\dot{\psi}(I_2 + M_3 l^2) = 0$$

where I_2 is the moment of inertia of the flywheel; l is the eccentricity of the mass M_3.

From equation (1.85) we find

$$\delta\dot{\psi} = \frac{M_3 l\cos\psi}{I_2 + M_3 l^2}\delta\dot{x}_1 \qquad (1.86)$$

$$J_x = \delta\dot{x}_1\left[M_1 + M_2 + M_3\frac{I_2 + M_3 l^2(\sin^2\psi - 0.5\theta|\sin 2\psi|\text{sign}\dot{x})}{I_2 + M_3 l^2}\right]. \qquad (1.87)$$

Noting that in the first phase of impact $\delta\dot{x}_1 = -\dot{x}_{1-}$, $J_x = J_{1x}$, while in the second, $\delta\dot{x}_1 = \dot{x}_{1+}$, $J_x = J_{2x} = RJ_{1x}$ and the change in the sign of velocity $\dot{x}_1$ occurs on the phase boundary, with the help of equation (1.87) we find the value of the post-impact velocity of the carriage as

$$\dot{x}_{1+} = -\mathcal{R}\dot{x}_{1-}\ , \qquad (1.88)$$

where

$$\mathcal{R} = R\frac{M_1 + M_2 + M_3\left[I_2 + M_3 l^2(\sin^2\psi - 0.5\theta|\sin 2\psi|)/(I_2 + M_3 l^2)\right]}{M_1 + M_2 + M_3\left[I_2 + M_3 l^2(\sin^2\psi + 0.5\theta|\sin 2\psi|)/(I_2 + M_3 l^2)\right]}.$$

The complete change of velocity as a result of collision is

$$\delta\dot{x}_+ = \dot{x}_{1+} - \dot{x}_{1-},\quad \delta\dot{\psi} = \dot{\psi}_+ - \dot{\psi}_-,$$

hence, in agreement with (1.86) and taking account of (1.88), we obtain

$$\dot{\psi}_+ = \dot{\psi}_- - \frac{M_3 l\cos\psi}{I_2 + M_3 l^2}(1 + \mathcal{R})\dot{x}_{1-}\ . \qquad (1.89)$$

Note that the normal impulse in (1.87), $J_x \geq 0$.As a result, in (1.88) $\mathcal{R} \geq 0$. The change of sign of these quantities shows that the mechanism becomes jammed in the first impact phase. Thus, initially, when $I_2 = 0$, $\psi = \pi/4$ the following condition will guarantee that this situation will not occur:

$$\theta < 1 + 2(M_1 + M_2)/M_3,$$

In conclusion let us estimate the effect of collision in a similar system with a moving boundary mass M_4, assuming $\theta = 0$ (Fig. 1.8, *c*). In this case, the corresponding equations for the impacting system (1.85) to (1.87) remain valid but for the boundary mass M_4 we find

$$-J_x = \delta\dot{x}_4 M_4 \qquad (1.90)$$

Noting that in the first phase of impact $\delta\dot{x}_1 = \dot{x}_0 - \dot{x}_{1-}$, $\delta\dot{x}_4 = \dot{x}_0 - \dot{x}_{4-}$, and in the second phase $\delta\dot{x}_1 = \dot{x}_{1+} - \dot{x}_0$, $\delta\dot{x}_4 = \dot{x}_{4+} - \dot{x}_0$ where $\dot{x}_0 = \dot{x}_{10} = \dot{x}_{40}$ common velocity of masses M_1 and M_4 on the phase boundary, and using equation (1.61), we obtain the following equations for the post-impact velocities analogous to (1.55) and (1.56)

$$\dot{x}_{1+} = \frac{(M - RM_4)\dot{x}_{1-} + M_4(1+R)\dot{x}_{4-}}{M + M_4},\quad \dot{x}_{4+} = \frac{M(1+R)\dot{x}_{1-}(M_4 - RM)\dot{x}_{4-}}{M + M_4}$$

Here $M = M_1 + M_2 + M_3(I_2 + M_3 l^2\sin^2\psi)/(I_2 + M_3 l^2)$ is a value, which represents an equivalent mass of the impacting systems

In agreement with equation (1.86), the post-impact angular velocity of the flywheel is defined as

$$\dot{\psi}_+ = \dot{\psi}_- + \frac{M_3 l\cos\psi}{I_2 + M_3 l^2}(\dot{x}_{1+} - \dot{x}_{1-})\ .$$

Problems involving constrained impact in systems with dry friction have received special consideration here because such problems are not adequately covered in the literature.

Hence, within areas the application of stereomechanics gives simple kinematic solutions during the study of collision.

§2 Oscillator with fixed stop

1. Using stereomechanics in impact theory makes it possible to find exact solutions for some dynamic models of vibro-impact systems. Finding such solutions is based on the stitching procedure using kinematic conditions of impact to stitch the solutions of differential equations describing motion in the intervals between collisions. Hence, in parallel with the some form of idealisation, characteristic of stereomechanics, the development of suitable models, requires the introduction of additional simplifying assumptions, which allow us to find solutions during the intervals. As a rule, this leads to the use of linear differential equations to describe the motion during the intervals.

Thus, exact solutions can be obtained by significant idealisation of real systems, at the price of results that do not match quantitatively the real situation and sometimes also give a distorted qualitative picture of the true motion. For example, it is characteristic of such idealised simplifications that dissipative effects in systems frequently leads to unrepresentative motion of the system.

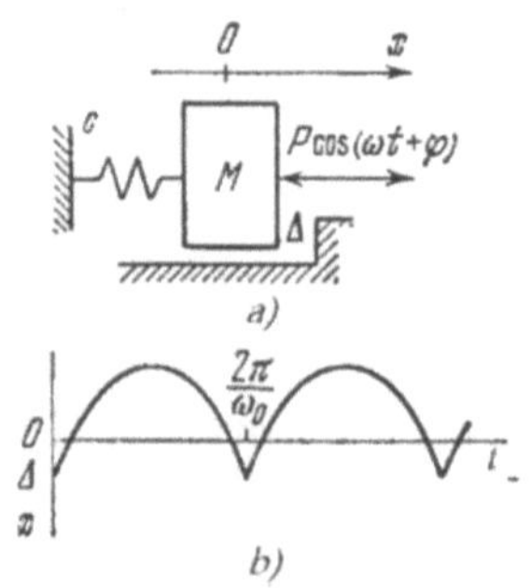

Fig. 2.1

Exact methods can generally only produce analytical results for models of small dimension. At the same time, exact methods can allow detailed analysis of motion of these models to be carried out, and these are important not only in themselves as areas of study, but also from the point of view of obtaining standard results with which various approximate methods can be compared. This latter point will be dealt with as we study approximate methods.

Using this approach, throughout this chapter we will examine the results of exact analysis for several types of dynamic models of vibro-impact systems, emphasising the qualitative side of the examined effects and studying the frequency behaviour of systems.

The case of simple restricted movement of the mass of a linear oscillator near a rigid stop (Fig. 2.1, *a*) reveals important features of the behaviour of such systems following collision. Many references in the literature [23, 48, 58, 146, 193] are devoted to the study of this dynamic model, and we shall use their results frequently in what follows.

We shall look initially at the natural vibration of this system, assumed to be a conservative oscillator, with totally elastic impact (coefficient of restitution $R = 1$). In the intervals between impacts, the motion of the mass M, measured from the state coinciding with the unstrained condition of the spring of stiffness c, satisfies the differential equation of an oscillator

$$M\ddot{x} + cx = 0 \tag{2.1}$$

At the instant when the mass impacts against the stop, which occurs when co-ordinate $x = \Delta$, its velocity instantaneously changes sign. We note that the positive value of Δ occurs when there is a gap between stops and mass, and becomes negative with initial interference.

Elements, which contact during impact form an *impact pair*. As we examine *one-dimensional* conditions, during which relative motion of the contacting elements can be written as a single co-ordinate, we will recognise *single-sided* and *double-sided* impact pairs depending on the number of contact surfaces for each of the interacting elements.

Rewriting equation (2.1) in the form

$$\ddot{x} + \Omega^2 x = 0, \tag{2.2}$$

where $\Omega = \sqrt{c/M}$ is the natural frequency of the linear oscillator, we obtain after integration

$$x = C_1 \cos\Omega t + C_2 \sin\Omega t, \tag{2.3}$$

where C_1 and C_2 are arbitrary constants.

Using the boundary conditions $t = 0, x = \Delta; t = 2\pi/\omega_0, x = \Delta$, where ω_0 is the natural frequency of the oscillator interacting with the stops, we find the motion of the system in the intervals between collisions as

$$x = \frac{\Delta}{\cos(\pi\Omega/\omega_0)} \cos\left(\Omega t - \pi\frac{\Omega}{\omega_0}\right). \tag{2.4}$$

Relationship (2.4) represents the vibration of the system only for one period $t \in [0, 2\pi/\omega_0]$ (Fig. 2.1, *b*) however, this information, naturally, is sufficient to represent the entire process. For example, expanding this solution as a Fourier series, using (2.4), we find for the coefficients a_{xi} $(i - 0, 1, \ldots)$ of the harmonics

$$a_{xi} = \frac{\omega_0}{\pi} \int_0^{2\pi/\omega_0} \frac{\Delta}{\cos(\pi\Omega/\omega_0)} \cos\left(\Omega t - \pi\frac{\Omega}{\omega_0}\right) \cos i\omega_0 t dt =$$
$$= \frac{2\Delta}{(\pi\Omega/\omega_0)[1 - (i\omega_0/\Omega)^2]} \tan\pi\frac{\Omega}{\omega_0} \tag{2.5}$$

and, as a result, the process examined can be represented along the entire time axis in the form

$$x = m_x + \sum_{i=1}^{\infty} \frac{2\Delta}{(\pi\Omega/\omega_0)[1 - (i\omega_0/\Omega)^2]} \tan\pi\frac{\Omega}{\omega_0} \cos i\omega_0 t, \tag{2.6}$$

where $m_x = a_{x0}/2$.

Figure 2.2 shows phase planes for the process examined for various stop arrangements. Since the motion of the representative points on the curved boundaries occurs uniformly with constant angular velocity Ω, but impact occurs instantaneously, the length of the curves for the corresponding closed cycles is proportional to the period of vibration of the system $T = 2\pi/\omega_0$ with collisions. From the diagram it follows directly that this period can, in general, depend on the amplitude of vibration and will not exceed the period of natural vibration of linear oscillator, i.e.

$$\Omega/\omega_0 \le 1, \tag{2.7}$$

while in the case when the stop and gap are arranged as in Fig. 2.2, *a,* as the

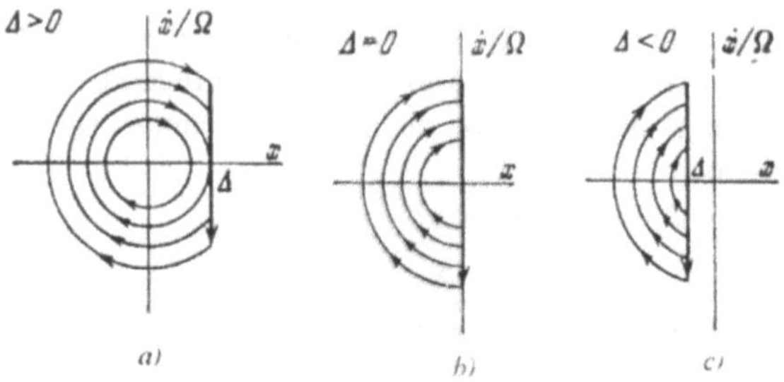

Fig. 2.2

amplitude of vibration increases, frequency increases and consequently, nonlinearity, introducing impact, takes a "hard" character; with interference, conversely, (Fig. 2.2, *c*), increasing amplitude causes a reduction in the vibration frequency, which we will call "soft" nonlinearity; finally, in the boundary case, $\Delta = 0$ (Fig. 2.2, *b*) corresponds to an isochronous system, although the nonlinear character of the vibration is retained. We note that

$$\begin{aligned} 0.5 < \Omega/\omega_0 \le 1 \quad &\text{when} \quad \Delta > 0 \\ \Omega/\omega_0 = 0.5 \quad &\text{when} \quad \Delta = 0 \\ \Omega/\omega_0 < 0.5 \quad &\text{when} \quad \Delta < 0 \end{aligned} \tag{2.8}$$

We will construct *backbone curves* for all three types of system, i.e. the relationship between amplitude and frequency of free vibration. When we refer to amplitude a_x of process $x(t)$, then we will mean the half-swing value (Fig. 2.1, *b*) as often used in non-harmonic vibration studies. Expressing the co-ordinate of the maximum deflection during process $x(t)$ as x_0, we find

$$a_x = \tfrac{1}{2}(\Delta - x_0) \qquad (a_x \ge 0). \tag{2.9}$$

According to (2.4)

$$x_0 = \Delta / \cos(\pi\Omega/\omega_0) \tag{2.10}$$

Substituting (2.10) into (2.9) we obtain the desired relationship

$$a_x = \frac{\Delta}{2}\left(1 - \frac{1}{\cos(\pi\Omega/\omega_0)}\right). \tag{2.11}$$

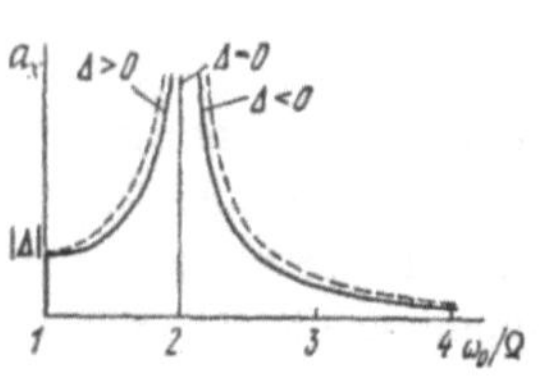

Fig. 2.3

In Fig. 2.3 backbone curves are presented for systems with various arrangements of stop (full lines). When the stop is installed with a gap, the vertical line of the backbone curve is added for isochronous vibration of a linear oscillator until they contact the stop. Later we will see that the type of nonlinearity obtained is characteristic for vibro-impact systems with single sided stops. Note that in all the examples examined (Fig. 2.3), as amplitude is increased, their frequencies approach more closely to double that of the linear oscillator.

In accordance with (2.7), series (2.6) converges fairly quickly and it is interesting to discover to what extent the described nonlinear behaviour of the backbone curve is shown by the first harmonic (*fundamental*) of vibration. According to (2.5) the amplitude of the fundamental is given by

$$a_{x1} = \frac{2\Delta}{\pi(\Omega/\omega_0 - \omega_0/\pi)} \tan \pi \frac{\Omega}{\omega_0} \qquad (a_{x1} \geq 0) \tag{2.12}$$

This relationship is shown dotted in Fig. 2.3. Comparison shows that the first harmonic of the process quite accurately represents the nonlinear behaviour of the system. The situation indicated is studied in Chapters 2 to 4 using approximate methods based on analysis of the fundamental.

2. Let us now consider forced vibration of a linear oscillators with stop, arising when a periodic force $P(t) = P\cos(\omega t + \varphi)$ (see Fig. 2.1, *a*) is applied to the vibrating mass. We shall find the loss of energy in the system using the coefficient of restitution *R*, assuming that $0 < R \leq 1$.

In the interval between collisions in the system, the following differential equation is valid

$$\ddot{x} + \Omega^2 x = a_p \cos(\omega t + \varphi), \tag{2.13}$$

where $a_p = P/M$; φ is the unknown initial phase of excitation. The general solution of the non-homogeneous linear equation (2.13) will be written in the form

$$x = C_1 \cos \Omega t + C_2 \sin \Omega t + a\cos(\omega t + \varphi), \tag{2.14}$$

where

$$a = \frac{a_p}{\Omega^2 \left|1 - \zeta^2\right|} \tag{2.15}$$

is the amplitude of forced vibration of the oscillator without a stop; $\zeta = \omega/\Omega$. We note that, in nonlinear systems, as well as the periodic motion with frequency ω, sub-periodic vibrations can also occur with a period that is a multiple of the exciting period . By taking the start time at the instant after impact, we write the boundary conditions for all combinations of periodic motion considered in the following form:

$$t = 0, x = \Delta, \dot{x} = \dot{x}_+; \quad t = 2\pi l/\omega, x = \Delta, \dot{x} = \dot{x}_-, \tag{2.16}$$

where l is regime multiple; $\dot{x}_-, \dot{x}_+$ are velocities before and after impact, related by (1.58)*. Using boundary conditions (2.16) to find the arbitrary constants C_1 and C_2 and phase φ in (2.14) we obtain

* As well as periodic regimes with a single impact per period of a mass against a stop as examined here, in vibro-impact systems it is also possible to find, similarly, periodic *multi-impact* processes, characterised by several impacts against the stop during the course of a single period. In general, single-impact regimes show the most intensive (resonant) character. They occur commonly in working processes and research is being carried out into their harmful effects. For this reason they are given special attention in this book.

$$C_1 = \Delta - a\cos\varphi, \quad C_2 = C_1 \tan\frac{\pi l}{\zeta}, \tag{2.17}$$

$$\sin\varphi = \frac{BD}{\Omega a}\dot{x}_-, \tag{2.18}$$

$$\cos\varphi = \left(1 - \frac{D}{\Omega\Delta}\dot{x}_-\right)\frac{\Delta}{a}, \tag{2.19}$$

where

$$B = \frac{1}{\zeta}\left(\frac{1-R}{1+R}\right)\tan\frac{\pi l}{\zeta}, \qquad D = -\frac{1}{2}(1+R)\cot\frac{\pi l}{\zeta} \tag{2.20}$$

Eliminating the trigonometric functions from equations (2.28) and (2.19), we can express the value of the velocity $\dot{x}_-$ by the following

$$\dot{x}_- = \frac{[1 \pm \sqrt{1-(1-a^2/\Delta^2)(1+B^2)}]\Omega\Delta}{D(1+B^2)} \tag{2.21}$$

Assuming in (2.21) that the expression under the radical $1-(1-a^2/\Delta^2)(1+B^2) \geq 0$, we obtain the condition defining the domain for real values of the velocity $\dot{x}_-$:

$$|\Delta|/a \leq \sqrt{1+1/B^2}\ . \tag{2.22}$$

According to (2.22) vibro-impact regimes can occur not only when the stops are arranged within the limits of the amplitude of vibration of linear oscillators $(0 < \Delta < a)$, but also outside this region $(|\Delta| > a)$. In the latter case, to develop into a vibro-impact regime the system must be supplied with sufficient energy for the process to start up, i.e. to receive what is called *hard excitation*. Later, when examining the motion of systems with double-sided stops, we will consider such features of vibro-impact systems in greater detail.

We note that when equation (2.21) and the values (2.17) to (2.19) obtained from it, are substituted into the motion law (2.14) we obtain double-valued functions for the system parameters. Hence, multivalued solutions that are characteristic of nonlinear vibration, also occur in vibro-impact systems. In addition, not all solutions satisfying (2.22) represent regimes that can exist. Searching for periodic solutions, we assumed that the mass received one collision per period of motion, making use of boundary conditions (2.16). From the structure of the model, it can be seen that the motion trajectory must not reach the stop in the interval between two calculated collisions, and this is not taken into account in the analysis carried out. By making motion satisfy conditions for the absence of additional intersections

$$x(t) \leq \Delta\,, \tag{2.23}$$

it is possible to isolate a narrower *existence domain* for the required solution from the domain of real values (2.22).

It is obligatory that the following inequality is satisfied in this case

$$\dot{x}_- \geq 0\,. \tag{2.24}$$

This obvious condition was not apparent either in the equations of motion or the boundary conditions and must be introduced additionally.

From (2.21), it follows that, if the term under the radical is greater than unity, then velocity $\dot{x}_-$ can take different signs and as a result, equation (2.24) is always satisfied for one and one only of them. Substituting into (2.21)

$$1-(1-a^2/\Delta^2)(1+b^2)>1$$

we find the following condition for a single solution $|\Delta| < a$. This solution corresponds to the positive sign in front of the radical in (2.21) which occurs when Δ/D > 0, or, using the notation of (2.20), in the following frequency bands:

$$l/m<\zeta<2l/(2m-1) \quad for \quad \Delta>0;$$
$$2l/(2m-1)<\zeta<l/(m-1) \quad for \quad \Delta<0 \tag{2.25}$$
$$(m=1, \ldots, 2, \ldots),$$

In the case when $\Delta/D < 0$, i.e. adjacent to frequency areas of (2.25), a regime occurs which corresponds to the negative sign of the radical.

Hence, for devices with gaps (interference),which do not exceed the amplitude of vibration of the linear oscillator, only one vibro-impact regime can exist for each multiple.

Equation (2.24) is satisfied for both values of velocity (2.21), if $|\Delta| > a$ and as a result $\Delta/D > 0$, i.e. in the frequency range found earlier (2.25). Here, in fact, equation (2.22) defines the upper boundary for the area of double-valued solutions, the lower boundary of which gives the ratio.

$$|\Delta|/a=1. \tag{2.26}$$

The inequality obtained still does not guarantee the absence of additional intersections. In general, it is difficult to find obvious analytical equations for the boundaries of existence domains as defined in equation (2.23). It would be necessary to find extreme values of deflection for the motion (2.14), using the sum of separate harmonic functions with different frequencies. Such proofs are usually obtained by checking the motion directly. This laborious procedure can be automated using computers.

We shall limit the examination of frequency effects, assuming $R = 1$, and neglecting energy dissipation during impact. For convenience in further analysis, we will change the law of motion (2.14) using (2.17) into the form

$$x=\frac{\Delta-a\cos\varphi}{\cos(\pi l/\zeta)}\cos\frac{\omega t-\pi l}{\zeta}+a\cos(\omega t+\varphi). \tag{2.27}$$

In agreement with equations (2.18) to (2.21), when R=1

$$\sin\varphi=0, \quad \cos\varphi=\pm 1,$$

and one of the extreme values (2.27) occurs at the instant $t=\pi l/\omega$. This value must satisfy condition (2.23)

$$x\left(\frac{\pi l}{\omega}\right)=\frac{\Delta-a\cos\varphi}{\cos(\pi l/\zeta)}+(-1)^l a\cos\varphi\le\Delta \tag{2.28}$$

Defining the value $\dot{x}_- = \frac{\Omega\Delta}{D}(1 \pm a/|\Delta|)$ from (2.21) and substituting this into (2.19) we find

$$a\cos\varphi = \mp\Delta / a|\Delta| . \tag{2.29}$$

Both now and later, the sign that is disposed like the similar sign before the radical in (2.21) must be taken.

Transforming inequality (2.28) using relationship (2.29), we eventually obtain the following conditions for the absence of the additional intersections

$$\frac{|\Delta|}{a} < \mp\cot^2\frac{\pi l}{2\zeta} \quad \text{when} \quad \frac{\Delta}{\cos(\pi l/\zeta)} > 0 \tag{2.30}$$

$$\frac{|\Delta|}{a} > \mp\cot^2\frac{\pi l}{2\zeta} \quad \text{when} \quad \frac{\Delta}{\cos(\pi l/\zeta)} < 0$$

for odd ratios ($l = 1, 3, \ldots$);

$$\frac{|\Delta|}{a} < \mp 1 \quad \text{when} \quad \frac{\Delta}{\cos(\pi l/\zeta)} > 0 \tag{2.31}$$

$$\frac{|\Delta|}{a} > \mp 1 \quad \text{when} \quad \frac{\Delta}{\cos(\pi l/\zeta)} < 0$$

for even ratios ($l = 2, 4, \ldots$).

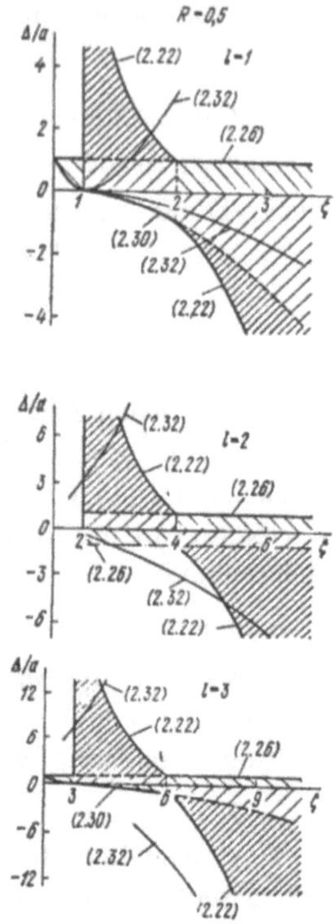

Fig. 2.4

Fig. 2.4 shows examples of the existence domains for regimes with various ratios ($l = 1, 2, 3$) constructed from inequalities, (2.22), (2.25), (2.26), (2.30), (2.31) and constant coefficient of restitution $R = 0.5$. Here the areas hatched with positive slope correspond to regimes with positive radicals in (2.21), while the negative shaded areas correspond to the negative radical. Areas where double-valued solutions occur are indicated by closer hatching.

The vibro-impact system considered occurs due to single-sided limitation of the vibration of the simplest form of linear oscillator, whose behaviour is well known. It is now simple to evaluate those effects that cause such limitations on the frequency range of vibro-impact regimes. Above all, we can see immediately (Fig. 2.4) that the domains in which vibro-impact regimes can exist mainly in the post-resonance range of the linear oscillator. ($\zeta > 1$), while for a system with interference ($\Delta < 0$) in the pre-resonance frequencies ($\zeta < 1$) such regimes do not exist in general.

Adding the curve of Δ/a to Fig. 2.4, using equation (2.15) for the amplitude-frequency characteristics of a linear system

$$\frac{\Delta}{a} = \frac{\Delta\Omega^2}{a_p}\left|1 - \zeta^2\right| \tag{2.32}$$

we can visualise the frequency range for the existence of different regimes. The intersection points of this curve with the boundary of the existence domains define

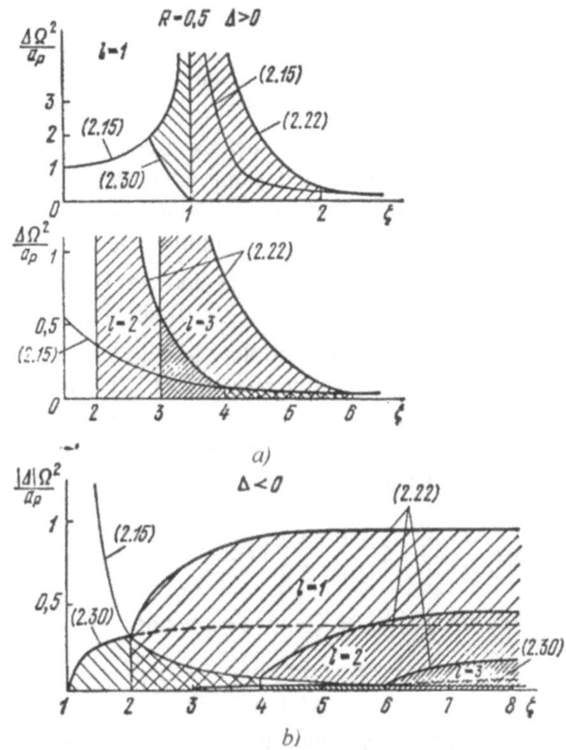

Fig. 2.5

the limiting frequencies for the occurrence of vibro-impact processes. Hence, for example, for systems with a gap $(\Delta > 0)$ near resonance $\zeta = 1$ and a unique vibro-impact regime will occur. At the same time, at post-resonance frequencies in the band $l < \zeta < 2l$ it is possible to cause dual regimes of ratio l.

In Fig. 2.5, *a*, for a system with an initial gap $(\Delta > 0)$ the existence domains are drawn against amplitude-frequency characteristics for a linear oscillator (2.15). The hatched areas, show the lower curves (2.15), which exist only for single-valued regimes, but in the areas included between the lines (2.15) and (2.22) - pairs can exist. Existence domains for regimes with ratios $l > 1$ are bounded on the left by frequency $\zeta = l$. The diagrams presented allow us to define the frequency range in which vibro-impact occurs for given values of a gap and to find the value of the gap for which such a process is possible at a given frequency. Fig. 2.5, *a* illustrates graphically the fact mentioned earlier of the existence of vibro-impact regimes for a device with a gap, as the amplitude of linear oscillation is increased $(\Delta > a)$. The value of such increase defines the boundaries (2.22). Notice that existence domains for various frequency ratios partially overlap. Obviously, in these zones development of one or other regime of existence depends initially on the way the transient process takes place.

It is interesting to note that neither increasing the gap nor reducing the exciting force will lead to the elimination of vibro-impact systems whatever the ratio, even in the presence of impact dissipation, which only contracts the frequency range. (As R increases, the limiting value (2.22) approaches more closely to the vertical line $\zeta = 2l$.) We must remember, however, that we have not taken account of other reasons for dissipation of energy, such as viscous damping, internal damping, etc. When we consider double-sided stops later, this aspect will be analysed. Jumping ahead, we will show that viscous losses influence the limits (2.22) of existence domains near frequencies $\zeta = l$, limiting the value of Δ at which vibro-impact regimes occur and reducing existence domains for regimes as l increases.

For a system with initial interference, $(\Delta < 0)$ the impact dissipation causes the elimination of regimes of higher multiples. In Fig. 2.4, such a situation occurs when $l = 3$, where the lower portion of curve (2.32) does not intersect with the boundary (2.22) of the area of existence. The limiting value of the related parameters for which a regime of multiple l occurs, is found by substituting equation (2.32) into condition (2.22), and writing it using (2.20) in the form

$$|\Delta| \le \frac{a_p}{\Omega^2}\left[1 + \zeta^2\left(\frac{1+R}{1-R}\right)^2\left(\cot\frac{\pi l}{\zeta}\right)^2\right]^{1/2} \Big/ \left|1 - \zeta^2\right|.$$

Simplifying this inequality, as $\zeta \to \infty$, we obtain

$$|\Delta| \le \frac{a_p}{\pi l \Omega^2}\frac{(1+R)}{(1-R)}. \tag{2.33}$$

Fig. 2.5, *b* shows existence domains of vibro-impact regimes for systems with initial interference. Note that existence domains for regimes with high ratios are completely enclosed within areas of lower ratio, while regimes of ratio l exist in the frequency range $\zeta \ge l$. For odd values of l, single valued domains are arranged below boundary (2.30), but for even values, below boundary (2.15). Double-valued domains occur above these lines, limited by (2.22). Equation (2.33) is shown as asymptotic to the upper boundary of the area of existence (2.22) (Fig. 2.5, *b*). The asymptotes, shown as dotted lines for the lower boundary (2.30) of the area of double values, are found similarly and defined by the relationship

$$\Delta = 4a_p/(\pi l \Omega)^2 \tag{2.34}$$

Note that relationship (2.33) can easily be used to choose the initial interference, guaranteeing the suppression of undesirable vibro-impact processes, occurring, for example, in kinematic coupled mechanisms when they are disconnected.

We shall examine regimes for various ratios at greater length and try to clarify the physical meaning. When the boundary conditions (2.16) were written, the ratio l was defined as the ratio of the period of system motion to the period of excitation. Hence a regime of ratio l is a process of frequency ω_l which is related to the excitation frequency ω by the expression

$$\omega_l = \omega / l. \tag{2.35}$$

Referring to Fig. 2.4 and Fig. 2.5, we note that frequency ω_l lies within the limits

$$\Omega \le \omega_l < 2\Omega \text{ when } \Delta > 0,$$

$$2\Omega \le \omega_l < \infty \text{ when } \Delta < 0.$$

Comparing this with the results obtained in section 1 during the examination of free vibration, we see that the frequency ω_l of the regime for any ratio coincides with one of the natural frequencies ω_0 (2.8) of the vibro-impact system, and hence the corresponding tuning of the system is called resonance.

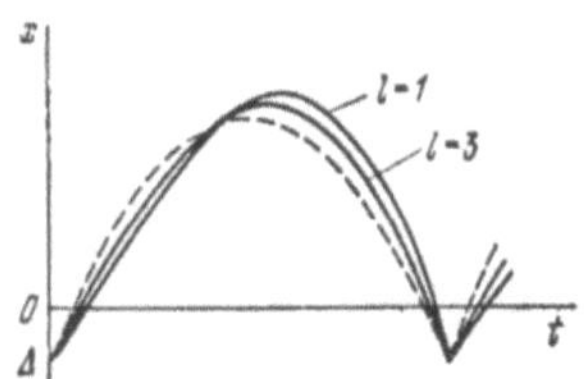

Fig. 2.6

Fig. 2.6 shows the motion for regimes with ratios $l = 1$ and $l = 3$, excited by an external force of frequency $\zeta = 1.75$, and $\zeta = 5.25$. For comparison the dotted lines show motion during free vibration with frequency $\zeta_0 = 1.75$. From the diagram it is obvious that the regimes for various ratio l are not very different from each other and at the same time they are quite close to natural vibration. It is also possible to establish this directly from the motion (2.27), rewriting it taking into account (2.35), in the form

$$x = \frac{\Delta - a\cos\varphi}{\cos(\pi\Omega/\omega_l)}\cos\left(\Omega t - \frac{\pi\Omega}{\omega_l}\right) + a\cos(\omega t + \varphi) \tag{2.36}$$

and comparing (2.36) to the free vibration (2.4). As the frequency of excitation ω increases, or the regime ratio increases, so the value $a \to 0$ and the law of motion (2.36) approaches more closely to (2.4).

Thus resonant motion of vibro-impact systems can be treated as free vibration sustained by external excitation. Such treatment is widely used in nonlinear mechanics and will be repeatedly made use of in what follows. The fact that resonant motion is close to free vibration can be used to estimate the amplitude of forced vibration using the backbone curve (2.11).

Fig. 2.7 shows backbone curves for various ratios $(l = 1, 2, 3)$, constructed using equation (2.11), using relationship (2.35). In this way we can approximate the branches of the resonance curves when $R = 1$. They are increasing infinite as $\zeta \to 2l$. Dissipation by impact $(R \ne 1)$ leads to the limiting value of vibration amplitude. Searching for the limiting frequency of vibro-impact regimes corresponding to the boundaries (2.22) of existence domains (Fig. 2.5), we find the higher points of the resonance curves, the achieved parts of which are shown in Fig. 2.7 by bold lines.

Later we will carry out detailed analysis of the amplitude-frequency characteristics of systems, and we will prove that the branches of the resonance curves, found from exact relationships, are arranged near the lines we have constructed, while the interpretation of resonance regimes given here, is convenient and sufficiently accurate, especially when estimating the maximum amplitude.

3. When the differential equations and boundary conditions were defined, only basic factors influencing the system were accounted for, based on the assumption that we could neglect small perturbations. Yet the external perturbations cause changes to the real motion from the solution found. For one period of motion, these small effects cause slight changes in the starting conditions of motion in the following interval from those that match the conditions of periodicity. However, it can be shown that as time elapses, small perturbations will accumulate and significantly distort the final result of motion of the system, so that it will differ considerably from the solutions found for undisturbed conditions. Such solutions are called *unstable*. Naturally, they are not produced in real systems. Thus the analysis of stability leading to the separation of physically possible solutions has a very important practical meaning.

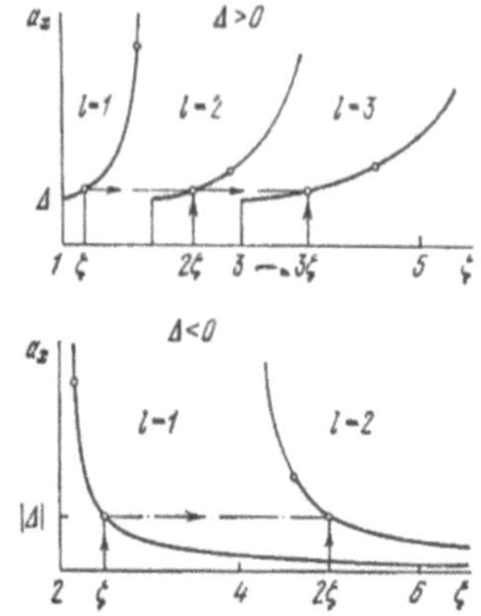

Fig. 2.7

A *stable* solution is one, which changes negligibly due to small perturbations of the initial conditions during any time interval. In this case, it is sufficient to find the tendency for perturbations to cause changes at the instant of a following collision. Stable motion occurs if, during continuous collisions, the perturbation either decreases or stays within limits; when unlimited growth of perturbation occurs, this represents instability. The procedure for such analysis was developed by R E Brunstein and A. E Kobrinsky [117].

Suppose the system realises a disturbing motion, such that at the instant of an arbitrary v-th collision the mass reaches the stops with a velocity $(\dot{x}_{-})_{v}$, differing from the undisturbed value $\dot{x}_{-}$ by a small quantity $\delta\dot{x}_{v}$ while the disturbing phase φ has increment $\delta\varphi_{v}$. This is equivalent to an increment δC_{1v} and δC_{2v} in the integration constants C_1 and C_2 for the solution (2.14). Assuming initial start time at the instant of impact, we can write the equation for the disturbed motion of the v-th interval as

$$\begin{aligned} x + \delta x_{v} = (C_1 + \delta C_{1v})\cos\Omega t_{v} + \\ + (C_2 + \delta C_{2v})\sin\Omega t_{v} + \\ + a\cos(\omega t_{v} + \varphi + \delta\varphi_{v}) \end{aligned} \tag{2.37}$$

The $(v+1)$-th collision occurs at the instant $t_{v} = 2\pi l/\omega + \delta T_{v}$ where δT_{v} is a small change in the period between collisions following the initial disturbance. Suppose that in (2.37) $t_{v} = t + \delta t_{v}$, where $\delta t_{v} = 0$ at the v-th instant and $\delta t_{v} = \delta T_{v}$ at the instant of the $(v+1)$-*th* impact, we can expand (2.37) in a Taylor series for small increments. From the equations obtained we subtract those for undisturbed motion (2.14) and we find the following equation for the disturbances in the v-*th* interval to second-order accuracy:

$$\begin{aligned} \delta x_\nu &= \delta C_{1\nu}\cos\Omega t + \delta C_{2\nu}\sin\Omega t - \\ &- \Omega\delta t_\nu (C_1 \sin\Omega t - C_2\cos\Omega t) - a(\omega\delta t_\nu + \delta\varphi_\nu)\sin(\omega t + \varphi) \end{aligned} \tag{2.38}$$

Writing the boundary conditions for the disturbed motion (2.38) at the instant after the ν-th and before the $(\nu+1)$th impact

$$\begin{aligned} &t = 0,\quad \delta t_\nu = 0,\quad \delta x_\nu = 0,\quad \delta\dot{x}_\nu = \delta\dot{x}_{\nu+};\\ &t = \frac{2\pi l}{\omega}, \delta t_\nu = \delta T_\nu,\ \delta x_\nu = 0,\quad \delta\dot{x}_\nu = \delta\dot{x}_{(\nu+1)-}, \end{aligned} \tag{2.39}$$

and the corresponding increment in the velocity during the $(\nu+1)th$ collision

$$\delta\dot{x}_{(\nu+1)+} = -R\delta\dot{x}_{(\nu+1)-}\,. \tag{2.40}$$

Applying conditions (2.39), (2.40), to (2.38) we obtain a recurrent relationship connecting the value of the disturbing initial conditions δC_1, δC_2 and phase $\delta\varphi$ to the two adjacent intervals of the disturbed motion

$$\begin{aligned} &\delta C_{1\nu} - a\delta\varphi_\nu \sin\varphi = 0,\\ &\delta C_{1(\nu+1)} - \delta C_{1\nu}\cos\frac{2\pi l}{\zeta} - \delta C_{2\nu}\sin\frac{2\pi l}{\zeta} + \\ &\qquad + (\delta\varphi_{\nu+1} - \delta\varphi_\nu)\frac{C_1}{\zeta}\tan\frac{\pi l}{\zeta} = 0,\\ &\delta C_{1\nu} R\sin\frac{2\pi l}{\zeta} - \delta C_{2(\nu+1)} - \delta C_{2\nu} R\cos\frac{2\pi l}{\zeta} + \\ &\qquad + \delta\varphi_{(\nu+1)}\left[\frac{RC_1}{\zeta} + a\zeta(1+R)\cos\varphi\right] - \delta\varphi_\nu \frac{RC_1}{\zeta} = 0 \\ &\qquad (\zeta = \omega/\Omega) \end{aligned} \tag{2.41}$$

In (2.41) and taking equation (2.17) into account as well, there is an obvious relationship between the incremental phase and period $\delta\varphi_{(\nu+1)} = \delta\varphi_\nu + \omega\delta T_\nu$, which reflects the fact of the accumulation of phase for ν cycles

$$\delta\varphi_\nu = \delta\varphi_0 + \sum_{i=1}^{\nu}\omega\delta T_\nu$$

The relationship obtained represents the linear finite difference equations [99], the solution of which take the form

$$\delta C_{1\nu} = h_1\beta^\nu,\quad \delta C_{2\nu} = h_2\beta^\nu,\quad \delta\varphi_\nu = h_3\beta^\nu, \tag{2.42}$$

where h_1, h_2, h_3 are coefficients independent from ν; β is a constant.

Substituting the assumed solution (2.42) into equation (2.41), we obtain the following system of linear ordinary equations for the relative co-efficient h_1, h_2, h_3:

$$\begin{aligned} &h_1 - h_3 a\sin\varphi = 0\,,\\ &h_1\left(\beta - \cos\frac{2\pi l}{\zeta}\right) - h_2\sin\frac{2\pi l}{\zeta} + h_3(\beta - 1)\frac{C_1}{\zeta}\tan\frac{\pi l}{\zeta} = 0\,, \end{aligned}$$

$$h_1 R\sin\frac{2\pi l}{\zeta} - h_2(\beta + R\cos\frac{2\pi l}{\zeta}) +$$

$$+h_3\left\{\beta\left[\frac{RC_1}{\zeta} + a\zeta(1+R)\cos\varphi\right] - \frac{RC_1}{\zeta}\right\} = 0$$

This set of equations has a non-zero solution only when the determinant of the coefficients is zero, i.e.

$$\begin{vmatrix} 1 & 0 & -a\sin\varphi \\ \beta - \cos\frac{2\pi l}{\zeta} & -\sin\frac{2\pi l}{\zeta} & (\beta - 1)\frac{C_1}{\zeta}\tan\frac{\pi l}{\zeta} \\ R\sin\frac{2\pi l}{\zeta} & -\left(\beta + R\cos\frac{2\pi l}{\zeta}\right) & \beta\left[\frac{RC_1}{\zeta} + a\zeta(1+R)\cos\varphi\right] - \frac{RC_1}{\zeta} \end{vmatrix} = 0 .$$

Expanding the determinant and using equation (2.17) to (2.19) we obtain the *characteristic equation* in the form

$$a_0\beta^2 + a_1\beta + a_2 = 0 \tag{2.43}$$

where

$$a_0 = \frac{\dot{x}_-}{\omega};\ a_1 = \Delta\zeta(1+R)\sin\frac{2\pi l}{\zeta} +$$

$$+\frac{\dot{x}_-}{2\omega}\left[(1+R)^2\left(2\zeta^2\cos^2\frac{\pi l}{\zeta} - 1\right) - (1-R)^2\cos\frac{2\pi l}{\zeta}\right];\ a_2 = \frac{R^2\dot{x}_-}{\omega},$$

Hence solving the stability problem leads to an estimate of the roots of the characteristic equation (2.43). If root $\beta_{1,2}$ is disposed within a unit circle, i.e.,

$$|\beta_{1,2}| < 1, \tag{2.44}$$

then, in accordance with (2.42), when $v \to \infty$, the value of the perturbations $\delta C_{1v} \to 0,\ \delta C_{2v} \to 0,\ \delta\varphi_v \to 0$, and the motion under investigation has *asymptotic* stability.

To estimate whether roots correspond to condition (2.44) it is not necessary to solve the characteristic equation. There exists for this aim Schur's criteria [160], which can be constructed directly from the coefficients of the equation or from the determinant of the characteristic equation [14]. For equations of the second degree, these criteria take the following form:

$$\left|\frac{a_2}{a_0}\right| < 1,\quad \left|\frac{a_1}{a_0 + a_2}\right| < 1. \tag{2.45}$$

The first of these gives the condition $R < 1$, which is always satisfied. The second is equivalent to the following two relationships

$$a_1 + a_0 + a_2 > 0, \tag{2.46}$$

$$a_1 - a_0 - a_2 < 0. \tag{2.47}$$

Substituting in (2.46) the values of the coefficients from (2.43) and making use of relationship (2.20) during any further conversion we have

$$\Delta\Omega/(D\dot{x}_{-}) < 1 + B^2 \tag{2.48}$$

Using the value given by equation (2.21) for $\dot{x}_{-}$, we obtain

$$\left[1 \pm \sqrt{1-(1-a^2/\Delta^2)(1+B^2)}\right]^{-1} < 1. \tag{2.49}$$

From this it follows immediately that the conditions for stability of the solution with the positive sign before the radical are satisfied when

$$|\Delta|/a < \sqrt{1+1/B^2}\,, \tag{2.50}$$

but solutions with the negative sign when

$$|\Delta|/a < 1. \tag{2.51}$$

Note that the stability inequalities found in (2.50), (2.51) coincide with the conditions for existence obtained earlier, (2.22), (2.26). Hence it appears that these conditions are always satisfied for those regimes corresponding to domains containing single valued solutions (2.4).

In domains with double-valued solutions, stability can only exist for a regime having a value of $\dot{x}_{-}$ together with the positive sign before the radical in equation (2.21).

Let us now examine the criterion (2.47). After conversion, similarly to that earlier, we obtain the following condition for stability:

$$\frac{1}{1 \pm \sqrt{1-(1-a^2/\Delta^2)(1-B^2)}} > \frac{1-C^2}{1+B^2}\,, \tag{2.52}$$

where

$$C^2 = \frac{(1+R^2)[1+\cos^2(\pi l/\zeta)] + 2R\sin^2(\pi l/\zeta)}{[\zeta(1+R)\cos(\pi l/\zeta)]^2}$$

When analysing (2.52) we will assume $R = 1$ for simplicity, taking into account that this equation depends uniformly on R. As a result inequality (2.52) takes the form

$$\frac{1}{1 \pm a/|\Delta|} > 1 - \frac{1}{\zeta^2\cos^2(\pi l/\zeta)}. \tag{2.53}$$

Using the positive sign, the stability conditions take the following form:

$$|\Delta|/a > \zeta^2\cos^2(\pi l/\zeta) - 1 \tag{2.54}$$

Using the negative sign, and taking account of the conditions for existence $|\Delta|/a < 1$ (Fig. 2.5), we obtain

$$|\Delta|/a < 1 - \zeta^2\cos^2(\pi l/\zeta)\,. \tag{2.55}$$

and in practice these all exhibit instability, excepting domains lying in the narrow zones close to frequency $\zeta = 2l$.

Fig. 2.8 shows boundaries for existence domains, hatched portions of which are stability domains, obtained according to (2.54, (2.55). Note that stability domains in general fall in the frequency range $l < \zeta < 2l$ when $\Delta > 0$ and $\zeta \geq 2l$ when

$\Delta < 0$, i.e., exactly where we constructed the backbone curves and discovered the possibility of large amplitude vibrations (Fig. 2.7).

4. Let us turn to the construction of amplitude-frequency characteristics for the system, taking the value given by (2.9) for amplitude, as before. Finding exact values for the amplitude is laborious and requires a search for the maximum deflection x_0 in equation (2.27). A simple solution occurs only when $R = 1$, and the maximum deflection is at the instant $t = \pi l / \omega$. However, the construction of the laws of motion for various values $R \neq 1$ (see, e.g., Fig 2.6) as discussed in section 2, and comparing the resonant regimes with free vibration shows that, even here the maximum deflection occurs close to the middle of period $t = \pi l / \omega$. Hence we will assume $x_0 \approx x(\pi l / \omega)$. Finding this value from (2.27) and substituting it into (2.9) we obtain for the amplitude a_x:

$$a_x = \frac{\Delta}{2}\left(1 - \frac{1}{\cos(\pi l / \zeta)}\right) - a\cos\varphi\left[(-1)^l - \frac{1}{\cos(\pi l / \zeta)}\right]. \tag{2.56}$$

Let us construct the resonance curves for regimes with ratios $l = 1$. In this case, the first term in (2.56) coincides with equation (2.11). According to (2.19), (2.31), the value $\cos\varphi$ is a double-valued function of the system parameters. Thus equation (2.56) defines two branches of the resonance curve, and (2.11) the backbone curve which separates them. The resonance curves for various stop arrangements are shown in Fig. 2.9. Here, item *1* indicates the backbone curves.

We note that in a system with initial gap $(\Delta > 0)$, linear harmonic vibration can also occur, of which the amplitude $a < |\Delta|$. It is easy to show that resonance curves (2.56) result from points where amplitude frequency characteristics of linear oscillators intersect the stop level $a = |\Delta|$. Really, with this condition, equation (2.21) gives a velocity value $\dot{x}_- = 0$, in agreement with (2.19), $a\cos\varphi = \Delta$ and as a result (2.56) gives $a_x = \Delta$. It is interesting that the same value of the amplitude a_x of linear resonance is reached when $\zeta = 1$. In this case, the second term (2.56) tends to zero as $\zeta \to 1$. Hence for pre-resonance frequencies near linear resonance, the amplitude-frequency characteristics of a nonlinear system show a characteristic fall, when amplitude a_x is less than the value of the initial gap Δ. Later, we will find that this effect leads to stability of systems with double-sided symmetrical stops when impact

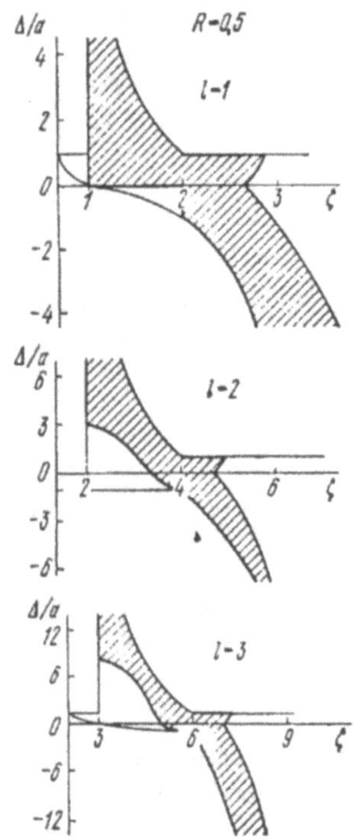

Fig. 2.8

occurs against one stop only (see §3, section 6).

Let us now find the maximum vibration amplitude. From equation (2.56) it is obvious that when $\cos\varphi = 0$ the resonance curve intersects the backbone curve. According to (2.19) and (2.21) one of the values of $\cos\varphi$ is zero, when $\Delta = a/B$. Multiplying both sides of the latter equation by $\frac{1}{2}\left(1-\frac{1}{\cos(\pi l/\zeta)}\right)$, using values (2.15), (2.20) we obtain

$$\frac{\Delta}{2}\left(1-\frac{1}{\cos(\pi l/\zeta)}\right) = a_x^*, \qquad (2.57)$$

where

$$a_x^* = \frac{a_p}{2\Omega^2}\frac{(1+R)}{(1-R)}\frac{\zeta\tan[\pi l/(2\zeta)]}{(\zeta^2-1)}. \qquad (2.58)$$

Comparing equations (2.56) and 2.57) it can be seen that equation (2.58) defines a curve in the (ζ, a_x) plane, which intersects the backbone curves exactly at resonance. This is the highest point on the resonance curve. When $R = 1$, branches of the resonance curves continue infinitely along the backbone curves and do not cross them anywhere (dotted lines in Fig.2.9, *c*). This follows directly from (2.56), taking into account that $\cos\varphi = \pm 1$, but also follows from equation (2.57).

For systems with interference ($\Delta < 0$) the resonance curve takes the form of a loop enclosing the backbone line and extending to high frequency areas $\zeta > 2l$. From Fig. 2.9 it is obvious that such behaviour exists when curve (2.58) intersects the backbone line higher. Substituting into (2.57) the limit when $\zeta \to \infty$, we obtain the limiting value for the conditions of existence of vibro-impact regimes

$$|\Delta| < \frac{a_p}{\pi/\Omega^2}\frac{1+R}{1-R}.$$

Earlier this equation was obtained from other considerations (see (2.33)).

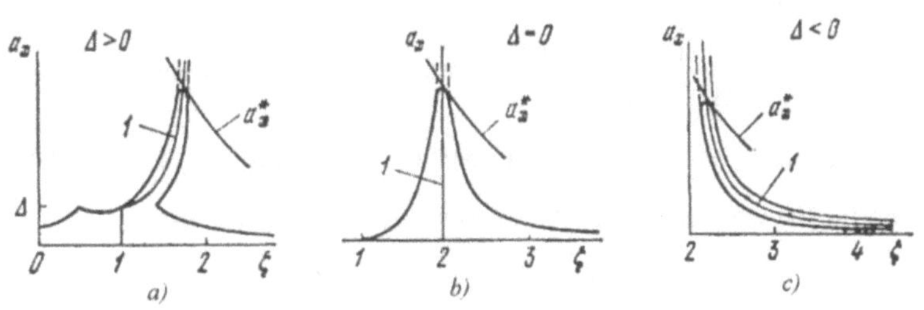

Fig. 2.9

Finally, systems with no initial gap ($\Delta = 0$), give an isochronous resonance curve for all ranges of existence, the maximum of which reaches frequency $\zeta = 2l$. At this value, equation (2.58) gives the following value for the resonance amplitude:

$$a_x^* = \frac{a_p}{\Omega^2}\frac{l}{(4l^2-1)}\frac{(1+R)}{(1-R)} \tag{2.59}$$

From this it can be seen that amplitude of resonant vibration dies away faster as the ratio l for the regime increases.

Let us look at limiting points of the resonance curves (Fig. 2.9, *a, c*) where the upper and lower branches merge. Such merging occurs when $\cos\varphi$ in (2.56) takes unit value, i.e., according to (2.19), (2.21) on the boundaries (2.22) of existence domains which coincide with the stability boundary (2.50). For these boundaries the value is

$$a\cos\varphi = \Delta/(1+1/B^2).$$

Substituting this in the equation for the resonance curves (2.56), we obtain the following equation for the stability bound:

$$a_x = \frac{\Delta}{2}\left(1-\frac{1}{\cos(\pi l/\zeta)}\right)-\frac{\Delta}{1+1/B^2}\left[(-1)^l-\frac{1}{\cos(\pi l/\zeta)}\right]. \tag{2.60}$$

Here the first term defines the backbone curve with which the stability boundaries coincide when $R = 1$. In general, the bound (2.60) occurs below the backbone curve, separating the branches of the resonance curve, from which it follows that the stable one in the double-valued existence domain is disposed above the backbone curve.

All that has been discussed here holds true for regimes of any ratio. It follows, however, that as l increases, the frequency range of regimes moves further away from the frequency of linear resonance, the second term in (2.56) becomes negligibly small, and branches of the resonance curves become closer to the backbone curve. However, higher points on the resonance curves approach the stability boundaries. Thus, the calculation of the amplitude of forced vibration set out in section 2 (Fig. 2.7) appears reasonably accurate.

To conclude this section we will check how the full analysis of resonance effects in systems can be represented by the first harmonic component of the process. With this aim, we shall present motion on all time axes, expanding (2.27) as a Fourier series:

$$x = m_x + \sum_{i=1}^{\infty}\frac{2(\Delta - a\cos\varphi)}{(\pi\Omega/\omega)([1-(i\omega/\Omega)^2}\tan\frac{\pi\Omega}{\omega}\cos i\omega t + a\cos(\omega t+\varphi), \tag{2.61}$$

where $m_x = (\Delta - a\cos\varphi)\frac{\omega}{\pi\Omega}\tan\frac{\pi\Omega}{\omega}$ is a constant for the process. Using relationships (2.19) and (2.20) we can modify the expression for m_x to the form

$$m_x = -\frac{(1+R)\dot{x}_- M}{Tc} = \frac{J}{Tc} = \frac{G}{c}. \tag{2.62}$$

In (2.62) $J = -M(1+R)\dot{x}_-$ is the impact impulse acting on mass M; $T = 2\pi/\omega$ is the period of the process; $G = J/T$ is the average value of the elastic force in the spring during the period. Hence, in the vibro-impact process examined, displacement of the middle of vibration occurs due to unsymmetrical single-sided collisions. In practice, similar effects are revealed as the development of static components of deflection in the operation of various measuring devices such as gyroscopes, indicators etc., and the effect has been given the name *drift*. The numerical value of drift is measured by m_x.

Substituting $a = 0$ into (2.51), we obtain a series, which is identical to that given by the analysis of free vibration (2.6). We have seen that for free vibration the backbone curves constructed for the fundamental coincide with the exact solution (Fig. 2.3). From (2.61) it follows that the amplitude of the fundamental ($i = 1$) of forced vibration is double-valued as a result of the double value of $\cos\varphi$, hence the branches of the resonance curve are grouped close to the backbone curve, and intersect it when $\cos\varphi = 0$.

§3. Oscillator with symmetrical stops

1. In the preceding section, we came across various effects, which occur during mechanical vibration of simple systems when collisions develop with a stop. A stop is regarded as a structural element of the system, which impart a nonlinear character to the motion. This is characterised by ambiguous behaviour of the system at the same values of the parameters, and considerably complicates the

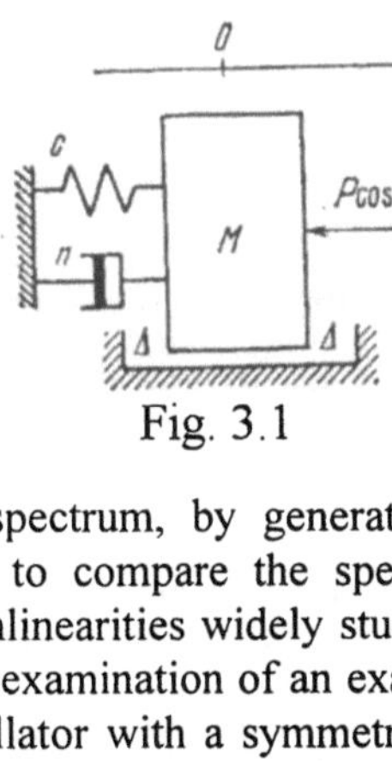

Fig. 3.1

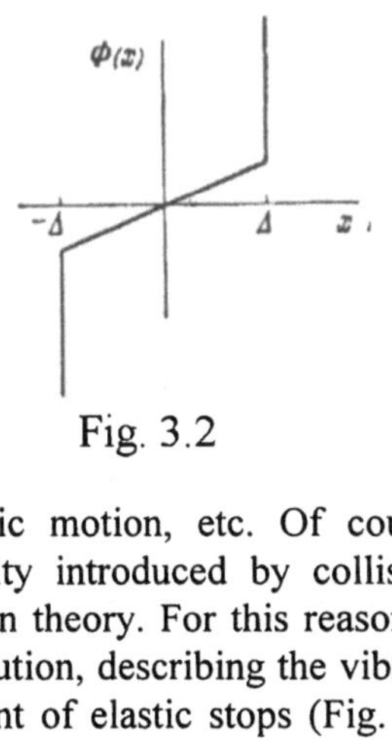

Fig. 3.2

vibration spectrum, by generating sub-periodic motion, etc. Of course, it is interesting to compare the specific nonlinearity introduced by collisions with similar nonlinearities widely studied in vibration theory. For this reason we shall turn to the examination of an exact periodic solution, describing the vibration of a linear oscillator with a symmetrical arrangement of elastic stops (Fig. 3.1) [40], and retaining the basic definitions of §2.

Fig. 3.2 shows the static characteristics of the system studied, $\Phi(x)$. To some extent, it limits characteristics to the usual representation of the elastic behaviour of a material with hardening. This case serves as the basis for studying the evolution of a typical dynamic model. At the same time, the specific structure of the system described allows us to overcome the inherent difficulty of exact analysis of the amplitude-frequency characteristics of vibro-impact systems.

We begin by studying free vibration, assuming the system as conservative. Assuming zero time at the instant after impact on the right hand stop, we obtain for equation (2.3) the following boundary conditions:

$$t = 0,\ x = \Delta;\quad t = \pi/\omega_0,\ x = -\Delta. \tag{3.1}$$

Defining an arbitrary constant, with the help of (3.1) we obtain the motion of the system in the interval between collisions

$$x = -\frac{\Delta}{\sin[\pi/(2\zeta_0)]}\sin\left(\Omega t - \frac{\pi}{2\zeta_0}\right) \qquad \left(t \in \left[0, \frac{\pi}{\omega_0}\right]\right), \tag{3.2}$$

where $\zeta = \omega_0/\Omega$. Differentiating (3.2) we find an expression for the velocity change

$$\dot{x} = -\frac{\Omega\Delta}{\sin[\pi/(2\zeta_0)]}\cos\left(\Omega t - \frac{\pi}{2\zeta_0}\right) \qquad \left(t \in \left[0, \frac{\pi}{\omega_0}\right]\right) \tag{3.3}$$

From the condition $\dot{x}(0) = \dot{x}_+ \le 0$ we obtain $\cot[\pi/2\zeta_0\} \ge 0$ and consequently,

$$\zeta_0 \ge 1. \tag{3.4}$$

This result also follows immediately from a study on of the phase plane for the system (Fig. 3.3).

Thus the free vibration frequency of the

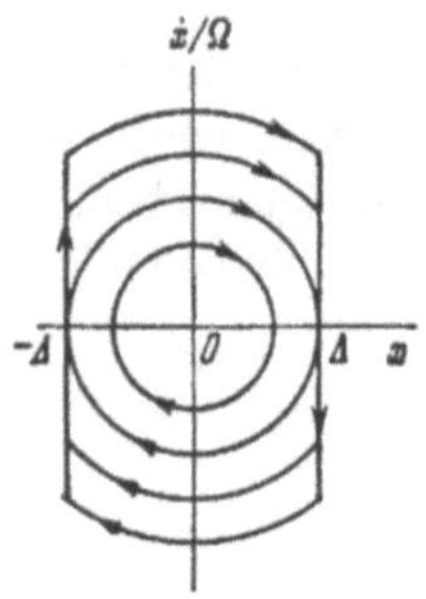

Fig. 3.3

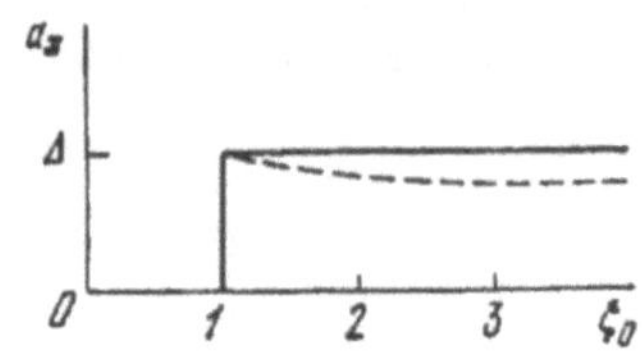

Fig. 3.4

system cannot be less than the natural frequency of the linear oscillator. Since the amplitude of vibration is constant and equal to Δ, the backbone curve of the system takes the form shown as a solid line in Fig. 3.4. The vertical portion of the backbone curve represents linear vibration of the oscillator with no impact against the stops.

Let us analyse this process as a Fourier series. For the amplitude of the i-th harmonic a_{xi} we find ($i = 1, 3, \ldots$)

$$a_{xi} = -\frac{2\omega_0}{\pi}\int_0^{\pi/\omega_0}\frac{\Delta}{\sin[\pi/(2\zeta_0)]}\sin\left(\Omega t - \frac{\pi}{2\zeta_0}\right)\cos i\omega_0 t\,dt = \frac{4\Delta\zeta_0\cot[\pi/(2\zeta_0)]}{\pi(i^2\zeta_0^2 - 1)}. \tag{3.5}$$

As a result, free vibration can be represented by the following series:

$$x = \sum_{i=1}^{\infty}\frac{4\Delta\zeta_0\cot[\pi/(2\zeta_0)]}{\pi(i^2\zeta_0^2 - 1)}\cos i\omega_0 t \quad (i = 1, 3, \ldots). \tag{3.6}$$

Using (3.4) we can see that the series converges rapidly, hence we can look at the expression for the amplitude of the fundamental

$$a_{x1} = \frac{4\Delta\zeta_0\cot[\pi/2\zeta_0)]}{\pi(\zeta_0^2 - 1)} \quad (a_{x1} \ge 0). \tag{3.7}$$

This relationship is shown dotted in Fig. 3.4. As in the case of a system with one stop, the backbone curve for the fundamental is very similar to the full characteristics of the system. In particular, as ζ_0 increases, the value of a_{x1} rapidly approaches to the constant value $8/\pi^2$.

2. Let us turn to the examination of forced vibration of systems due to a periodic force $P\cos\omega t$ taking identical coefficients of restitution for both stops and describing the continuous energy dissipation of the system by means of a viscous damper whose action is proportional to the velocity of a reduced mass, through b (Fig.3.1).

In the interval between collisions, the motion of the system can be described by the differential equation

$$\ddot{x} + b\dot{x} + \Omega^2 x = a_p \cos(\omega t + \varphi_0)\,, \tag{3.8}$$

where φ_0 is the unknown initial phase of excitation, $b = n/M$, $a_p = P/M$, we find as in the case of the single-sided stop, the periodic solution

$$x = a\cos\omega t \tag{3.9}$$

Here

$$a = \frac{a_p}{\sqrt{(\Omega^2-\omega^2)^2 + b^2\omega^2}} = \frac{a_p}{\Omega^2\sqrt{(1-\zeta^2)^2 + 4r^2\zeta^2}} \quad (r = b/(2\Omega))\,, \tag{3.10}$$

$$\sin\varphi_0 = \frac{2r\zeta}{\sqrt{(1-\zeta^2)^2 + 4r^2\zeta^2}}, \quad \cos\varphi_0 = \frac{1-\zeta^2}{\sqrt{(1-\zeta^2)^2 + 4r^2\zeta^2}}.$$

This solution corresponds to the motion of the system without impact occurring naturally when $a < \Delta$.

To find the vibro-impact behaviour, we can write the general solution for equation (3.8) as

$$x = \exp(-r\Omega t)(C_1\cos\Omega_1 t + C_2\sin\Omega_1 t) + a\cos(\omega t + \varphi)\,. \tag{3.11}$$

Here $\Omega_1 = \Omega\sqrt{1-r^2}$, C_1, C_2 are arbitrary constants: φ is the phase of the force and depends on the choice of starting condition. The general solution written in the form (3.11) assumes that $r < 1$. The basic effects of vibro-impact behaviour, which we shall study for the remainder of this book, become more obvious as the result of small values of the viscous damping coefficient. This allows us to simplify the representation by taking $\Omega_1 = \Omega$.

We will study the periodic regimes characterised by alternate impacts of the mass against the stops. Because of symmetry of the system, the period of motion consists of the sum of two identical intervals, each of which covers the time intervals between two consecutive collisions. Consequently, to get a full picture of the motion, it is sufficient to find the law of motion for one of these intervals. For analysis we will look at the interval beginning after the impact on the right hand stop (Fig. 3.1). Then the boundary condition of the regimes for various ratios l (l = 1, 3, ...) takes the form

$$t = 0, \qquad x = \Delta, \quad \dot{x} = -R\dot{x}_-;$$
$$t = \pi l / \omega, x = -\Delta, \quad \dot{x} = -\dot{x}_-. \tag{3.12}$$

Substituting boundary conditions (3.12) into (3.11) we can express the arbitrary constants and the phase, in terms of system parameters. As a result the motion in the regimes examined will be as follows:

$$x = \exp(-r\Omega t)\left(\cos\Omega t - \frac{\exp r\vartheta + \cos\vartheta}{\sin\vartheta}\sin\Omega t\right)\frac{DR}{\Omega}\dot{x}_- + \tag{3.13}$$
$$+ a\cos(\omega t + \varphi) \qquad (t \in [0, \pi l / \omega],)$$

where, similar to §2

$$\sin\varphi = \frac{BD}{a\Omega}\dot{x}_-, \quad \cos\varphi = \left(1 - \frac{D}{\Omega\Delta}\dot{x}_-\right)\frac{\Delta}{a} \tag{3.14}$$

$$\dot{x}_- = \frac{[1 \pm \sqrt{1 - (1 - a^2/\Delta^2)(1 + B^2)}]\Omega\Delta}{D(1 + B^2)} \tag{3.15}$$

Here we make use of the following non-dimensional terms:

$$B = \left(\frac{1}{D} - \frac{\exp r\vartheta + \cos\vartheta + r\sin\vartheta}{\sin\vartheta}\right)\frac{1}{\zeta},$$
$$D = \frac{(1 + R)\sin\vartheta}{\exp r\vartheta + 2\cos\vartheta + \exp(-r\vartheta)}, \quad \vartheta = \frac{\pi l}{\zeta}. \tag{3.16}$$

3. As long as the amplitude of the vibro-impact systems studied is constant, $a_x = \Delta$, the structure of the amplitude-frequency characteristics is defined by existence domains for regimes in the ζ, Δ plane. We shall study the boundaries of the areas of existence. To begin with, we shall isolate the areas of validity for the value of $\dot{x}_-$. Assuming that the term under the radical in (3.15) is non-negative, we obtain similarly to (2.22)

$$\Delta / a \le \sqrt{1 + 1/B^2}\,. \tag{3.17}$$

As we found already in §2 areas of validity are wider than existence domains since the structure of the model is based on the fact that in between collisions the mass must be situated between the stops, not exceeding the level $|x| = \Delta$ (i.e., with no additional intersections) which was not reflected in the previous analysis and consequently in (3.17). Meanwhile, this limitation is not satisfied with the given model for all values of the system parameters, hence later we shall study at length the effect of this additional limit on the results of the analysis, but now we shall examine some basic dynamic features, resulting from inequality (3.17).

Since in the basic case $B \neq 0$, then, similarly to the situation described in §2, vibro-impact regimes of systems can exist not only when the inter-stop distance is less than the amplitude of the linear oscillator $(\Delta < a)$ but also when the gap is expanded to a value higher than the linear vibration of the system $(\Delta > a)$ up to the value identified with the equals sign in (3.17). Naturally, when $\Delta > a$ the same tuning results in harmonic vibration (3.9) of the linear system, hence development into a vibro-impact regime can occur as the result of an additional impulse.

Within the domains of real values, we shall separate existence domains for this study, i.e., by combining the system parameters during which the law of motion in these regimes satisfies the conditions for the absence of additional intersections

$$|x(t)| \leq \Delta . \tag{3.18}$$

In the general case, it is not possible to find the analytic bounds of the areas of existence from the general condition (3.18) for the given model, since the law of motion is written as the sum of components of harmonic functions with non-multiple frequency. Instead, as was shown in §2, for systems with reasonably simple structures, analytic methods can expose the approximate boundaries of the existence domains, and we will use these here.

According to (3.12) for the model examined

$$\dot{x}_{-} \geq 0 . \tag{3.19}$$

Turning to expression (3.15) we find that when $\Delta < a$ one, and only one of the two values of satisfies condition (3.19) $\dot{x}_{-}$. Hence, for this model due to the arrangement of stops, within the limits of vibration of the linear system only one regime can exist for each ratio studied. Noting that the term D (see(3.16)) depends on the sign of $\sin\vartheta$ we obtain that when

$$\zeta \geq l \tag{3.20}$$

these regimes correspond to the positive sign, and when $\zeta < l$ on the minus sign.

We will now look at how condition (3.19) influences the areas of existence of vibro-impact regimes for which $\Delta \geq a$. According to (3.15) in this case the expression below the radical is less than or equal to unity and consequently both values of $\dot{x}_{-}$ take the sign of the coefficient in the denominator, in other words the sign of the value of $\sin\vartheta$. With the requirement that $\sin\vartheta \geq 0$, we again come to the condition (3.20).

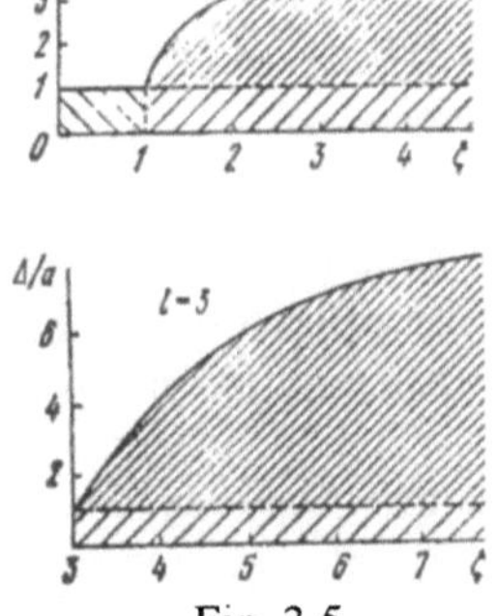

Fig. 3.5

Consequently, in the frequency range (3.20) for every point in the area of real values, satisfying the condition $\Delta \geq a$ two periodic vibro-impact regimes can exist with ratios from 1 to l inclusive. Note that when $\Delta = a$, one of the values of $\dot{x}_{-}$ is equal to zero independently of the other system parameters. This represents linear vibration of the oscillator just touching the stops but without impact.

From the backbone curves for the system (Fig. 3.4) we can convince ourselves that inequality (3.20) represents the condition when the excitation frequency coincides with one of the natural frequencies of the vibro-impact system, and this correspondence can be interpreted as resonance. Later, such a treatment will receive a fuller grounding and will be developed effectively.

Fig. 3.5 shows examples of cross-sections of the parameter space with the existence domains hatched. The sections are constructed using (3.17) for constant values of R, r, and using $l = 1$ and $l = 3$. Here the area hatched with positive slope lines denotes the existence domains with the positive sign in (3.15), with opposite hatching for the negative sign. The closer spaced hatching when $\Delta > a$ shows the existence of coupled regimes.

4. Interest is naturally aroused as to the extent to which the qualitative analysis obtained is preserved when constructing additional boundaries for existence domains on the basis of (3.18). We will study these boundaries for the case $R = 1$, $r = 0$.

Based on (3.13) to (3.15), the laws of motion will be written for this case in the form

$$x = -\frac{\Delta \pm a}{\sin[\pi l/(2\zeta)]}\sin\left(\Omega t - \frac{\pi l}{2\zeta}\right) + a\cos(\omega t + \varphi) \qquad (t \in [0, \pi l/\omega]) \tag{3.21}$$

in which $\varphi = \pi$ with the positive sign and $\varphi = 0$ with the negative sign. Hence, the movement of the system in the intervals between impacts is represented by the sum of portions of two harmonic functions, which are anti-symmetrical values relative to $t = \pi l/(2\omega)$.

Initially, let us look at motion when $\zeta < l$. In the range of t examined , both harmonic functions reach extremes. In general, these extremes are shifted in time, however, we will assume for convenience in the situation of absence of additional intersections, the case when both extremes occur simultaneously and with opposite phase. At this instant

$$|x| = \left|\frac{\Delta \pm a}{\sin[\pi l/(2\zeta)]} - a\right|. \tag{3.22}$$

Earlier, it was shown that in the frequency range examined only regimes with the negative sign can exist satisfying the condition $\Delta < a$. Then, from (3.22) we find $|x| > \Delta$. Hence, when $\zeta < l$ regimes with alternate impacts on the stops do not occur.

At the same time, when $\Delta < a$, the stops present an obstacle to the development of harmonic vibration, which leads to the development of collisions, meaning that the system develops vibro-impact regimes of a different type. From Fig.2.9, *a*, it can be seen that these can be regimes with excitation frequency and characterised by impacts on one stop only.

We shall now study the effects of equation (3.18) on the frequency domains satisfying (3.20). Initially we note that, studying the stability of these regimes as carried out in the previous section, for the given model the requirements of stability are also satisfied by regimes with the plus sign before the radical in (3.15) and properly in (3.21), hence, we shall now only study such regimes. This is a typical feature of nonlinear resonance, which we will regularly come across.

When $\zeta < l$ in the examined time interval t, the first harmonic term of (3.21) decreases monotonically (Fig. 3.6) and the second term depends on l. When $l = 1$, this term grows monotonically which can mean violation of the condition $x \geq -\Delta$. Making the situation worse, let us replace the harmonic terms *1, 2 by* the linear terms *3, 4*. Obviously, if the conditions for the absence of additional intersections are satisfied for the linear approximation when $t = 0$, then they are known to be satisfied for the harmonic terms for the arbitrary value t considered. Consequently, we can put $\Delta + a - \pi a/2 \geq -\Delta$, whence

$$\Delta / a \geq (\pi - 2)/4 \approx 0.29 . \tag{3.23}$$

Hence, even for such artificial narrowing, the existence domains for basic vibro-impact regimes found ($l = 1$) cover almost all pre-resonance areas of the parameter range. Note that when condition (3.23) is not satisfied and the stops are arranged with relatively small gap, the character of motion is preserved; however, the methods used here are of little use for general estimation, since fluctuating effects occur according to the value of the basic displacement. For this reason, inequality (3.23), which guarantees the existence of the regimes examined, will be taken conditionally as the boundary of an existence regime.

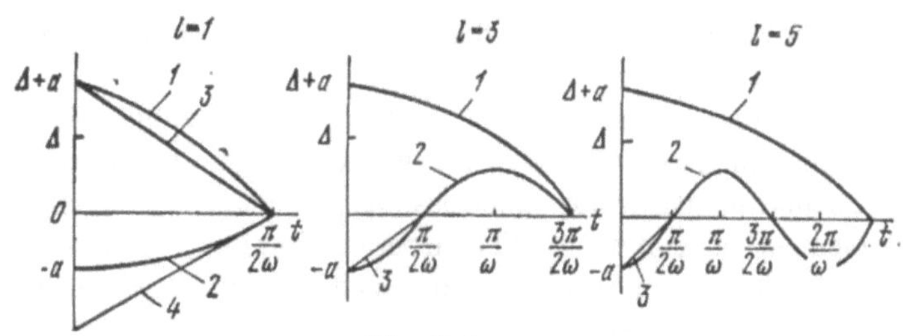

Fig. 3.6

When l = 3, 5, ... the first harmonic component does not change, while the second oscillates, and achieves a positive value of amplitude (Fig. 3.6). As a result, when $\Delta < a$, condition (3.18) is violated and for this reason sub-periodic regimes do not occur here. When $\Delta \geq a$ in the case when condition (3.18) is violated, this happens at the boundary $x = \Delta$. We can limit our study to the segment $t \in [0, \pi/\omega]$. As long as condition (3.18) is satisfied when t = 0, it is sufficient that in the examined segment the second term of the harmonic does not grow so fast that it extinguishes the first. Such a requirement can be majorised by two conditions, the modulus of the derivative of the first harmonic term when $t = 0$, must not be less than the modulus of the slope of line *3* (this guarantees condition (3.19) in the portion $t \in [0, \pi/\omega]$), but when $t = \pi/(2\omega)$ it must not be less than the derivative of the second term (this guarantees (3.18) when $t \in [\pi/(2\omega), \pi/\omega]$). The first condition gives

$$\frac{\Delta + a}{\tan[\pi l/(2\zeta)]} \Omega \geq \frac{2a\omega}{\pi},$$

or after conversion

$$\frac{\Delta}{a} \geq \frac{2\zeta}{\pi} \tan \frac{\pi l}{2\zeta} - 1 . \tag{3.24}$$

We will now examine the second condition at the instant $t = \pi/(2\omega)$. It takes the form

$$\frac{\Delta + a}{\sin[\pi l/(2\zeta)]} \Omega \cos \frac{\pi(l-1)}{2\zeta} \geq a\omega .$$

Assuming $\zeta >> l$ we obtain, putting $\cos(0) = 1$

$$\frac{\Delta}{a} \geq \zeta \sin\frac{\pi l}{2\zeta} - 1. \tag{3.25}$$

Using (3.20) conditions (3.24), (3.25) can be joined together in the majorising relationship:

$$\frac{\Delta}{a} \geq \zeta \tan\frac{\pi l}{2\zeta} - 1, \tag{3.26}$$

which we will take as a boundary for the existence of the sub-periodic vibro-impact regime. Beginning with some value $\zeta > l$, condition (3.26) is always satisfied, while the boundaries corresponding to various values of $l > 1$ are arranged one above the other, higher than the level $\Delta / a = 1$, and do not intersect each other..

5. Study of the existence domains on the basis of condition (3.18) allowed us to refine the boundaries and to detail the structure of the system parameters extensively, hence, no matter how laborious this procedure such investigation is an important element of analysis to find accurate solutions. From this springs a question, in what measure are the results obtained when $R = 1$, $r = 0$, pertinent to a real system that includes dissipative effects?

First of all we notice that if inequality (3.17) defining the dissipative effects of the system gave us the upper limit for the area of existence for frequency of excitation ζ and amplitude Δ, then condition (3.18) or its more particular form (3.19) lead to the appearance of the lower boundaries (3.20), (3.23), (3.26). This reflects various physical characteristics of the limits specified.

Inequality (3.17) defines the energy requirements of a system to cause vibration. Hence, any solutions satisfying the differential equation (3.8) and boundary conditions (3.13) and also with additional intersections satisfy (3.17) because they bring about the conditions of energy balance. Therefore, it is natural that increasing dissipation of energy in the system affects only the upper limits by the swing and rapidity of vibration. Y I Neimark showed that for the general system this boundary always coincides with one of the boundaries of stability (corresponding to the positive unit root when studying the roots of the characteristic equation inside a unit circle) [148]. In future we shall refer to this as the *energy boundary of vibration*.

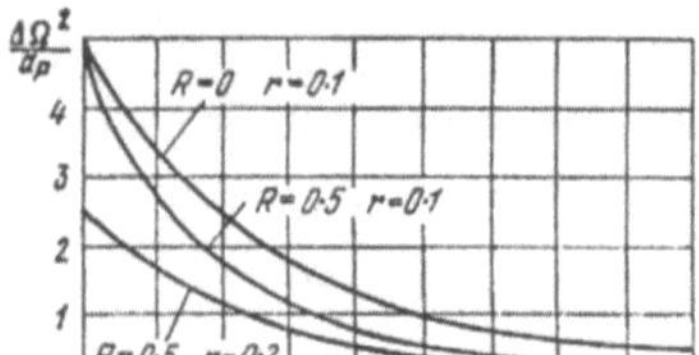

Fig. 3.7

Conversely, condition (3.18) produces the structural limitations on the defined vibration trajectory (with additional intersections) and does not, in principle, prevent an increase in the amplitude and frequency of vibration. The boundaries (3.20), (3.23), (3.26) obtained on the basis of (3.18) represent existence domains of vibro-impact regimes, and will be referred to as the *structural boundaries of vibration.*

To estimate the influence of dissipative factors on the structural boundaries we shall examine in more detail the specific law of motion of the system. Comparing (3.21) with (3.2) we see that the first harmonic term in (3.21) is similar to the free vibration of a conservative system with frequency $\zeta_0 = \zeta / l$, investigated in section 1, differing only in the value a. As ζ increases from unity, then, according to (3.10), the value of a rapidly decreases and tends to zero, hence the basic motion of the system (3.21) becomes more and more like a pure conservative system, but the law of motion becomes more and more saw-toothed. Analysing (3.13) it can be shown that there is a similar tendency also when taking account of the dissipation of energy in the system due to the compensating energy loss of the external excitation. This confirms the resonant character of the vibro-impact regimes studied in section 3. As a result, accounting for small dissipation during vibratory motion, cannot change the qualitative character of conditions (3.20), (3.23), (3.26), concerning high frequency motion. The specific result of (3.22) shows that similar conclusions can be obtained with regard to low frequency motion. Calculated results gives good confirmation of these conclusions.

Hence, dissipative effects of systems, in practice, influence only the energy boundaries of vibration (3.17). We will study this in more detail. Fig. 3.7 shows examples of these boundaries constructed for $l = 1$, for various values of the coefficients of viscous and impact dissipation of energy. It can be seen that both mechanisms for energy dissipation give numerically identical configurations to the energy boundary of vibration, decreasing monotonically as excitation frequency increases. This result allows us to make an equivalent substitution of one mechanism of dissipation for the other, a process we will practice widely when presenting approximate methods.

In addition, it is useful to notice that the influences of viscous and impact dissipation have several special features. In particular, as ζ / l increases towards unity, according to (3.16), $B \to \infty$, and consequently condition (3.17) gives

$$\Delta / a \leq 1 \tag{3.27}$$

which is independent of R. This follows from the fact that the regimes on the boundaries are impact free (see section 3). When we neglect viscous friction $(r = 0)$, we find the limit of B equal to zero, and from condition (3.17) notwithstanding (3.27) we obtain $\Delta / a < \infty$. Consequently, close to this boundary for the frequency range for the existence of vibro-impact regimes, accounting for viscous damping remains important but impact dissipation can be neglected. However, study of effects in this area is of little interest, because of the sluggish impact regimes in these areas.

For the more important high frequency area, tending to the limit when $\zeta \to \infty$ in (3.16), we find from (3.17)

$$\lim_{\zeta \to \infty} \frac{\Delta}{a} \leq \sqrt{1 + \frac{\pi^2 l^2 (1+R)^2}{4(1-R)^2}} \tag{3.28}$$

The right hand side is, of course, finite only when $R \neq 1$ and does not depend on r, i.e., impact dissipation at high frequency appears stronger than viscous dissipation.

For this reason, as we search for the exact solutions of resonant motion, we will concentrate mainly on impact dissipation of energy.

Expanding (3.16) as a series in terms of the small parameter r and limiting the expansion to linear terms we obtain directly the equation of condition (3.17) in terms of basic system parameters in the following form:

$$\frac{\Delta}{a} \le \sqrt{1+\left(\frac{\zeta \sin(\pi l/\zeta)}{(1-R)[1+\cos(\pi l/\zeta)]/(1+R)+\varepsilon}\right)^2}, \tag{3.29}$$

where

$$\varepsilon = r\left(\frac{\pi l}{\zeta} + \sin\frac{\pi l}{\zeta}\right) \Big/ \left(1+\cos\frac{\pi l}{2\zeta}\right). \tag{3.30}$$

From (3.29) it can be seen that as the energy dissipation in the system increases, so the energy boundary of vibration reduces steadily. However, even for the limiting value of impact dissipation corresponding to $R=0$ this boundary exceeds the amplitude of non-impact vibration a for all values of the ratio l for all frequency ranges in existing regimes.

Comparing (3.29) with (3.23), (3.26) and using the fact that

$$\tan\frac{\pi l}{2\zeta} = \sin\frac{\pi l}{\zeta} \Big/ \left(1+\cos\frac{\pi l}{\zeta}\right), \tag{3.31}$$

we can establish that for domains of any ratio, beginning with some value $\zeta > l$, the energy boundary must be above the structural boundary which leads to the existence of a regime.

In section 4, we have already mentioned that the lower boundaries for existence domains of sub-periodic regimes for various ratios (see (3.26)) are disposed above one another with increase of l, and do not intersect one another for all frequency ranges (3.20) while in the whole of this range $\Delta/a > 1$ for any $l > 1$. This implies that it is not feasible to transfer directly from harmonic vibration to the sub-periodic regimes, whatever the frequency ratio, and requires the use of additional energy for the starting process (hard excitation). Soft transfer from harmonic vibration to vibro-impact can occur only for the basic regime (l = 1) according to (3.23), because their structural boundaries are very close to zero. However, in agreement with (3.29), as R increases the existence domains of vibro-impact regimes for various ratios overlap and consequently, due to the small changes in the system parameters the regimes can transfer continuously from one

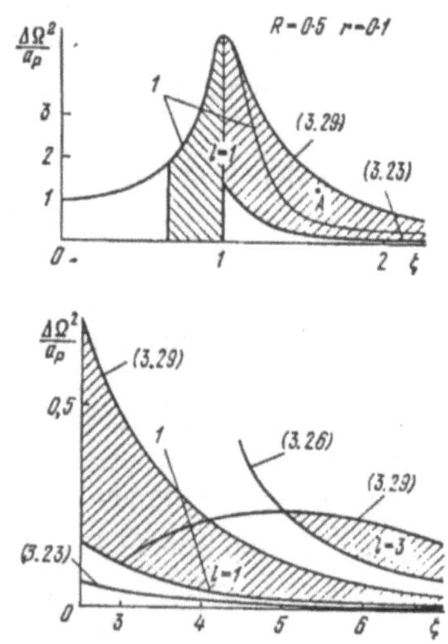

Fig. 3.8

to another. In areas of overlap, establishment of one or other regime depends on the character of the preceding transfer process.

Hence, an investigation of the influence of dissipation shows that it considerably decreases the size of existence domains for vibro-impact regimes, but it cannot completely eliminate them. This explains why it is so difficult to combat collisions in machines, which occurs in systems of vibration isolation, bearing assemblies and various other types of mechanical structures. In such cases, when it is occurs in actual assemblies, then according to (3.30), an effective measure to eliminate collisions can be to increase the gap Δ beyond the limits of the existence domains.

Fig. 3.8 shows examples of existence domains for regimes on the basis of equations (3.20), (3.23), (3.26) and (3.29). It is convenient to show them with the help of the basic amplitude-frequency characteristic corresponding to linear system *1*. These areas are shown hatched with positive slope, while those with negative slope hatching are the existence domains for vibro-impact regimes having the same frequency as the excitation and characterised by impact on one stop only with amplitude $a_x < \Delta$ (see section 4). The frequency range for existence of this area is plotted from the results obtained in §2. Naturally, depending on the transfer process such a regime remains on one or other of the stops.

6. Transferring to the $\zeta, a_x\Omega^2/a_p$ plane, and drawing a horizontal straight line corresponding with the level Δ of the stops, together with the existence domains for vibro-impact (Fig. 3.9, *a*), we produce the amplitude-frequency characteristics of the system. In Fig. 3.9, *a*, this is indicated with a solid line corresponding to the basic regime. On the horizontal portion *1* a symmetrical vibro-impact regime exists having alternate collisions of the mass against each

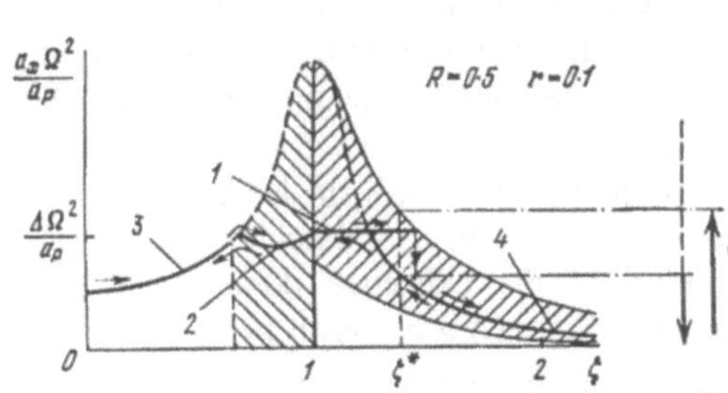

Fig. 3.9

stop. In the pre-resonance frequency zone ($\zeta < 1$), branch *2* of the frequency-amplitude characteristics is constructed from the results of §2, for an actual system of vibro-impact regimes with frequency equal to that of the excitation and colliding against one stop. Moreover, the added branches *3*, *4*, representing the amplitude characteristics for harmonic vibration of a linear oscillator, lie below the level of the stops.

From Fig. 3.9, *a*, it can be seen that as the exciting frequency is gradually increased (see arrows) from the zero regime of the system, passing through the frequency zone corresponding to harmonic vibration, then after the first collisions, the regime changes initially to one with single- sided impact. Subsequently, when it passes through the resonant frequency of a linear oscillator, a symmetrical vibro-impact regime is established, with alternate collisions on each stop, which can be prolonged in high frequency domains until the system reaches the existence boundary. When the system frequency intersects the boundary values, the vibrator "breaks-away" from the stops and once again harmonic vibration becomes established for all subsequent frequency ranges.

A different process occurs as the frequency is decreased from a high value Here the vibro-impact regime is established at a frequency lower than the breakaway frequency and then the system develops in the opposite direction to its amplitude-frequency characteristics. Note that in this case vibro-impact regimes can exist only within the limits of the frequency range in which the stops obstruct harmonic vibration of a linear oscillator.

Thus, during run-up to the high frequency zone, basic vibro-impact regimes occur more widely than during run-down, i.e., as was to be expected, a similar pulling occurs as for systems of the nonlinear hard type. At the same time, in contrast to systems with smooth nonlinearity, in vibro-impact systems, linear and nonlinear vibration is clearly separated. Transfer from regime to regime in a pulling field can only happen due to special additional excitation.

As well as the description of a pulling effect in regimes *due to frequency*, in vibro-impact systems it is also possible to obtain pulling effects *due to amplitude*. In the system studied this can occur for every excitation frequency which exceeds the natural frequency of the linear oscillator (e.g., ζ^* in Fig. 3.9, *a*) by means of the introduction of stops in the harmonic vibration zone of the oscillator (Fig. 3.9, *b*) before a vibro-impact regime becomes established and then slowly expanding the gap. As a result the vibro-impact regime is maintained up to the value of the amplitude at breakaway, when the stops reach the level defining the boundaries for the existence domains. After breakaway in the system, the original harmonic vibration is re-established.

Specifically, the structure of the system leads to the fact that its amplitude-frequency characteristics (Fig. 3.9, *a*), coincide with the backbone curves (Fig. 3.4), within the limits of the existence domains of symmetrical regimes . Moreover, in zones of pulling $(\Delta > a)$, each point of the characteristic exists in two regimes with different collision velocities, in agreement with the results of section 3. We have already mentioned that, as is usual for nonlinear systems in double-valued areas, the amplitude-frequency characteristics corresponding to lesser velocity are unstable. In §14, we will show that this condition is typical for a large number of basic systems.

Let us consider the behaviour of the fundamental for systems in the symmetrical regimes studied. For simplification, we will ignore energy dissipation in the system. Expanding (3.21) as a Fourier series, we obtain, similarly to (3.5), (3.6)

$$x = \left[\frac{4\zeta(\Delta \pm a)\cot[\pi l/(2\zeta)]}{\pi l(\zeta^2 - 1)} \mp a\right]\cos\omega t + \\ + \sum_{\substack{i=1 \\ i \neq l}}^{\infty} \frac{4\zeta(\Delta \pm a)\cot[\pi l/(2\zeta)]}{\pi l[(i\zeta/l)^2 - 1]} \cos\frac{i\omega}{l} t \qquad (i = 1, 3, \ldots). \tag{3.32}$$

When $\zeta > l$ the value a tends to zero, as does $1/\zeta^2$, and the series (3.32)

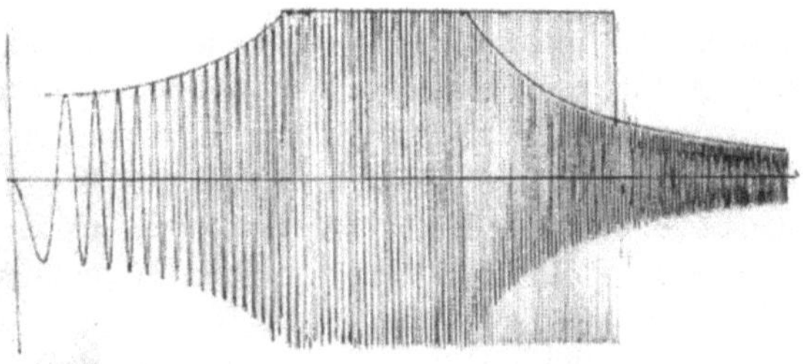

Fig. 3.10

converges as rapidly as (3.6) in practice. Consequently, independent of the existence of the nonlinearity of the system when the periodic regime occurs it causes a filtering effect on the fundamental $(i = 1)$.

The amplitude of the first harmonic when $l = 1$ is found from (3.32) after rearrangement as

$$a_{x1} = \frac{4\Delta\zeta\cot[\pi/(2\zeta)]}{\pi(\zeta^2 - 1)} \pm a\left[\frac{4\zeta\cot[\pi/(2\zeta)]}{\pi(\zeta^2 - 1)} - 1\right]. \tag{3.33}$$

Comparing (3.33) with (3.7) we see that the two branches of the amplitude-frequency characteristics for the fundamental are located on either side close to the backbone curve, and rapidly converge to it as ζ is increased (see (3.10)). Since existence domains for the fundamental coincide naturally with the calculated values for the existence domain already determined, then the study of the first harmonic gives all the required qualitative information relevant to the process studied. In the next chapters we shall try to obtain this same information using approximate methods.

Let us find the equation for the gain coefficient of the fundamental due to harmonic excitation. Denoting this by *k*, we find from (3.32) using (3.10)

$$k = \frac{a_{xl}}{a_p} = \frac{4\Delta\zeta\cot[\pi l/(2\zeta)]}{\pi l a_p(\zeta^2 - l^2)} \pm \frac{1}{\omega^2 - \Omega^2}\left[\frac{4\zeta\cot[\pi l/(2\zeta)]}{\pi l(\zeta^2 - l^2)} - 1\right] \tag{3.34}$$

As is usual in nonlinear systems, this coefficient depends on both amplitude and frequency of vibration. From (3.34) it follows that in the frequency range of existence for the solution $\zeta > 1$, the gain coefficient can be reasonably well approximated by its first term. Thus, when studying vibro-impact processes of the type discussed, the value of the gain coefficient due to harmonic excitation in the high frequency range can be obtained with sufficient accuracy by analysing free vibration of the system. This will be explained in detail in §9, during the presentation of suitable approximate procedures.

To examine the behaviour of solutions during transient processes, a study was made of a simulation of the system under consideration [43]. Fig. 3.10 shows vibration traces from the model as the frequency of excitation of the system was gradually increased and, superposed on it, the vibration during subsequent decreasing frequency. The pulling in the high frequency areas can be seen clearly as well as the single sided character of collisions in pre-resonant zones. Calculation of the collision velocity confirmed that when $\omega > \Omega$, vibration occurs in all regions corresponding to the positive sign in (3.15).

To study the transition from the harmonic regime to vibro-impact, the domains of attraction were constructed on the phase plane inside the pulling zone. Fig. 3.11 shows such a construction for point *A*, of the plane of parameters, (see Fig. 3.8). In this case, harmonic vibration and basic vibro-impact regimes exist, the limiting cycle of each being marked in Fig.3.11 by the numbers *1* and *2*. The procedure for constructing the zones of attraction was to establish the limiting cycle of a transition process during start-up of the system from various starting conditions corresponding to points on the scaled grid applied to the phase plane. The initial phase of excitation during all start-ups was identical (zero in the given example). From Fig. 3.11 it can be seen that, when the starting impulse exceeds certain

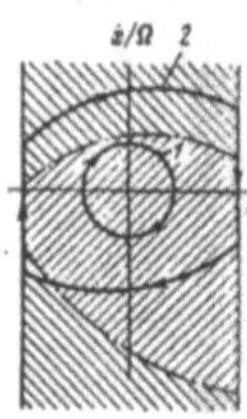

Fig. 3.11

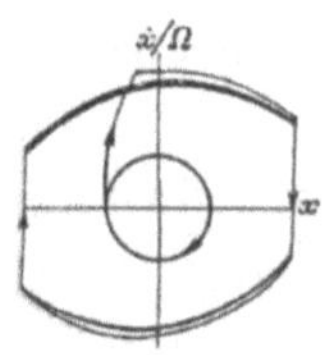

Fig. 3.12

Fig. 3.13

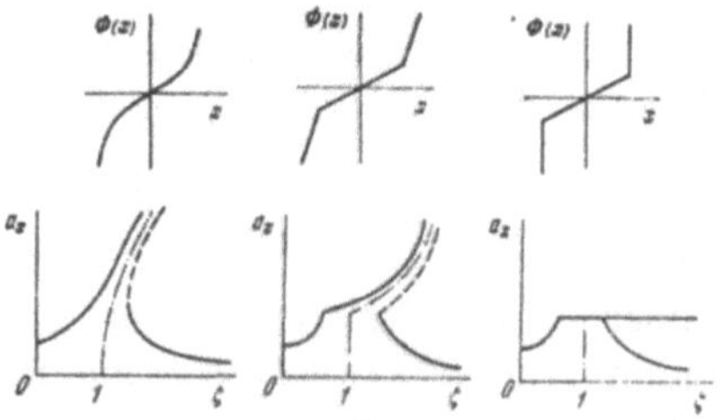

a) b) c)

Fig. 3.14

threshold values, it always lead to the establishment of a vibro-impact regime. Fig. 3.12 shows an example of the phase trajectory for such a transition.

Fig. 3.13 shows examples recorded for a model of various actual regimes.

These all correspond to identical frequency of excitation. Here *1* represents harmonic vibration, *2* represents the basic vibro-impact regime, *3* represents the sub-periodic regime with frequency ratio 3. The transition of systems into vibro-impact regimes occurs as the result of a starting impulse.

Summarising the analysis carried out, it can be said that the dynamic model studied is the limiting case of a hardening type of nonlinear system not only in terms of structure but also in function. For comparison, Fig. 3.14 shows various static elastic characteristics of the hardening type, and the amplitude-frequency characteristics resulting from them. In this figure, unstable branches are shown dotted, i.e., as already mentioned, , the resonance with the highest velocity of the two possible resonances is the one that becomes established. Similar results were obtained in the systems studied. The cause of such similarity can be understood if one takes into account the change from elastic characteristics of Fig. 14, *b* to the characteristics in Fig. 3.14, *c* corresponds to the smooth increase in the stiffness of the edge portions . As the analysis shows, the sloping portions of the backbone curves gradually lengthen, the steepening portions move away in the direction of higher frequency, in the limit tending to infinity, but double-valued branches of the resonance curves approach more closely to the backbone line. It is significant that the appearance of nonlinearity beyond the limits of small effects does not lead to a qualitative transformation of the system behaviour. Hence, no matter what expressions are specified, the basic character of nonlinear behaviour for vibro-impact systems, obtained using exact methods, are similar to "basic" nonlinear hardening systems, showing the success possible with approximate methods. This analogy leads to the conclusion about the usefulness of applying approximate methods for the analysis of vibro-impact systems. In §9, the model studied will be investigated using a harmonic linearisation technique and the results obtained will be compared with the results of exact analysis.

7. The effects of pulling of vibro-impact regimes described here were studied in an experimental device designed for the high frequency system of photoelectric scanning [26]. Fig. 3.15 shows the experimental arrangement. The beam *1*, fixed at *4*, passes through the gap between two electromagnets *2*, which is supplied with alternating current from a generator. On the end of the beam a fixed ball *3* is installed in the gap between stops *5*. The stops were designed as a slot narrowing in the direction perpendicular to the picture plane. The beam and the excitation system were assembled on a single body, arranged so that they could be moved along the slot through an angle formed by the stops to change the size of the gap. The vibration of the ball was registered by a wire strain gauge whose signal was recorded using an oscillograph.

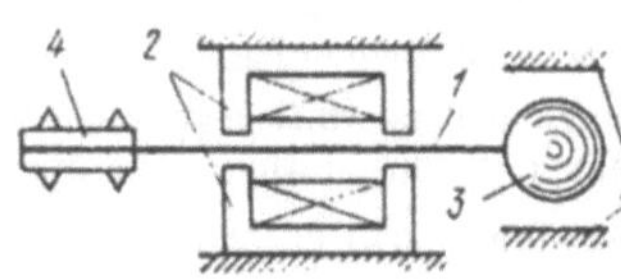

Fig. 3.15

Fig. 3.16 shows several typical oscillograms obtained from the vibrating system with a natural linear frequency $\Omega/2\pi \approx 750\,\text{Hz}$. On the oscillograms the central line shows the signal from the strain gauge, which is proportional to the displacement of the ball from central symmetry, and the lower curve shows a timebase with a frequency of 500 Hz.

The two antinodes in Fig. 3.16, *a* represent the uniform increase of the exciting frequency and the subsequent uniform decrease. An example of the frequency scale of the excitation is given above the oscillogram. The frequency curve was recorded with a stationary gap $\Delta = 5 \times 10^{-4}$ m . Various lengths of antinode demonstrate various pulling of the vibro-impact regime as the frequency of excitation was increased. The beat at the end of the oscillogram occurs after the breakdown of the vibro-impact regime into a mixture of damped free vibration and forced vibration at the fundamental. At the ends of the oscillogram, vibration without collision is recorded.

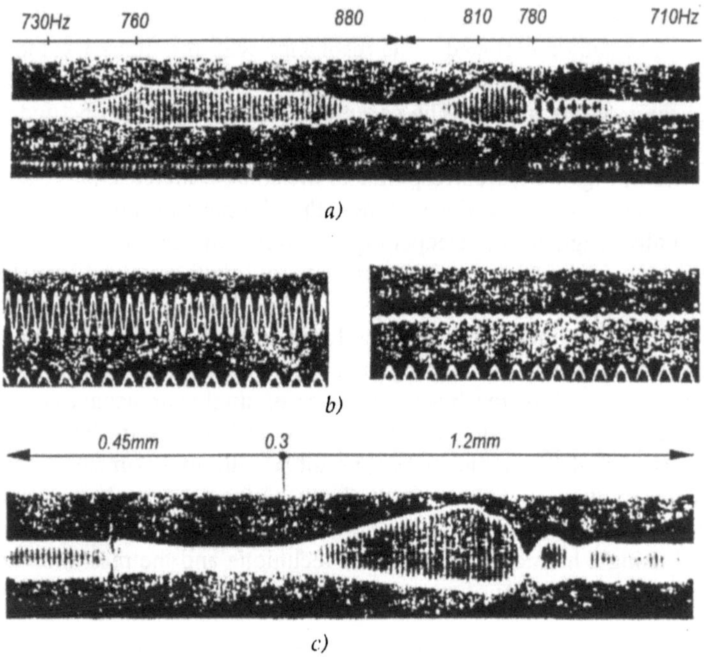

Fig. 3.16

Fig. 3.16 *b* shows the results near the breakdown point at a frequency $\omega/(2\pi) = 880\text{Hz}$ in the vibro-impact and harmonic regimes, into which the system changes as a result of breakdown. Comparison with the oscillogram in Fig. 3.16 *b* shows that the breakdown from the vibro-impact regime to harmonic represents a fivefold decrease in amplitude.

The oscillogram in Fig. 3.16 *c* was recorded at a fixed frequency $\omega/(2\pi) = 880\text{Hz}$ and represents amplitude pulling which happens in the system when the gap between the stops initially starts to decrease to a value at which vibro-impact is triggered. This instant is characterised by a sudden peak in the left-hand part of the oscillogram as the gap is further decreased to $\Delta = 3 \times 10^{-4}$ m and then increased up to $\Delta = 12 \times 10^{-4}$ m . A vibro-impact regime continues to hold in the system. After breakdown of this regime as it crosses the boundaries of the existence domains (Fig. 3.7, *a*) the original form of harmonic vibration reappears. Thus, as a result of a gradual increase in the distance between the stops, the system is pulled to a regime whose amplitude is several times greater than the regime without collisions.

By measuring the frequency of breakdown of vibro-impact regimes for various values of gap it is possible to construct a boundary for their existence domains (Fig. 3.17). Its configuration corresponds to the energy boundary of vibration (3.29) (see Fig. 3.7).

§4. Principles of percussion machine dynamics

1. Percussion machines are widely used in various fields of technology, particularly for processes that include demolition or deformation of workpiece or materials. Their principal advantage is the simple production of large forces arising from the collision of a machine striker with workpiece either directly onto the workpiece or onto a tool which transforms impact energy into a working process.

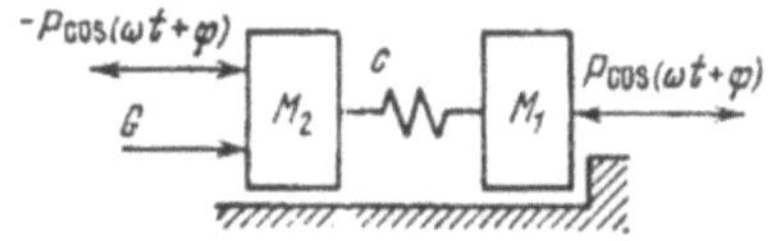

Fig. 4.1

The dynamic system for percussion must include a drive arrangement, the models of the vibro-impact mechanism and the working process. The model of impact interaction, various examples of which were studied in §1, describes the character of the work process. The character of the drive produces a significant influence on the dynamics of the system, interactions between the vibrator and its drive will be studied in §12 (see sections 3 to 7). Here we will examine the basic dynamics of a typical two-mass model of a percussion machine (Fig. 4.1) taking into account the mass of the striker and body with connecting elements isolated from the striker by an elastic spring. For simplicity we will assume that the drive provides periodic excitation of the system independent of its vibration, while the working process (impact) can be described kinematically using stereomechanic theory. This allows us to develop the analysis with exact methods and to evaluate the effect introduced to take account of the mass of the body on the dynamics of the vibro-impact mechanism by comparison to the more highly idealised model examined in §2 (Fig. 2.1), also used to model percussion machines.

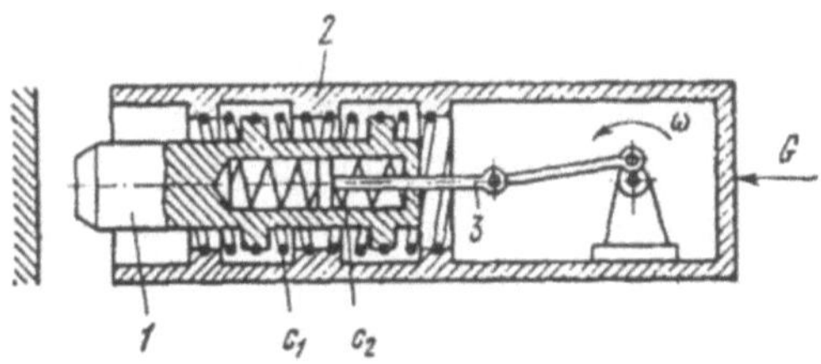

Fig. 4.2

As an example, Fig. 4.2 shows schematically an actual example of a percussion machine. Striker *1* assisted by spring c_1 connected to body 2, while spring c_2 is attached to the piston rod of a simple crankshaft driven mechanism. The whole device is pressed against a stop by static force G (the feed). We shall assume that the driving device for the crank shaft supplies it with a constant angular velocity ω, while the motion of the piston rod *3* relative to the body is simple harmonic motion of the form $x_3 = r\cos(\omega t + \varphi)$, where r is the crank radius. Neglecting dissipative forces we can write the differential equations of motion in the form

$$M_1\ddot{x}_1 = -c_1(x_1 - x_2) - c_2[x_1 - (x_2 + x_3)],$$
$$M_2\ddot{x}_2 = -c_1(x_2 - x_1) - c_2[(x_2 + x_3) - x_1] + G,$$

where x_1, x_2 are coordinates of the striker and body: M_1, M_2, - their masses.

Writing $c = c_1 + c_2$, $P = c_2 r$, we obtain

$$M_1\ddot{x}_1 = c_1(x_2 - x_1) + P\cos(\omega t + \varphi), \qquad (4.1)$$
$$M_2\ddot{x}_2 = c_1(x_1 - x_2) - P\cos(\omega t + \varphi) + G. \qquad (4.2)$$

Thus, performance analysis of this machine leads to the study of the dynamic model shown in Fig. 4.1, in which vibration is excited by harmonic forces acting with opposite phase on both masses. The study of similar models leads to the modelling of a wide range of percussion machines.

If mass M_2 is sufficiently large, then its vibration can be neglected. Then according to (4.2) $\ddot{x}_2 \to 0$ when $M_2 \to \infty$. It would appear that the model described is identical to that of an oscillator with a stop studied earlier in §2. However, this is not so.

The characteristic feature of the model consists of the following. In the static condition under the influence of force G, mass M_2 can take up some equilibrium position and creates a strain $\Delta = -G/c$ in the system. Let us assume that under the influence of an excitation force, the system undergoes vibration with period $T = 2\pi l/\omega$, during which time mass M_1 collides once with the stop. In this case, equilibrium must be satisfied between an impulse of force G for the period and the increment of momentum, i.e.

$$GT = M_1(\dot{x}_{1-} - \dot{x}_{1+}). \qquad (4.3)$$

From this, using the relationship between the velocities $\dot{x}_{1+} = -R\dot{x}_{1-}$, we can immediately find the value of the pre-impact velocity

$$\dot{x}_{1-} = GT/[M_1(1+R)]. \qquad (4.4)$$

From this expression it can be seen that velocity $\dot{x}_{1-}$ does not depend on the force of excitation. Thus, if mass M_2 is large, it takes some position of dynamic equilibrium, differing from the static value m_{x2} and of value

$$\Delta = -(m_{x2} + G/c) \qquad (4.5)$$

similar to the gap (interference) in the system studied in §2. Thus, here the gap does not have a fixed value, but changes according to the vibration regime. This predetermines a series of specific features for the behaviour of systems. We shall examine several of these for the case $M_2 \to \infty$, using the solutions of §2.

During natural vibration of an oscillator impacting on a stop, according to (2.4) the velocity is

$$\dot{x}_{1-} = -\Delta\Omega\tan(\pi\Omega/\omega_0). \qquad (4.6)$$

Assuming in (4.4) that $T = 2\pi/\omega_0$ and $R = 1$, after substitution in (4.6) we obtain the following value for the equivalent gap:

$$\Delta = -\pi\frac{G}{c}\frac{\Omega}{\omega_0}\cot\left(\pi\frac{\Omega}{\omega_0}\right). \qquad (4.7)$$

Substituting (4.7) in equation (2.11) we obtain the equation of the backbone curve

$$a_{x_1} = \frac{G}{2c}\frac{\Omega}{\omega_0}\tan\left(\pi\frac{\Omega}{2\omega_0}\right). \tag{4.8}$$

This is shown as a solid line in Fig. 4.3.

The equivalent gap for forced vibration is found by solving equation (2.21) for the value of Δ:

$$\Delta = \frac{\dot{x}_{1-}D}{\Omega} \pm \frac{1}{2\omega}\sqrt{(2a\omega)^2 - (1-R)^2\dot{x}_{1-}^2}\,. \tag{4.9}$$

Assuming for the sub-radical quantity in (4.9)

$$2a\omega - (1-R)\dot{x}_{1-} \geq 0$$

and using (2.15) and (4.4) when $T = 2\pi l/\omega$ we obtain the condition defining the existence domains for the value of Δ, as

$$\frac{G}{P} \leq \frac{1}{\pi l}\frac{(1+R)}{(1-R)}\frac{\omega^2}{\left|\Omega^2 - \omega^2\right|}. \tag{4.10}$$

Obviously, of all the solutions satisfying (4.10) the system can achieve only those when the gap (interference) $\Delta \geq G/c$. Making (4.9) comply with this condition and limiting the value of R to $R = 1$, we obtain

$$\pm a \geq -G/c - \dot{x}_{1-}D/\Omega. \tag{4.11}$$

The analysis shows that for regimes with frequency ratio $l = 1$, both values of (4.9) satisfy condition (4.11) only in the range $\omega > \Omega$, and consequently

$$G/P > 1/2, \tag{4.12}$$

but only one of the solutions lies in the remaining parts of the area (4.10). Regimes with high ratios always exist in the frequency range

$$\omega > l\Omega. \tag{4.13}$$

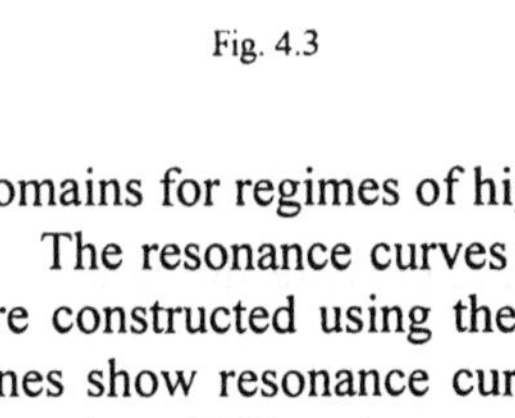

Fig. 4.3

Fig. 4.4 shows existence domains of the solution for frequency ratios l = 1, 2, corresponding to the inequalities (4.10), (4.13). Note that existence domains for regimes of high ratios completely overlap those of lower ratio.

The resonance curves for values of the amplitude of forced vibration, Fig. 4.5, are constructed using the same method used to construct Fig. 2.7. Here the thin lines show resonance curves when $R = 1$, constructed using the backbone curve equation (4.8) using expression (2.35). The portions of the resonance curves shown by thick lines arise when energy dissipation occurs during collisions $(R \neq 1)$. The limiting frequency for the development of vibro-impact behaviour is found from the boundaries of the existence domains (Fig. 4.4). Here, we meet an interesting feature of the systems examined: it appears that neither the increase in dissipation during impact, nor increase in clamping force can eliminate vibration with infinite amplitude at the fundamental resonance frequency $\omega = \Omega$. (Later we will show that the limit arises because of viscous damping.) The vibrations shown can also occur at multiple frequencies $\omega = l\Omega$, however, they can be is suppressed.

In agreement with (4.10), sub-periodic resonances occur when

$$\frac{G}{P} \le \frac{(1+R)}{(1-R)} \frac{1}{\pi(l^2-1)} . \tag{4.14}$$

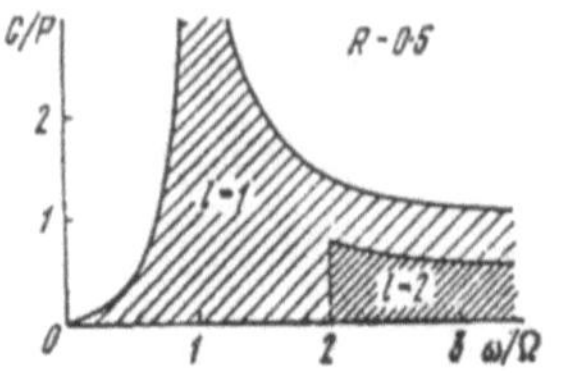

Fig. 4.4

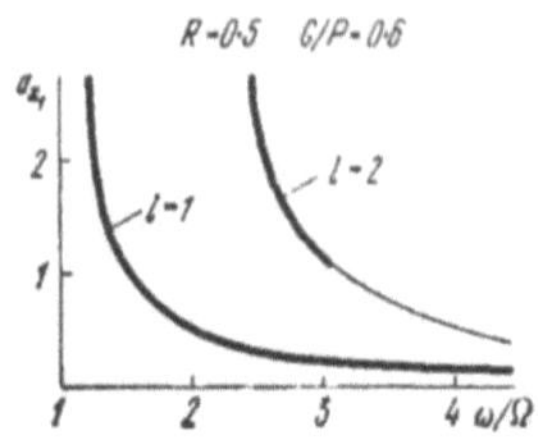

Fig. 4.5

2. Let us now consider the general case of vibration of the system under investigation (4.1) described by the complete set of equations (4.1), (4.2). We shall limit this study to the basic periodic regimes when ratio $l = 1$. Introducing relative coordinates $x = x_1 - x_2$, from (4.1), (4.2) we obtain the equations for relative motion

$$\ddot{x} + \Omega^2 x = a_P \cos(\omega t + \varphi) - \frac{G}{M_2}, \tag{4.15}$$

where $\Omega = \sqrt{cM/(M_1 M_2)}$; $a_P = PM/(M_1M_2)$; $M = M_1 + M_2$.

The general solution of (4.15) takes the form

$$x = C_1 \cos\Omega t + C_2 \sin\Omega t + a\cos(\omega t + \varphi) - G/(M_2\Omega^2), \tag{4.16}$$

where $a = a_P / |\Omega^2 - \omega^2|$.

Substituting (4.16) in the right hand side of equations (4.1), (4.2) we can write the general solution for a system of non-homogeneous linear equations in the following form:

$$x_1 = \frac{M_2}{M}[C_1 \cos\Omega t + C_2 \sin\Omega t + a\cos(\omega t + \varphi)] + C_3 + C_4 t + \frac{G}{2M}t^2$$

$$x_2 = -\frac{M_1}{M}\left[C_1 \cos\Omega t + C_2 \sin\Omega t + a\cos(\omega t + \varphi) - \frac{G}{c}\right] + C_3 + C_4 t + \frac{G}{2M}t^2 \tag{4.17}$$

Taking the time origin at the instant after impact we can write the boundary conditions for the desired periodic motion

$$t = 0, \quad x_1 = 0, \quad \dot{x}_1 = \dot{x}_{1+}, \quad x_2 = x_{20}, \quad \dot{x}_2 = \dot{x}_{20};$$

$$t = 2\pi/\omega, \; x_1 = 0, \quad \dot{x}_1 = \dot{x}_{1-}, \quad x_2 = x_{20}, \quad \dot{x}_2 = \dot{x}_{20},$$

where x_{20}, $\dot{x}_{20}$ are the co-ordinates and velocity of mass M_2 at the moment of collision.

Using the boundary conditions to find the constants C_1, C_2, C_3, C_4 and phase φ in solution (4.17) we obtain

$$C_1 = -\frac{\pi G}{M_1 \Omega \omega} - \cot\frac{\pi}{\zeta}, \qquad C_2 = -\frac{\pi G}{M_1 \Omega \omega},$$

$$C_3 = \frac{\pi G M_2}{M M_1 \Omega \omega} \cot\frac{\pi}{\zeta} - a\cos\varphi, \qquad C_4 = -\frac{\pi G}{M\omega}, \tag{4.18}$$

$$\sin\varphi = -\frac{(1-R)M\dot{x}_{1-}}{2M_2 a\omega} \tag{4.19}$$

where $\zeta = \omega/\Omega$,

$$\dot{x}_{1-} = \frac{2\pi G}{(1+R)M_1\omega} \tag{4.20}$$

Using (4.20) and the notation of (4.15), equation (4.19) can be modified to the form

$$\sin\varphi = -\frac{\pi G}{P}\frac{(1-R)}{(1+R)}\frac{\left|1-\zeta^2\right|}{\zeta^2} \tag{4.21}$$

Assuming $|\sin\varphi| \le 1$ in (4.21), we obtain the condition for existence of the required periodic solution as

$$\frac{G}{P} \le \frac{1}{\pi}\frac{(1+R)}{(1-R)}\frac{\zeta^2}{\left|1-\zeta^2\right|},$$

which coincides with the inequality (4.10) found earlier. Hence, regardless of the ratio of mass M_1 to mass M_2, all regimes correspond to points in the existence domains as shown in Fig. 4.4

Substituting the constants given by (4.18) in the motion (4.17) we obtain

$$x_1 = \frac{G}{M}\left(\frac{t^2}{2} - \frac{\pi}{\omega}t\right) + \frac{M_2}{M}\{a[\cos(\omega t + \varphi) - \cos\varphi] - \\ - \frac{\pi G}{M_1 \Omega\omega \sin(\pi/\zeta)}\left[\cos\left(\Omega t - \frac{\pi}{\zeta}\right) - \cos\frac{\pi}{\zeta}\right]\} \tag{4.22}$$

$$x_2 = \frac{G}{M}\left(\frac{t^2}{2} - \frac{\pi}{\omega}t\right) + \frac{M_2}{M}\{a\left[\frac{M_1}{M_2}\cos(\omega t + \varphi) - \cos\varphi\right] - \\ - \frac{\pi G}{M_1 \Omega\omega \sin(\pi/\zeta)}\left[\frac{M_1}{M_2}\cos\left(\Omega t - \frac{\pi}{\zeta}\right) + \cos\frac{\pi}{\zeta}\right]\} + \frac{M_1}{M}\frac{G}{c} \tag{4.23}$$

As previously, by the term 'amplitude of vibration' we mean the value of half the sweep. It can be shown that when $R = 1$, functions (4.22), (4.23) take extreme values at the instants $t = 0$ and $t = \pi/\omega$. Bearing in mind that masses M_1 and M_2 reach their extreme displacements at these instants, we will take the value

$$a_{x_i} = \tfrac{1}{2}[x_i(0) - x_i(\pi/\omega)] \qquad (i = 1, 2)$$

as the amplitude of their vibration.

As a result, using (4.22), (4.23) we obtain the following equations for the amplitude-frequency characteristics:

$$a_{x_1} = \frac{\pi^2 G}{4M\omega^2} + \left(\frac{M_2}{M}\right)^2 \frac{G}{c}\frac{\pi}{2\zeta}\tan\frac{\pi}{2\zeta} + \frac{M_2}{M} a\cos\varphi, \tag{4.24}$$

$$a_{x_2} = -\frac{\pi^2 G}{4M\omega^2} + \frac{M_1 M_2}{M^2}\frac{G}{c}\frac{\pi}{2\zeta}\tan\frac{\pi}{2\zeta} + \frac{M_1}{M} a\cos\varphi. \tag{4.25}$$

Initially, let us examine the special case mentioned earlier when mass M_2 is considerably greater than mass M_1. Converting (4.24), (4.25), to the limit when $M_2 \to \infty$ and as a result $M_2 / M \to 1$, we find

$$a_{x_1} = \frac{G}{c}\frac{\pi}{2\zeta}\tan\frac{\pi}{2\zeta} + a\cos\varphi, \qquad a_{x_2} \equiv 0. \tag{4.26}$$

Here the first term is identical to the equation of the backbone curve (4.8). Since the value of $\cos\varphi$ is a double-valued function of the system parameters, equation (4.26) defines two branches of the resonance curves separated by the backbone lines. The amplitude-frequency characteristics are shown in Fig. 4.6, where the thin lines show the backbone curves. When $R = 1$, according to (4.21), $|\cos\varphi| = 1$. Substituting into (4.26), the obvious requirement

$$a_{x_1} \geq 0, \tag{4.27}$$

the apparent necessary conditions for the absence of additional intersections, we find that both branches of the resonance curve exist for all frequencies in the range $\zeta > 1$, when inequality (4.12) obtained earlier as the result of other reasons, is satisfied.

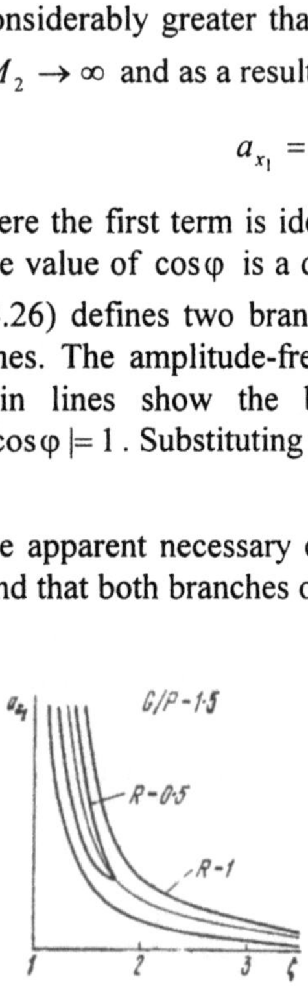

Fig. 4.6

When $R \neq 1$ the value of the initial phase φ is defined by equation (4.21). For this case, the branches of the resonance curve are enclosed between the proper branches of the system without dissipation. From (4.26), it can be seen that when the value of $\cos\varphi = 0$, the resonance curves intersect the backbone curves at points which lie on the boundary of the existence domain.

In general, for arbitrary ratio of the mass of the system, the mass M_2 is moving. The resonance curves for both masses look similar as in Fig. 4.6. This can be confirmed by comparing equations (4.24), (4.25) with expression (4.26). They differ because of the appearance of the term $\pi^2 G/(4M\omega^2)$, which represents the possibility of vibration for all systems as a unit mass M, impacting against stops. The additional multiplying factors before both of the other two terms represent a combined mass term, representing a degree of coupling between the masses due to their relative vibration. In this case, inequality (4.27) gives the conditions for existence of both branches of the resonance curves in the form

$$\frac{G}{P} \geq \frac{1}{2}\frac{(M_1 + M_2)}{M_2} \tag{4.28}$$

a particular case of which appears in condition (4.12). We note that even here the loss of energy during collision reduces the frequency range of the vibro-impact regime and does not lead to the elimination of resonance with infinite amplitude at frequency $\omega = \Omega$. In this case at the amplitude of resonance, a considerable influence is shown by other forms of damping, inherent always in real systems.

The amplitude of vibration of mass M_2 also influences the character of the system considerably, since it defines the vibration activity of the body of the machine. The tendency to obtain maximum amplitude of the striker naturally leads to an increase in the vibration of the body, the allowable level of which, especially for hand machines, is limited by rigid standards. Examining the equations for amplitude (4.24), (4.25), we see that in the neighbourhood of resonance, $\zeta \to 1$, the first term is negligibly small by comparison with the two others and the amplitude of vibration of mass M_2 becomes

$$a_{x_2} \approx (M_1 / M_2) a_{x_1}.$$

Hence, the level of vibration in machines depends mostly on the ratio of the mass of the body to that of the striker.

3. In §3 we found that viscous damping changes the boundaries of the existence domains in the vicinity of linear resonance. In the models examined earlier, the impact regime had a weak character in this frequency range. Hence a study of these regions was not of particular interest, and special attention was given to impact dissipation and its influence on vibro-impact resonances. However, in the two-mass system examined regimes with maximum amplitude occur particularly in the vicinity of the linear resonance frequency and it can be assumed that viscous damping shows an important influence here. To calculate this influence we will use an approximate process based on the establishment of the equations for energy balance during motion, displayed by the first harmonic of natural vibration. We shall examine the limiting case as $M_2 \to \infty$.

The equations for the amplitude of the fundamental are obtained with the help of expression (2.12) and equation (4.17), which defines the equivalent gap. We have

$$a_{x_1} = \frac{2G}{c[(\omega_0^2 / \Omega^2 - 1)]} \tag{4.29}$$

In Fig. 4.3, this expression is shown as a dotted line for comparison with the exact backbone curve (4.8) drawn with a solid line.

Suppose under the action of a harmonic excitation $P(t) = P\cos(\omega t + \varphi)$ and the force of viscous friction $n\dot{x}$, mass M_1 achieves harmonic vibration $x_1 = a_{x1}\cos\omega t$. During this motion, the work of the excitation force for one period is equal to

$$E_e = \int_0^{2\pi/\omega} P(t)\dot{x}_1 dt = \pi a_{x_1} P \sin\varphi \tag{4.30}$$

and the work of the viscous friction force is

$$E_{d1} = \int_0^{2\pi/\omega} n\dot{x}_1 dt = \pi\omega a_{x_1}^2 n. \tag{4.31}$$

From the condition of balance between the energy entering the system due to excitation, E_e and the energy E_{d1} dispersed by the dissipative forces during the period we obtain

$$\sin\varphi = a_{x_1} n\omega / P.$$

Since $|\sin\varphi| \le 1|$, we obtain the following condition of existence:

$$a_{x_1} \le P/(n\omega). \tag{4.32}$$

Hence, viscous damping limits the maximum vibration amplitude. It is

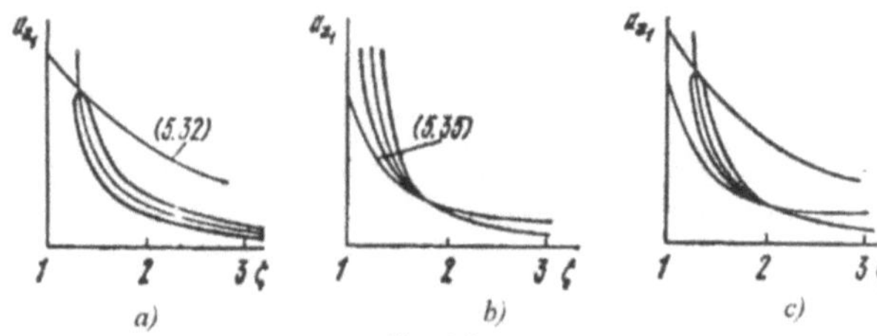

Fig. 4.7

convenient to find the limiting amplitude graphically as the points where the backbone curves (4.29) intersect the energy boundary of vibration, defined exactly by equation (4.32). The resulting resonance curves take the form shown in Fig. 4.7 *a*.

By similar methods we can find the conditions for existence when impact dissipation only is present. The loss of energy during impact takes the form

$$E_{d2} = \frac{M_1}{2}(\dot{x}_{1-}^2 - \dot{x}_{1+}^2) = \frac{M_1 \dot{x}_{1-}^2}{2}(1 - R^2),$$

or using (4.20)

$$E_{d2} = \frac{2\pi^2 G^2 \Omega^2}{c\omega^2} \frac{(1-R)}{(1+R)}. \tag{4.33}$$

From the conditions of energy balance, (4.31) and (4.33) we obtain

$$\sin\varphi = \frac{2\pi G^2 \Omega^2}{c\omega^2 a_{x_1} P} \frac{(1-R)}{(1+R)} \tag{4.34}$$

Demanding that $|\sin\varphi| \le 1$ we obtain the following condition for existence:

$$a_{x_1} \ge \frac{2\pi G^2 \Omega^2}{c\omega^2 P} \frac{(1-R)}{(1+R)} \tag{4.35}$$

From this it follows that impact dissipation provides the lower limit of the resonance curve. When calculating this form of dissipation only, the resonance curves takes the form shown in Fig. 4.7 *b*. The minimum amplitude is defined by the points where the backbone curve intersects the line given by expression (4.35).

When combining the effects of viscous and impact dissipation, the energy balances (4.30), (4.31), (4.33) give the following condition for existence:

$$\frac{P}{2\pi\omega}\left(1-\sqrt{1-\frac{4G^2nC}{\omega P^2}}\right)\le a_{x_1}\le\frac{P}{2\pi\omega}\left(1+\sqrt{1-\frac{4G^2nC}{\omega P^2}}\right), \tag{4.36}$$

where $C = 2\pi(1-R)/[M_1(1+R)]$.

In this case the resonance curves (Fig. 4.7 *b*) take the peculiar form of loops, the frequency range of which becomes compressed as the dissipative condition of the system increases.

Using the fact that the term under the radical in (4.36) is positive we find the ratio between the parameters at which it is possible to obtain resonance

$$\frac{G}{P}\le\sqrt{\frac{\omega M_1(1+R)}{8\pi n(1-R)}}. \tag{4.37}$$

When equality applies in (4.37), the resonance curves degenerate to a point defining the intersection of the backbone curve with the line $a_{x_1} = P/(2n\omega)$.

Percussion machines often try to perform as resonant ones. It is obvious that regimes that conform to the maximum resonance curves are suitable for optimum productivity as well as energy consumption. However, these conditions require that the relationship between the system parameters is fully maintained. Hence, the frequency of excitation ω^*, corresponding to the maximum resonance curve, depends on the value of the excitation force *P*, the applied force *G* and dissipative factors. Assuming that in the vicinity of ω^*, the influence of impact damping is negligible, from (4.29) and (4.32) we find the relationship of the parameters during tuning of resonance.

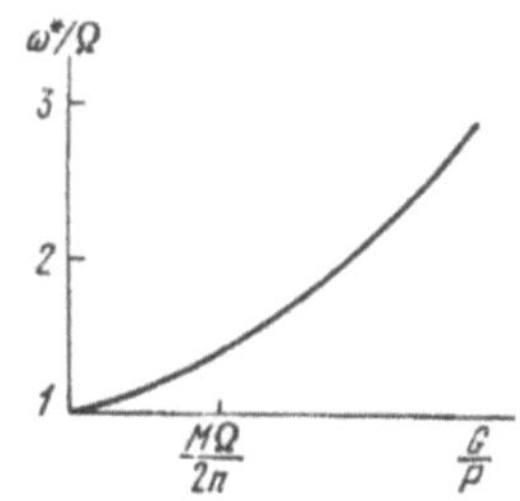

Fig. 4.8

$$\frac{G}{P}=\frac{M_1}{2\pi\omega^*}(\omega^{*2}-\Omega^2). \tag{4.38}$$

Fig. 4.8 shows the dependence of the resonant frequency ω^* on the feed force *G*, according to equation (4.38). From this, the resonance amplitude changes in accordance with expression (4.32).

In §20 we will concentrate especially on excitation and stabilisation of resonance in vibro-impact systems.

4. Up to this point, during the study of a range of vibro-impact systems, it has been assumed that the excitation is applied in the prescribed periodic form. In percussion machines, such effects are created by a various range of vibration exciters. Amongst them, and described above, are drives which convert mechanical rotation to reciprocating motion (Fig. 4.2) and numerous electromagnetic and electrodynamic exciters, pneumatic and hydraulic vibration exciters.

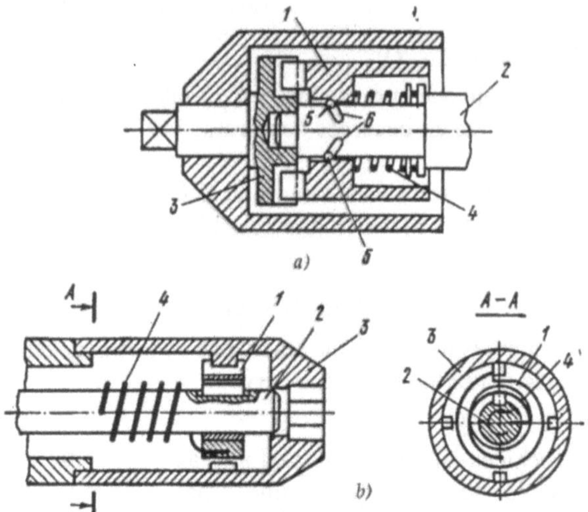

Fig. 4.9

However, there is a wide class of machine typically using vibro-impact principles of excitation, based on systematic interaction of an oscillating mass moving relative to a series of stops [17]. An obvious examples of such motion is often seen when boys run a stick along a fence.

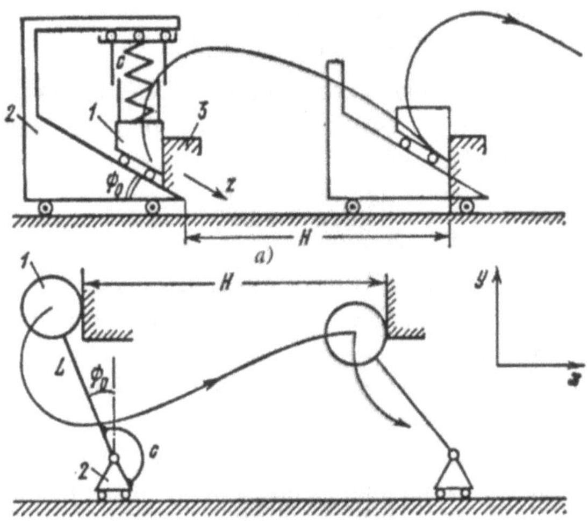

Fig. 4.10

An example of such a method appears as an impact rotating mechanism (Fig.4.9, *a*) widely used in the construction of screwthread wrenches, screwjacks, clamping devices etc. The mechanism consists of a striker *1*, surrounding a driving shaft *2*, with a helical v-shaped groove on the shaft surface, *6*. The ball *5* is mounted in the groove of the shaft, and lies in a similar v-shaped groove on the internal surface of the striker. With the help of a spring *4*, the lugs on the striker mesh with those on the anvil *3*. The mechanism operates in the following manner. During the time when the external force on the anvil is less than the value of the moment created as a result of the preliminary force of compression, the anvil and the striker move as one body. During braking of the anvil *3*, rod *2* continues to rotate and exerts a screwing effect on the striker *1* and disconnects the lugs, which after jumping once again, are put into mesh under the influence of spring 4.

In Fig.4.10 *a*, is shown a plane analogue of this mechanism (in Fig. 4.9 and 4.10 similar elements are denoted by identical figures). The driving shaft is modelled by a moving carrier which has a sloping plane *2* and the lug of the anvil by the stop *3* against which rides the striker *1*, elastically suspended by a spring of stiffness *c*, moves along the sloping plane.

This system represents one of the dynamic models of a double mass system studied earlier. Its features, including superposition of additional constraints, is related to the vibrating body itself, which does not have a straight line trajectory but requires plane motion to be considered. At the same time, the arrangement of the striker on the carrying body leads to a geometric summation of its relative (vibratory) and its translational motion, significantly increasing the impact efficiency.

Superposing various constraints allows us to put into practice various types of mechanism producing vibration from the possibilities described. Fig, 4.9 *b* shows yet another different construction for a vibro-rotation mechanism. The striker *1*, eccentrically mounted on the driving shaft *2* and connected to it by spring *4*,

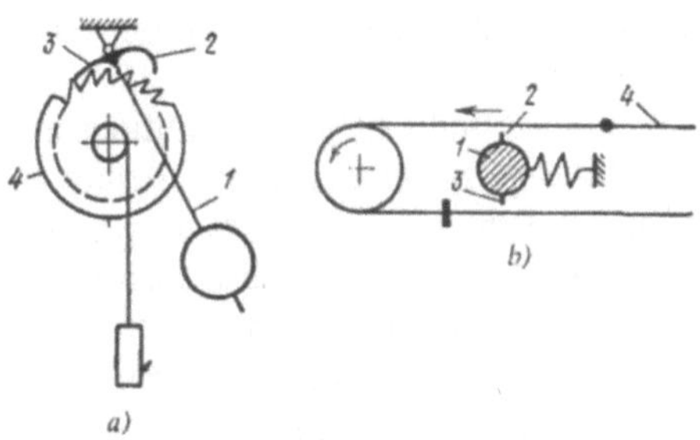

Fig. 4.11

meshes with the lugs on the anvil *3* through its own teeth. During braking the lugs of the rotating shaft *2*, because of its eccentricity, take the striker out of mesh with the anvil. The striker rotates the spring *4*, which on release, gives the striker an additional rotation relative to the eccentric. As a result the striker applies an impact to the following lug of the anvil and the whole process repeats. Fig. 4.11 *b* shows the plane analogy for this mechanism. Here, striker *1* rotates in conjunction with straight-line motion of the carriage *2* and the spring of stiffness *c* associated with it. Later it will be shown that the character of additional constraints between vibrating and supporting bodies significantly influences the vibration characteristics.

Note that in this manner we can produce excitation of many different mechanisms. One of the earliest to be produced practically was the watch mechanism known as the escapement (Fig. 4.11 *a*). During vibration of a pendulum *1*, one of the fingers attached to it, *2*, *3*, becomes clear of the teeth on the moving wheel *4*, which rotates through one step for one swing of the pendulum while the other finger gradually meshes with a tooth on the moving wheel. At the same time the impulse from the moving wheel is transmitted through the fingers to the pendulum and in this fashion maintains its vibration. A simplified dynamic model of a similar mechanism (Fig. 4.11, *b*) was described in [9] and examined in detail in [9, 51 to 53]. The moving wheel was modelled as an endless belt *4*, fitted with teeth. A spring-mounted mass *1* imitates the pendulum, reciprocating between the branches of the belt and supplying the fingers *2*, *3* with delicate impacts. In the interval between impacts, the belt can only move forward through one tooth, which instantaneously retracts after impact, so that the point of its attachment can freely move past the finger of the oscillator. Since only the tooth on the lower branch retracts itself, the tooth on the upper branch moves accordingly and so on, in turn.

Further refinement of the models requires calculation of the geometric constraints superposed on the vibrating body and provision for unimpeded bending of the fingers and the corresponding teeth of the rotating wheel after collision, similar to that shown in the model in Fig. 4.10 *b*. The models of Fig. 4.10 can be used also to explain spurious vibro-impact effects which happen when operating a rotating clutch mechanism, or rotating a ratchet mechanism etc.

We shall study the dynamic model described, (Fig. 4.10) assuming that the main body *2* moves uniformly with velocity v_0. Such an assumption allows us to simplify the study considerably, reducing it to the analysis of a system with a single degree of freedom, whose vibrations are synchronised by a periodic change with time of its kinematic structure.

Using the stitching method we find existence conditions and the law of motion for regimes characterised by regular impact of the striker onto each stop. The motion of the striker relative to the carrying body has a periodic character and completely defines the regime sought. We shall use z as the relative coordinate of the point mass M_1 of the striker, measured from the free position of the elastic element. For the second model (Fig. 4.10, *b*) coordinate $z = (\psi - \alpha)L$ introduces a condition to identify the situation. Here ψ is the current angular coordinate: α is the angle characterising the free position of the elastic element; both angles are measured from the vertical axis. Neglecting the loss of energy in the intervals between impacts we can write the differential equation for the relative motion in the form

$$\ddot{z} + \Omega^2 z = 0 . \qquad (4.39)$$

For the first model $\Omega^2 = c\sin^2\psi_0 / M_1$ and for the second $\Omega^2 = c/(M_1 L^2)$. The boundary conditions for the regimes examined take the form

$$\begin{aligned} &t = 0, && z = z_0, \quad \dot{z} = \dot{z}_+; \\ &t = T = H / v_0, && z = z_0, \quad \dot{z} = \dot{z}_- . \end{aligned} \qquad (4.40)$$

Using the boundary condition when $t = 0$, the solution of equation (4.39) is as follows:

$$z = z_0 \cos \Omega t + \frac{\dot{z}_+}{\Omega} \sin \Omega t \qquad (4.41)$$

After substitution of the boundary conditions when $t = T$ we find

$$z_0 = -\frac{\dot{z}_-}{\Omega} \cot \frac{\Omega t}{2} \qquad (4.42)$$

$$\dot{z}_+ = -\dot{z}_- \qquad (4.43)$$

The kinematic relationships of impact interactions with constraints between the striker and the supporting body (constrained impact) have already been obtained in §1, section 7. Using relationship (1.83) connecting the velocity before and after impact and using equations (4.42) and (4.43) we obtain

$$z_0 = -\frac{(1+R)(\cos\psi_0 + \theta \sin\psi_0) v_0 \cot(\Omega t/2)}{[(1-R)\cos\psi_0 + \theta(1+R)\sin\psi_0]\Omega \cos\psi_0}. \qquad (4.44)$$

For the first model (Fig. 4.10, *a*) ψ_0 is a fixed parameter and hence formula (4.44) defines a unique solution for the problem. For the second model (Fig. 4.10, b) $z_0 = (\psi_0 - \alpha)L$, where ψ_0 is the unknown angular coordinate at collision defined by the transcendental equation (4.44). In this case, the solution is not unique, which corresponds with the multiplicity of periodic regimes for the system.

For both models, relationship (4.44) gives as a condition for existence of the solutions found, $\Omega T \neq 2\pi n$ $(n = 1, 2, \ldots)$ which has an obvious physical meaning. For the second model, moreover, the solution ψ_0 must comply with inequality (1.81) the condition developed to ensure the absence of jamming during restricted impact.

The relative motion of the striker (4.41), with the help of equations (4.42), (4.43) can be converted to the form

$$z = z_0\left(\cos\Omega t + \tan\frac{\Omega t}{2}\sin\Omega t\right) = \frac{z_0}{\cos(\Omega t/2)}\cos\Omega\left(t - \frac{T}{2}\right) \qquad (4.45)$$

$$(t \in [0, T]).$$

An impression of the character of the periodic regimes for the second model can be found easily by substituting $\theta = 0$ in (4.44).

The result can be written as

$$1/(\psi_0 - \alpha) = K \cos\psi_0, \qquad (4.46)$$

where

$$K = -\frac{\Omega L(1-R)}{v_0(1+R)} \tan\frac{\Omega t}{2}. \qquad (4.47)$$

Fig. 4.12 shows examples for graphic solutions of equation (4.46). These solutions produce multiple values, the elements of which are related to the number of revolutions of the angular coordinate of the striker at the moment of impact. Rearranging (4.42) in the form

$$\psi_- = -\Omega(\psi_0 - \alpha)\tan(\Omega T/2), \qquad (4.48)$$

when $\tan(\Omega T/2) > 0$, coordinate $\psi_0 - \alpha$ and velocity ψ take opposite signs at the moment of impact, when $\tan(\Omega T/2) < 0$, they take the same sign. The first case represents impact during conditions of pre-arranged strain, the second represents conditions for a gap. This situation was similarly obtained earlier for the double mass model (Fig. 4.1) where the value of the equivalent gap (interference) (4.9) changed depending on the vibration regime.

From (4.46) it can be seen that as the solution of equation (4.46) moves further away from the asymptote $\psi_0 = \alpha$ (Fig. 4.12) then it rapidly approaches the value $\pm\pi n/2$ $(n = 1, 3, \ldots)$. Hence, practical interest is aroused only by those few values of ψ_0 nearest the asymptote which can represent real processes satisfying the conditions for the absence of jamming (1.81).

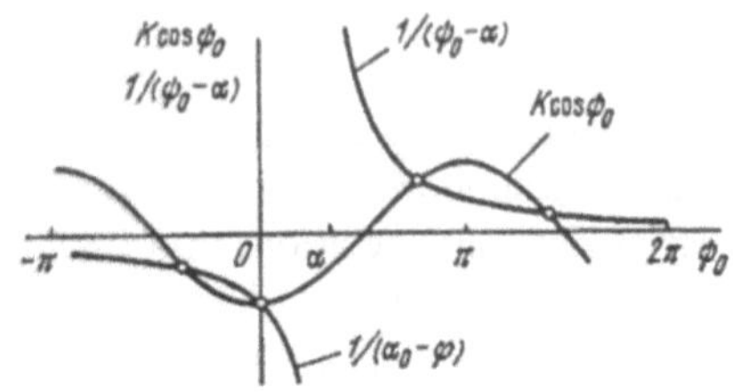

Fig. 4.12

5. Let us study the stability of solutions due to small perturbations of the phase coordinates. For simplification we shall limit the presentation to the case $\theta = 0$, which will preserve all the basic features of the qualitative picture of stability. The study will be carried out simultaneously for both models, using the finite difference scheme presented in §2 for the study of asymptotic behaviour near stationary solutions.

In agreement with (4.41) we will write the equation for the disturbed motion after the ν–*th* impact

$$z = (z_0 + \delta z_\nu)\cos\Omega t + \frac{\dot{z}_+ + \delta\dot{z}_{\nu+}}{\Omega}\sin\Omega t$$
$$\dot{z} = -(z_0 + \delta z_\nu)\Omega\sin\Omega t + (\dot{z}_+ + \delta\dot{z}_{\nu+})\cos\Omega t \tag{4.49}$$

where δz_ν and $\delta\dot{z}_{\nu+}$ are small increases in coordinate and velocity at the instant after the ν–*th* impact. We will find the increment in the phase coordinate due to the following $(\nu+1)$–*th* impact. Substituting in (4.49)

$$t = T + \delta T_\nu, \qquad z = z_0 + \delta z_{\nu+1}, \qquad \dot{z} = \dot{z} + \delta\dot{z}_{(\nu+1)+}$$

we obtain after linearisation the equations for disturbances

$$\delta z_{\nu+1} = \delta z_\nu\cos\Omega T + \frac{\delta\dot{z}_\nu}{\Omega}\sin\Omega T + \delta T_\nu\dot{z}_-,$$
$$\delta\dot{z}_{(\nu+1)-} = -\delta z_\nu\Omega\sin\Omega T + \delta\dot{z}_{\nu+}\cos\Omega T + \delta T_\nu\Omega^2 z_0 \tag{4.50}$$

The increment in the period during the ν–*th* interval δT_ν will be expressed through the increment of the phase coordinates (Fig. 4.10). Taking into account that for the first model, the impact condition takes the form

$$t = T + \delta T_v, \quad x = v_0 t + z\cos\psi_0 = H,$$

we find, incrementing this equation

$$\delta x_\nu = v_0\delta T_\nu + (\delta z_{\nu+1} - \delta z_\nu)\cos\psi_0 = 0,$$

whence

$$\delta T_\nu = -\frac{\cos\psi_0}{v_0}(\delta z_{\nu+1} - \delta z_\nu)\,. \tag{4.51}$$

A similar relationship is obtained for the second model by incrementing $x = v_0 t + L\sin\psi$ in the vicinity of $\psi = \psi_0$.

Incrementing equation (1.83) through velocity $\dot z$ and angle ψ in the vicinity of the stationary value when $\theta = 0$, we find the relationship between the increment of velocity before and after impact for the second model

$$\delta\dot z_{\nu+} = -R\delta z_{\nu-} - D\delta z_\nu \tag{4.52}$$

where

$$D = \frac{(1+R)v_0\sin\psi_0}{L\cos^2\psi_0}; \quad \delta z_\nu = L\delta\psi_\nu$$

Note that as ψ_0 nears the value $\pm\pi n/2$ $(n = 1, 3, \ldots)$ the condition for small value of $\delta\dot z_{\nu+}$ is violated, hence the stability analysis for these values does not provide true results. However, as was shown earlier, a periodic regime has no practical meaning in this case.

For the first model, the relationship between the increment of velocity before and after impact can be obtained formally from(4.52) when $D = 0$.

Substituting (4.51. (4.52) into (4.50), and substituting for the value $\dot z_-$ from (4.42) we obtain a system of linear homogeneous equations in finite-difference form

$$\frac{2\delta z_{\nu+1}}{1-R} - \delta z_\nu\left(\frac{1+R}{1-R} + \cos\Omega T - \frac{D}{\Omega}\sin\Omega T\right) + \delta\dot z_{\nu-}\frac{R}{\Omega}\sin\Omega T = 0,$$

$$\delta z_{\nu+1}\frac{1+R}{1-R}\Omega\cot\frac{\Omega T}{2} - \delta z_\nu\left(\frac{1+R}{1-R}\Omega\cot\frac{\Omega T}{2} - \Omega\sin\Omega T - D\cos\Omega T\right) +$$
$$+\delta\dot z_{(\nu+1)-} + \delta\dot z_{\nu-}R\cos\Omega T = 0.$$

Seeking their solution in a form similar to (2.42), we find the following characteristic equation after simple transformation:

$$2\beta^2 - \beta\left[(1+R)^2 + (1-R)^2\cos\Omega T - \frac{D}{\Omega}(1-R)\sin\Omega T\right] + 2R^2 = 0\,.$$

Eventually, using Schur's criterion (2.45) we find the conditions for stability

$$\left|R^2\right| < 1, \qquad \left|\frac{(1+R)^2 + (1-R)^2\cos\Omega T - \dfrac{D}{\Omega}(1-R)\sin\Omega T}{2(1+R^2)}\right| < 1\,. \tag{4.53}$$

Substituting $D = 0$ in (4.53) we obtain for the first model $|\cos\Omega T| < 1$, i.e., the condition for stability is satisfied for all the corresponding parameters for which the periodic regimes exist.

Substituting in (4.53) the value of D from (4.52) we obtain, using (4.46), (4.47) conditions for stability for the second model in the form

$$\left|(1+R)^2+(1-R)^2\cos\Omega T+(1-R)^2(1-\cos\Omega T)(\psi_0-\alpha)\tan\psi_0\right|< \\ <2(1+R^2) \tag{4.54}$$

Equation (4.54) is equivalent to two inequalities

$$(\psi_0-\alpha)\tan\psi_0<1, \qquad (\psi_0-\alpha)\tan\psi_0>-d\,, \tag{4.55}$$

where

$$d=\frac{1}{1-\cos\Omega T}\left[\frac{2(1+R^2)+(1+R)^2}{(1-R)^2}+\cos\Omega T\right].$$

Thus, depending on the sign of the quantity $\psi_0-\alpha$ the conditions for stability are as follows:

$$\frac{1}{\psi_0-\alpha}\gtrless\tan\psi_0, \qquad \frac{1}{\psi_0-\alpha}\gtrless-\frac{\tan\psi_0}{d}. \tag{4.56}$$

Here the upper value corresponds to positive value and the lower to a negative value of $\psi_0-\alpha$. In Fig. 4.13, stability domains constructed in agreement with (4.56) are shown hatched. The dotted line shows the graphic solution for equation (4.46) (Fig. 4.12).

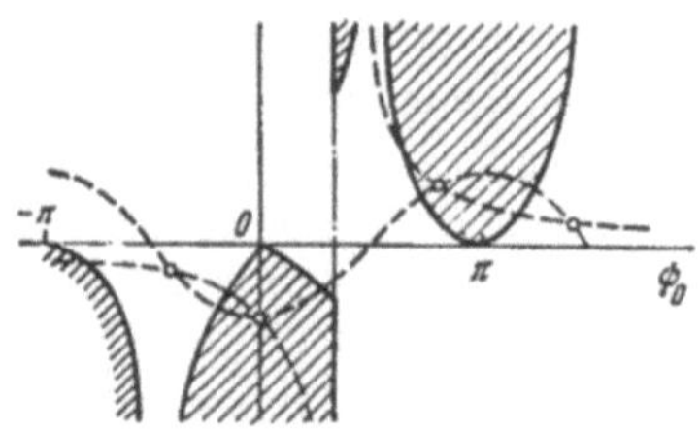

Fig. 4.13

6. The formulas obtained make possible complete design calculations for impact rotary mechanisms similar to those shown in Fig. 4.9, and to synthesise their structural elements and to tune the parameters of vibratory systems. Fig. 4.14 shows examples of a comparison between relative, *1*, and absolute, *2*, solutions (full line), and the practical (dotted line) trajectory for a striker taken from a real machine manufactured according to the design of Fig. 4.9, *a*. The calculated trajectory can be constructed either using the parametric equations

$$x=v_0t+z\cos\psi_0, \quad y=-z\sin\psi_0 \quad (t\in[0,H/v_0])$$

or with the help of graphical construction (Fig. 4.14) offsetting relative coordinate z along the direction of the inclined plane from point $x=v_0t$ along the abscissa. Since the striker performs complex helical motion in the real mechanism, for calculation it is necessary to take into account that in the model the mass of the striker M_1 has a corrected value related to the actual inertia parameters of the striker through the relationship

$$M_1=\frac{I}{r^2}\cos^2\psi_0+M\sin^2\psi_0,$$

where I, M are the moment of inertia and the mass of the striker; r is the radius of the trajectory of the centre of the ball (Fig. 4.9, *a*). It is not difficult to use the known absolute trajectory, to choose such a configuration of colliding elements, so that the striker will bend the teeth of the anvil without touching in the interval between calculated impacts.

An important parameter of the machine is the power of its drive mechanism. By analogy with (4.3) we will find the average force G developed by the drive. It is related to the value of the normal impulse J appearing on the striker during impact, by the expression

$$G=J/T=Jv_0/H. \tag{4.57}$$

Using equation (1.82) we find when $\theta=0$

$$J=(1+R)J_1=M_1(1+R)[(v_0/\cos^2\psi_0)-(\dot z_-/\cos\psi_0)] \tag{4.58}$$

Substituting in equation (4.57) the value (4.58) and the value of $\dot z_-$ from 4.42) we find using

(4.44)

$$G = \frac{2M_1 v_0^2}{H\cos^2\psi_0}\frac{(1+R)}{(1-R)}. \tag{4.59}$$

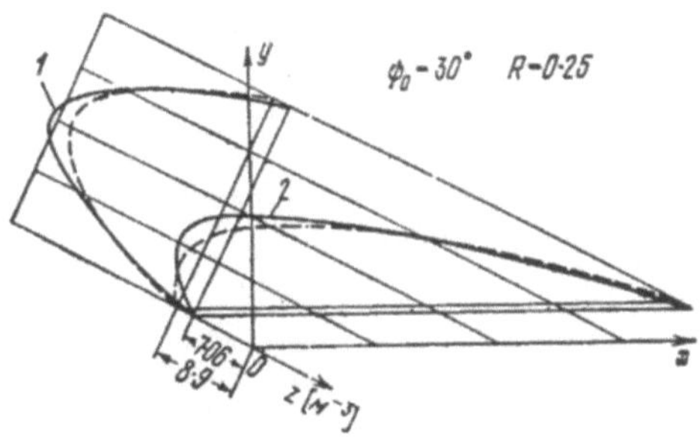

Fig. 4.14

The nominal power of the drive mechanism is found from the formula

$$N = Gv_0.$$

Note that equation (4.59) also gives the feed, i.e., the static force which must be applied to the body of the machine during operation.

§5. Stitching method

1. To conclude the presentation of the results of analysis of vibro-impact systems using exact methods, we will give a short generalisation of the calculation procedures used for the analysis of existence conditions and stability of periodic solutions in systems with an arbitrary finite number of degrees of freedom, vibrating with several impacts per period. For an explanation of the procedure, it is convenient to illustrate it by dealing with the phase space of the system.

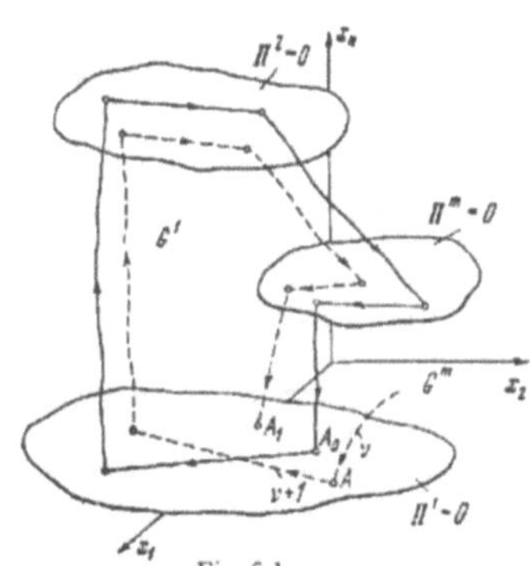

Fig. 5.1

We will look at a vibro-impact system described by a set of differential equations of the n-th degree, i.e., having $n/2$ degrees of freedom, which describe the system behaviour in the interval between collisions. The system described has n phase coordinates, $x_1, \ldots, x_n$ as a measure of which we can take the generalised coordinates and velocities of its point masses.

The study of the structure of any vibro-impact system reveals the functional interrelations between its generalised coordinates corresponding to all possible system configurations in which collisions of its elements occur. We shall call these unilateral constraints *impact constraints*.

In the general case, each impact constraint is described by a function, expressed through phase coordinates, time t, and time shift (phase) t_0, between external periodic force and system motion, definable from the chosen starting time point and written as:

$$\prod^k (x_1, \ldots, x_n, t_0, t) = 0 \quad (k = 1, 2, \ldots, m) \tag{5.1}$$

It is necessary to take into account the dependence of impact constraints on time when elements of the system are lead kinematically to the assigned periodic law. That is why, it is assumed that function (5.1) is periodic in t and t_0 with excitation period T.

As we study each time interval of the moving system between alternate collisions, it is convenient to combine the start of the interval with the instant of impact. As long as impact constraints given by (5.1) are applied to the system only at this instant ($t = 0$) then the expression can be contracted:

$$\Pi^k(x_q, t_0^k) = 0 \quad (q = 1,2,\ldots,m) . \tag{5.2}$$

Expression (5.2) originates in the phase space surfaces (Fig. 5.1), called *surfaces of discontinuity*. The number of discontinuity surfaces depends on the structure of the system and in the general case is an arbitrary finite quantity. The appearance of the representative point (end of the phase vector) on the discontinuity surface corresponds with the collision of some of the system elements, as the result of which their velocity instantaneously changes in agreement with the kinematic relationship of stereomechanical theory and the phase point moves along these surfaces to a new position. Hence, for each discontinuity surface $\Pi^k = 0$ some functional transformations of the phase coordinate occur which in the general form can be written as follows:

$$x_i^- = \Phi_i^k(x_i^+, \ldots, x_n^+, t_0,) \quad (i = 1,2,\ldots,n) . \tag{5.3}$$

Here, the negative and positive signs relate to values before and after impact. The time dependency in (5.3) arises in the same way following kinematic excitation of some elements and for this reason can be considered as T - periodic. Taking into account that the transformation (5.3) occurs when $t = 0$ it can be contracted and written in the form

$$x_i^- = \Phi_i^k(x_q^+, t_0^k) . \tag{5.4}$$

Noting that the transformation can exist on the discontinuity surface only if the following condition is satisfied

$$\sum_{q=1}^{n} \frac{\partial \Pi^k}{\partial x_q} \frac{\partial x_q}{\partial t_0^k} + \frac{\partial \Pi^k}{\partial t_0^k} \neq 0 \tag{5.5}$$

which denotes that at the moment of intersection with a discontinuity surface the points shown possess non-zero velocity relative to them.

Surfaces $\Pi^k = 0$ divide the phase space of the system into the zones $G^1, \ldots, G^s$, in the limits of which the point described fulfils continuous motion, corresponding to the movement of the system in the interval between impacts. This movement can be written as the solution of some systems of differential equations. Just suppose G^s is a zone between surfaces $\Pi^s = 0$ and $\Pi^{s+1} = 0$. The solution of the system of differential equations can be written provisionally as:

$$x_i^s = x_i^s(C_1^s, \ldots, C_n^s, t_0^s, t) = x_i^s(C_j^s, t_0^s, t) \quad (j = 1,\ldots,n) , \tag{5.6}$$

where C_j^s is a set of integration constants.

Using the accepted meanings, each type of periodical movement of the system with an arbitrary finite number, m, of collisions for a period of movement T, can be written as a sequence of values of index k defining a series of collision elements. For a study of regimes of the defined type it is expedient to arrange these sequences in order of the values

$$1, 2, \ldots, m \tag{5.7}$$

by means of a corresponding relabelling of the discontinuity surfaces (it is possible that various numbers may be repeated). This sequence characterises also types of non-periodic motions to which the series of colliding elements correspond (5.7). The interval of motion during which the sequence (5.7) is realised once will be called the *cycle* of non-periodic motion. For convenience in our study the beginning and end of the cycle will be arranged on the surface $\Pi^1 = 0$.

In phase space (Fig. 5.1), periodic motion of the form (5.7) will correspond with the closed trajectory shown, intersecting the discontinuity surfaces and the areas between them in accordance with the given sequence (5.7) and in agreement with expressions (5.4) and (5.6). With small perturbations, the system motion, deviating from the periodic regime (shown dotted in Fig. 5.1), undergoes $\Pi^1 = 0$ at point A, different from A_0, corresponding to a periodic trajectory, and, passing in sequence (5.7) to the remaining discontinuity surfaces, arrives again at $\Pi^1 = 0$ in the new point A_1. In this case also, the trajectory of the phase point in phase space also follows the same sequence of transformations of its phase coordinates. Hence, the aggregate motion of the system characterised by the defined series of collisions between its elements is adequately represented by several groups of sequential transformations. Denoting the group of sequential transformations by the letter Q, the motion of form (5.7) can be interpreted as a form of a point mapping for some surfaces (for example, $\Pi^1 = 0$) onto itself after transformation Q. Conditionally this can be written as $QA = A_1$.

Periodic motion of the type studied will be correlated with stationary point A_0 by transformation Q, i.e., $QA_0 = A_0$. If, as a result of multiple application of transformation Q to the points, we find ourselves in the low ε-neighbourhood of static point A_0 , as the points continuously approach it, then the point A_0 is called *asymptotically stable*.

The geometric interpretation of the examined combination of transformation Q in the form of the mapping of the point of some surfaces on themselves, has been called the *method of point mapping*. Such an interpretation allows the use of topological representation and methods to establish many important effects of the transformations described. The idea for point mapping which arose from the work of H. Poincaré, and G. Birkhoff, has found wide applications in the theory of nonlinear vibrations, thanks to the researches of A. A. Andronov and his school [10]. Contemporary methods for point mapping and a detailed bibliography, are more fully explained in the monograph by Y. I. Neimark [171]. In particular, a number of works on the application of this approach in the theory of vibro-impact systems is presented (see also [57]).

The analytical relationship between the phase coordinate of the points constructed before and after transformation Q produce expressions in the form of *difference equations*. This means we may use the methods and theorems of difference equations for the study of the type of transformation described [99]. Most important here is the distribution of Lyapunov's method for stability for a system of difference equations. The contemporary state of this problem and its applied aspects are set out in the book by P V Bromberg [68]. A E Kobrinsky and A A Kobrinsky illuminate problems and provide a bibliography for using this method in the theory of vibro-impact systems in the monograph [118].

Both approaches are adequate since they are different mathematical interpretations exist for one and the same transformation based on the idea of stitching associated with names of L LeCornu, A D Papaleksy and A Sommerfeld [8, 141].

We have already met examples of the application of difference equations for the analysis of stability in vibro-impact systems (see §2, section 3, §4, section 4), and in Chapter 5, when synthesising vibro-impact systems we will make use of the method of point mapping.

2. We will find the conditions for existence of a T-periodic regime of type (5.7) (see Fig. 5.1). We will denote the passage of time of the phase point of area G^s as t^s . We will take $\Pi^1 = 0$ as the original surface for examination . At the moment of leaving it, the point described has coordinates $\{x_i^1(C_j^1 . t_0^1 . 0)\}$ .The next entry of the point described on the surface $\Pi^1 = 0$ occurs during its motion from surface $\Pi^m = 0$ when $t = t_m$. The values of the phase coordinates at this point will be as follows: $\{x_i^m(C_j^m . t_0^m . t^m)\}$. Using (5.4) we obtain the conditions for periodicity in the form

$$x_i^m(C_j^m, t_0^m, t^m) = \Phi_i^1[x_q^1(C_j^1, t_0^1, 0), t_0^1] \quad (i = 1,\dots,n) . \tag{5.8}$$

These equations must be supplemented with the conditions connecting the coordinates on the remaining discontinuity surfaces

$$\begin{gathered} x_i^{k-1}(C_j^{k-1}, t_0^{k-1}, t^{k-1}) = \Phi_i^k[x_q^k(C_j^k, t_0^k, 0), t_0^k] \\ (i = 1,\dots,n; k = 2,\dots,m) \end{gathered} \tag{5.9}$$

and the conditions attached to the phase coordinates converted according to (5.8), (5.9) on to the discontinuity surfaces, i.e.,

$$\Pi^k\left[x_q^k(C_j^k, t_0^k, 0), t_0^k\right] = 0 \quad (k = 1,\dots,m) . \tag{5.10}$$

Since

$$\begin{gathered} t^s = t_0^{s+1} - t_0^s \quad \text{when} \quad s = 1,\dots,m-1, \\ t^m = T - \sum_{s=1}^{m-1} t^s \end{gathered} \tag{5.11}$$

expressions (5.8) to (5.10) provide sufficient conditions for determining the arbitrary constants $C_j^k (j = 1,\ldots,n; k = 1,\ldots,m)$ and phase $t_0^k (k = 1,\ldots,m)$.

Let us look at the stability of the periodic regimes found. As the result of small perturbations applied to the system, the value of the phase coordinates for the point describing a stable solution at the instant it appears on the discontinuity surface and its departure therefrom, will be different from the corresponding value for periodic motion by a small increment. These increments can be equated through the incremental values of C_j^k, t_0^k and t^k. The question of stability is decided by studying recurrent relationships coupling the variation of these values belonging to various cycles of excitation, for some discontinuity surfaces, for example, $\Pi^m = 0$.

Let us follow the motion of the representative point on surface $\Pi^1 = 0$ (Fig. 5.1). On this surface it appears at point *A*, after finishing the *v-th* cycle of disturbed motion. At this time its coordinates will be as follows: $x_i[C_j^m + (\delta C_j^m)_v, t_0^m + (\delta t_0^m)_v, t^m + (\delta t^m)_v]$, where δ means a small variation in the corresponding value. At the instant of departure of the point from surface $\Pi^1 = 0$ its coordinate will take the form $[x_i^1(C_j^1 + \delta C_j^1, t_0^1 + \delta t_0^1, 0), 0]$. Comparing variations pertaining to various cycles produced on surface $\Pi^m = 0$, those which have no index *m* will not be indicated by number of the cycle, since they are under consideration for the one cycle *v*+1 only.

Coupling the phase coordinates in the system on surface $\Pi^1 = 0$ before and after collision with the help of equation (5.4) we find

$$x_i^m[C_j^m + (\delta C_j^m)_v, t_0^m + (\delta t_0^m)_v, t^m + (\delta t^m)_v] = \\ = \Phi_i^1[x_q^1(C_j^1 + \delta C_j^1, t_0^1 + \delta t_0^1, 0), t_0^1 + \delta t_0^1]. \tag{5.12}$$

Expanding (5.12) as a Taylor series for small variations and noting that $(\delta t^m)_v = \delta t_0^1 - (\delta t_0^m)_v$ we obtain to second degree accuracy

$$\sum_{j=1}^{n} \frac{\partial x_i^m}{\partial C_j^m}(\delta C_j^m)_v + \left(\frac{\partial x_i^m}{\partial t_0^m} - \frac{\partial x_i^m}{\partial t^m}\right)(\delta t_0^m)_v - \sum_{q=1}^{n} \frac{\partial \Phi_i^1}{\partial x_q^k} \sum_{j=1}^{n} \frac{\partial x_q^1}{\partial C_j^1}\delta C_j^1 + \\ + \left[\frac{\partial x_i^m}{\partial t^m} - \frac{\partial \Phi_i^1}{\partial t_0^1} - \sum_{q=1}^{n} \frac{\partial \Phi_i^1}{\partial x_q^k}\frac{\partial x_q^1}{\partial t_0^1}\right]\delta t_0^1 = 0. \tag{5.13}$$

It is not difficult to see that equation (5.13) can be obtained by means of formal variation of equation (5.8). Carrying out this operation on equation (5.9), (5.10) and noting that according to (5.11) $\delta t^s = \delta t_0^{s+1} - \delta t_0^s$, we obtain

$$\sum_{j=1}^{n} \frac{\partial x_i^{k-1}}{\partial C_j^{k-1}}\delta C_j^{k-1} + \left(\frac{\partial x_i^{k-1}}{\partial t_0^{k-1}} - \frac{\partial x_i^{k-1}}{\partial t^{k-1}}\right)\delta t_0^{k-1} - \sum_{q=1}^{n} \frac{\partial \Phi_i^k}{\partial x_q^k} \sum_{j=1}^{n} \frac{\partial x_q^k}{\partial C_j^k}\delta C_j^k + \\ + \left[\frac{\partial x_i^{k-1}}{\partial t^k} - \frac{\partial \Phi_i^k}{\partial t_0^k} - \sum_{q=1}^{n} \frac{\partial \Phi_i^k}{\partial x_q^k}\frac{\partial x_q^k}{\partial t_0^k}\right]\delta t_0^k = 0 \quad (k = 2,\ldots,m), \tag{5.14}$$

$$\sum_{l=1}^{n} \frac{\partial \Pi^k}{\partial x_q^k} \sum_{j=1}^{n} \frac{\partial x_q^k}{\partial C_j^k}\delta C_j^k + \left(\frac{\partial \Pi^k}{\partial t_0^k} + \sum_{l=1}^{n} \frac{\partial \Pi^k}{\partial x_q^k}\frac{\partial x_q^k}{\partial t_0^k}\right)\delta t_0^k = 0 \quad (k = 1,\ldots,m). \tag{5.15}$$

Note that in equations (5.14), (5.15) when $k = m$ variations $\delta C_j^m, \delta t_0^m$ do arise, however, in contrast to the similar variation in equation (5.13) they belong to the following *v*+1-th cycle of disturbed motion. Thus equations (5.13) to (5.15) cause recurrent relationships between the variations of the arbitrary constants $\delta C_j^m, \delta t_0^m$ belonging to various cycles of the disturbed motion in the form of a system of difference equations. Taking into account the character of the solutions of these equations (see §2, section. 3), we can put in equations (5.14), (5.15) $\delta C_j^m = \beta(\delta C_j^m)_v, \delta t_0^m = \beta(\delta t_0^m)_v$ (β is a numerical constant). In such a way, equations (5.13) to (5.15) converge to a system of linear ordinary algebraic equations for the variations. Setting the determinant of this system to zero, we obtain the characteristic determinant, the roots of which represent stability for the motion studied. It is not difficult to convince oneself that maximum order of the determinant for finding the constant β is equal to n, since when using equation (5.15) the phase variations in equations (5.13), (5.14) can be linear expressions through corresponding variations in the constants of integration.

The stability conditions for the periodic solution considered exist by finding the roots of the characteristic determinant inside a unit circle $|\beta_i| < 1 \, (i = 1,\dots,n)$. If any of the roots lie outside the limits of this area of the complex plane, then it represents an unstable solution. Estimating the roots of the characteristic determinant is possible using Schur's criterion, or equivalent (see §2, section 3).Using the substitution $\beta = (1+s)/(1-s)$ it is possible to transform the unit circle on the complex plane for root β onto the left-hand half-space of the complex plane of root s, and to use the Rouse-Hurwitz criterion for stability [24, 161].

The stitching method appears convenient for studying low order systems, examples of which have been presented in this chapter, or for systems having a regular structure [116, 166, 168]. In references [125, 127, 172] this procedure is simplified to find approximate solutions for periodic regimes of vibro-impact systems, with small nonlinearities.

In the following section we will acquaint the reader with another method to find exact periodic solutions, which is more convenient for dealing with systems of large dimension since it allows the wide use of linear analysis of the structures composing vibro-impact systems.

Chapter 2. Equivalent linearisation of vibro-impact systems

For this relief much thanks.

William Shakespeare: Hamlet.

§6. Equations of vibro-impact systems

1. When we constructed exact solutions we took one mathematical model to describe motion between impacts and another for the instant of collision. The first we described by linear differential equations; and the second using finite expressions from stereomechanical impact theory, used as the boundary conditions of the process. Such a description hampers the general analysis of vibro-impact systems, largely because it is impossible to predict the successive instants when the bodies will collide. Using the procedure of finding a periodic solution, which makes it possible to postulate regularity of collisions, does, in some measure, make it possible to overcome these difficulties, but, even in these cases, the solution constructed does not appear completely adequate due to the additional possibility of the body trajectory crossing the boundary levels. Separate descriptions of inter-impact motion and the impact process are even more problematic when using a more general concept of collisions between solids. For this reason, we will turn to the construction and analysis of single forms to record the equation of motion for the colliding elements of a vibro-impact system on all time axes, describing the fully combined motion occurring. This is possible by introducing nonlinear expressions reflecting the process of force interaction of colliding bodies or their elements.

The required expressions are constructed on the basis of force characteristics of contact interaction $\Psi(x)$ or $\Psi(x,\dot{x})$ (see §1), and in the case of single sided impact interaction when relative coordinate $x=\Delta$ take the form

$$\Phi(x)=\Psi(x-\Delta)\eta(x-\Delta), \tag{6.1}$$

$$\Phi(x,\dot{x})=\Psi(x-\Delta,\dot{x})\eta(x-\Delta) \tag{6.2}$$

where $\eta(x)$ is Heaviside's unit step function

$$\eta(x)=\begin{cases}0 & \text{when} \quad x<0,\\ 1 & \text{when} \quad x\geq 0.\end{cases}$$

In the case of a symmetrical double sided interaction, the elements of the impact pairs fixed at the basic gap 2Δ when $x=0$, the similar characteristic will be:

$$\Phi(x)=\Psi(|x|-\Delta)\eta(|x|-\Delta)\operatorname{sign}x \tag{6.3}$$

$$\Phi(x,\dot{x}) = \Psi(|x|-\Delta,\dot{x})\eta(|x|-\Delta)\mathrm{sign}x \tag{6.4}$$

Expressions of the type (6.1), (6.3) will be called *static force characteristics of the impact pair*, while type (6.2), (6.4) will be called the *dynamic force characteristics*

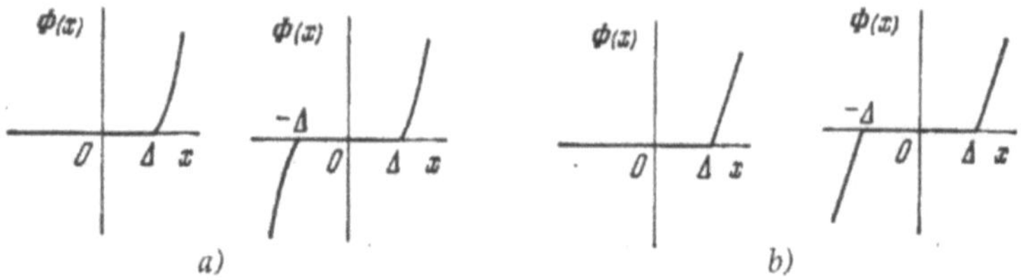

Fig. 6.1

of the impact pair. The dynamic characteristics are frequently presented in the form of the sum of elastic components $\Phi_E(x)$ or $\Phi_E(x,\dot{x})$ and dissipative constituents $\Phi_D(x)$ or $\Phi_D(x,\dot{x})$.

In particular, when describing contact interaction using Hertzian theory (see (1.35), (1.36)) the static characteristics take the form (Fig. 6.1, *a*)

$$\Phi(x) = c_1(x-\Delta)^{3/2}\eta(x-\Delta) \tag{6.5}$$

in the case of single sided interaction and

$$\Phi(x) = c_1(|x|-\Delta)^{3/2}\eta(|x|-\Delta)\mathrm{sign}x \tag{6.6}$$

for double sided collisions. To approximate the force characteristics of contact interaction linear expressions take the form (Fig. 6.1, *b*)

$$\Phi(x) = c_2(x-\Delta)\eta(x-\Delta), \tag{6.7}$$

$$\Phi(x) = c_2(|x|-\Delta)\eta(|x|-\Delta)\mathrm{sign}x. \tag{6.8}$$

Similar characteristics can also be constructed using the hypotheses of stereomechanical theory of impact [37]. Let us look at the direct central impact of two forward moving bodies with masses M_1 and M_2. Let us assume that collision occurs when the relative coordinate $x = \Delta$. According to (1.46) the equation for relative motion of the bodies in this case takes the form

$$M\ddot{x} = -\Phi \qquad (M = M_1M_2/(M_1+M_2)), \tag{6.9}$$

where Φ is the principal vector of the interaction force. An effect equivalent to collision can be obtained by introducing an "instantaneous" force Φ acting only in the narrow layer $\Delta \le x < \Delta+\varepsilon$, where ε is a small quantity, but, as it approaches the limit, by decreasing the thickness of the layer to zero. Let us substitute this force in the sum of two components: the elastic component

$$\Phi_E = \Phi_1\{\eta(x-\Delta)-\eta[x-(\Delta+\varepsilon)]\} \tag{6.10}$$

and the dissipative

$$\Phi_D = \Phi_2\{\eta(x-\Delta)-\eta[x-(\Delta+\varepsilon)]\}\mathrm{sign}\dot{x}. \tag{6.11}$$

The constant coefficients Φ_1 and Φ_2 will be defined so that as x changes from Δ to $\Delta+\varepsilon$, the value of the relative velocity $\dot{x}$ changes from $\dot{x}_-$ to zero, but during the change of x from $\Delta+\varepsilon$ to Δ the velocity changes from zero to $-R\dot{x}_-$, where $\dot{x}_-$ is the relative velocity of the bodies at the start of contact. As a result the work of the force during impact will be as follows:

$$(\Phi_1+\Phi_2)\varepsilon = M\dot{x}_-^2/2, \quad (\Phi_1-\Phi_2)\varepsilon = M\dot{x}_-^2R^2/2.$$

Solving this system of equations for the coefficients Φ_1 and Φ_2 we find

$$\Phi_1 = M(1+R^2)\dot{x}_-^2/(4\varepsilon), \qquad \Phi_2 = M(1-R^2)\dot{x}_-^2/(4\varepsilon). \tag{6.12}$$

Substituting the value of the coefficients Φ_1 and Φ_2 in (6.10), (6.11) and noting that the usual limit when $\varepsilon\to 0$ is not valid here, we apply the full transition in the generalised sense [98, 198]. As a result we find

$$\Phi_E = \frac{M\dot{x}_-^2(1+R^2)}{4}\delta_+(x-\Delta), \tag{6.13}$$

$$\Phi_D = \frac{M\dot{x}_-^2(1-R^2)}{4}\delta_+(x-\Delta)\mathrm{sign}\dot{x}. \tag{6.14}$$

Here $\delta_+(x)$ is the right-hand Dirac function since its singularity is arranged to the right from the coordinates of the triggering of the $\delta-$function. The symbol $\delta_+(x-\Delta)\mathrm{sign}x$ is understood in the sense

$$\delta_+(x-\Delta)\mathrm{sign}\dot{x} = \begin{cases} \delta_+(x-\Delta) & \text{when} \quad \dot{x}>0 \\ -\delta_+(x-\Delta) & \text{when} \quad \dot{x}<0 \end{cases}$$

Due to symmetrical double-sided interaction, elastic and dissipative constants take the form

$$\Phi_E = \frac{M\dot{x}_-^2(1+R^2)}{4}\delta(|x|-\Delta)\mathrm{sign}x, \tag{6.15}$$

$$\Phi_D = \frac{M\dot{x}_-^2(1-R^2)}{4}\delta(|x|-\Delta)\mathrm{sign}\dot{x}. \tag{6.16}$$

The symbols used here take the following meanings:

$$\delta(|x|-\Delta)\mathrm{sign}x = \begin{cases} \delta_+(x-\Delta) & \text{when} \quad x>0, \\ -\delta_-(x+\Delta) & \text{when} \quad x<0, \end{cases}$$

$$\delta(|x|-\Delta)\mathrm{sign}\dot{x} = \begin{cases} \delta_+(x-\Delta) & \text{when} \quad \dot{x}>0, \\ -\delta_-(x+\Delta) & \text{when} \quad \dot{x}<0, \end{cases}$$

where $\delta_-(x)$ is the left-hand Dirac function. Remember that dimensions of the Dirac function are $[x]^{-1}$. The resulting outcome assumes that the impact process is limited to the level $x=\pm\Delta$ in the forward and reverse directions and consequently the double triggering of the $\delta-$function corresponds to the two phases of impact. In this way,

$$\int_{\Delta}^{\Delta+\varepsilon}\delta_+(x-\Delta)dx = \int_{-\Delta}^{-\Delta-\varepsilon}\delta_-(x+\Delta)dx = 1. \tag{6.17}$$

Thus, when using stereomechanical impact theory, the elastic and dissipative components of the dynamic characteristics of impact pairs appear to be dependent on the coordinates and velocity of impact, i.e.

$$\Phi(x,\dot{x}) = \Phi_E(x,\dot{x}) + \Phi_D(x,\dot{x})\,. \tag{6.18}$$

When $R = 1$, equations (6.14), (6.16) change to zero, and from (6.13) to (6.15) we obtain the force characteristic for the case of an totally elastic impact

$$\Phi_E(x,\dot{x}) = \frac{M\dot{x}_-^2}{2}\delta_+(x-\Delta)\,, \tag{6.19}$$

$$\Phi_E(x,\dot{x}) = \frac{M\dot{x}_-^2}{2}\delta(|x|-\Delta)\mathrm{sign}x\,. \tag{6.20}$$

The force characteristics of the impact pair can also be obtained by considering the direct central impact of two forward moving bodies as the result of the effects of a nonlinear elastic force of type (6.5) or (6.7) with an unlimited increase in the coefficients c_1 or c_2.

For single-sided and double-sided interactions, the limiting cases conform to the static characteristic drawn in Fig. 6.2 *a*, *b*. Due to the presence of elastic bonds of stiffness c between the interacting bodies, these characteristics acquire the form shown in Fig, 6.2 *c*, *d*. It is natural that in this case, the extra elasticity can be taken into account as the effective parallel to the nonlinear characteristic shown in Fig. 6.2 *a*, *b*:

$$\Phi(x) = cx + \Phi_E(x)\,. \tag{6.21}$$

In the approach to the study of vibro-impact systems using the elastic interaction characteristics shown in Fig. 6.2, energy losses during impact can be calculated independently by introducing an equivalent dissipative force $\Phi_D(\dot{x})$, convenient for analysis, found from the conditions for an energy balance. In addition, it is often expedient to take into account a specific regime of an actual system, therefore the practical possibilities of such calculation will be presented during the investigation of real systems. For us at the moment it is important that with the help of different nonlinear expressions of the type studied, all the basic forms of force interactions for colliding bodies or their contacting elements can be written.

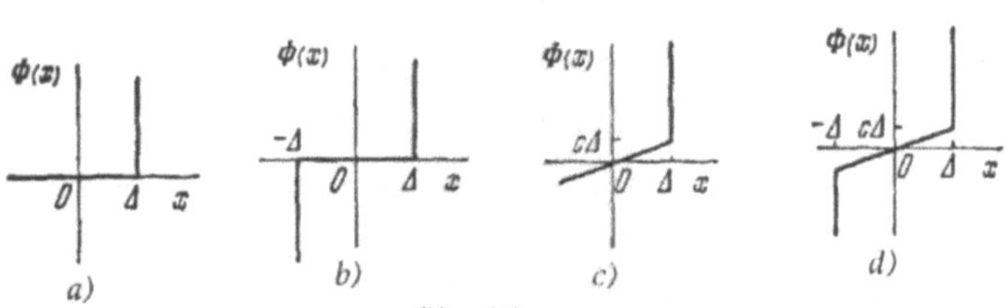

Fig. 6.2

2. No matter which of the nonlinear force characteristics, described above, we substitute into equation (6.9), we will obtain the differential equation of relative motion of the colliding bodies describing the total actual motion on all time axes

$$M\ddot{x} + \Phi(x,\dot{x}) = 0\,. \tag{6.22}$$

The effects of the external force on the collision of the bodies and also the additional elastic and dissipative connections are taken into account by the introduction into equation (6.22) of corresponding additive terms.

Substituting $M_2 \to \infty$ in (6.9) we obtain $M = M_1$, and equation (6.22) then represents the impact of a body against a limiter. In the event, if the latter moves according to the law $x_0(t)$ we must substitute $x = x_1 - x_0(t)$ in (6.22), where x_1 is the absolute coordinate of the body. Using the impact pair characteristic $\Phi(x,\dot{x}))$ we can build-up, for example, the equation for oblique sliding impact of the material point mass M on a fixed obstruction (Fig. 1.5). Based on the hypothesis of dry friction we take for force G acting in the tangential direction,

$$G(x,\dot{x},\dot{y}) = -\theta\Phi(x,\dot{x})\mathrm{sign}\dot{y}, \tag{6.23}$$

where θ is the coefficient of dry friction.

The equation for the normal coordinate x takes the form (6.22). For the tangential coordinate y using (6.23)

$$M\ddot{y} + \theta\Phi(x,\dot{x})\mathrm{sign}\dot{y} = 0\,. \tag{6.24}$$

The methods used to describe impact interactions can naturally be extended into more general cases for two-dimensional and three-dimensional interactions. Such cases are not of particular interest here since the purpose of this book is to deal specifically with a study of vibro-impact systems, hence, later on we will limit the systems considered to those with one dimensional impact interactions. By introducing to the structure of the dynamic model for such systems, a nonlinear elastic dissipative element $\Phi(x,\dot{x})$ to characterise the one-dimensional interaction forces in impact pairs, it is possible to construct the differential equations for vibro-impact systems with an arbitrary number of degrees of freedom by the usual methods.

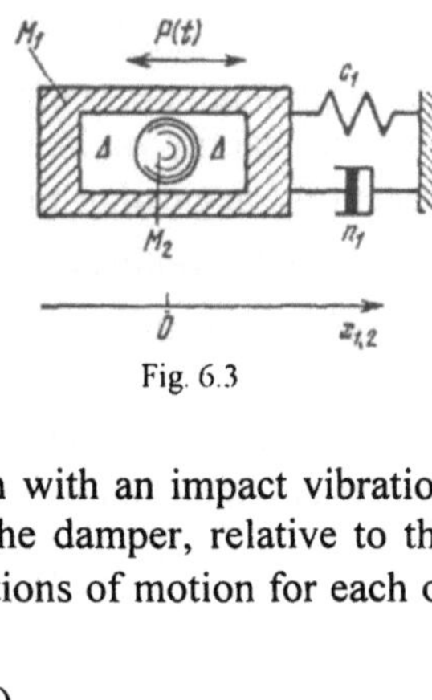

Fig. 6.3

We shall examine several examples. Fig. 6.3 shows the dynamic model of a simple vibrating system with an impact vibration absorber. Writing the coordinates of displacement of the damper, relative to the damped mass $x_2 - x_1$ through x we will write the equations of motion for each of the masses in the form

$$\begin{aligned} M_1\ddot{x}_1 &= -n_1\dot{x}_1 - c_1x_1 - \Phi(x,\dot{x}) + P(t), \\ M_2\ddot{x}_2 &= \Phi(x,\dot{x}). \end{aligned} \tag{6.25}$$

Writing the differential operator as $s = d/dt$, we can rewrite the equations of the system (6.25) in the following form:

$$\begin{aligned} (M_1s^2 + n_1s + c_1)x_1 + \Phi(x,sx) &= P(t) \\ M_2s^2x_2 - \Phi(x,sx) &= 0. \end{aligned} \tag{6.26}$$

Solving equations (6.26) accordingly relative to x_1 and x_2 and subtracting the first from the second, after reorganisation we obtain the equation for the relative coordinate x:

$$\begin{gathered}(M_1M_2s^4+M_2n_1s^3+M_2c_1s^2)x+\\+[(M_1+M_2)s^2+n_1s+c_1]\Phi(x,sx)=-M_2s^2P(t).\end{gathered}\tag{6.27}$$

In general, for a stationary holonomic mechanical system with n degrees of freedom in the presence of a single nonlinearity in the form of a one-dimensional impact pair between the *k-th* and the *m-th* masses, the equations can be arranged in the form

$$\begin{aligned}&\mathscr{D}_{11}(s)x_1+\ldots+\mathscr{D}_{1k}(s)x_k+\ldots+\mathscr{D}_{1m}(s)x_m+\ldots+\mathscr{D}_{1n}(s)x_n=P_1(t),\\&\ldots\ldots\ldots\ldots\ldots\ldots\ldots\ldots\ldots\\&\mathscr{D}_{k1}(s)x_1+\ldots+\mathscr{D}_{kk}(s)x_k-\Phi(x,sx)+\ldots+\mathscr{D}_{km}(s)x_m+\\&\qquad\qquad+\ldots+\mathscr{D}_{kn}(s)x_n=P_k(t),\\&\ldots\ldots\ldots\ldots\ldots\ldots\ldots\ldots\ldots\\&\mathscr{D}_{m1}(s)x_1+\ldots+\mathscr{D}_{mk}(s)x_k+\ldots+\mathscr{D}_{mm}(s)x_m+\Phi(x,sx)+\\&\qquad\qquad+\ldots+\mathscr{D}_{mn}(s)x_n=P_m(t),\\&\ldots\ldots\ldots\ldots\ldots\ldots\ldots\ldots\ldots\\&\mathscr{D}_{n1}(s)x_1+\ldots+\mathscr{D}_{nk}(s)x_k+\ldots+\mathscr{D}_{nm}(s)x_m+\ldots+\mathscr{D}_{nn}(s)x_n=P_n(t),\end{aligned}\tag{6.28}$$

where $\mathscr{D}_{pq}(s)=M_{pq}s^2+n_{pq}s+c_{pq}$; $M_{pq},n_{pq},c_{pq}\;(p=1,\ldots,n;\,q=1,\ldots,n)$ are coefficients of positive definite quadratic form

$$T=\tfrac{1}{2}\sum_{p,q=1}^{n}M_{pq}\dot{x}_p\dot{x}_q,\quad B=\tfrac{1}{2}\sum_{p,q=1}^{n}n_{pq}\dot{x}_p\dot{x}_q,\quad \Pi=\tfrac{1}{2}\sum_{p,q=1}^{n}c_{pq}x_px_q$$

characterising the corresponding kinetic energy, dissipative functions and potential energy of linear parts of the system [24], obtained from the original system by neglecting the impacting interacting mass $\Phi(x,sx)\equiv 0$; $x_1,\ldots,x_n$ are generalised coordinates of the system; $x=x_m-x_k$ are coordinates of the relative disposition of the colliding masses; $P_1(t),\ldots,P_n(t)$ are generalised external forces.

Transferring the nonlinear terms in (6.28) to the right-hand side and solving the system of equations as linear in terms of coordinates x_k and x_m, we find

$$\begin{aligned}x_k(t)&=\left[\frac{\mathscr{R}_{kk}(s)}{\mathscr{D}(s)}-\frac{\mathscr{R}_{mk}(s)}{\mathscr{D}(s)}\right]\Phi[x(t),sx(t)]+\sum_{i=1}^{n}\frac{\mathscr{R}_{ik}(s)}{\mathscr{D}(s)}P_i(t),\\x_m(t)&=\left[-\frac{\mathscr{R}_{mm}(s)}{\mathscr{D}(s)}+\frac{\mathscr{R}_{km}(s)}{\mathscr{D}(s)}\right]\Phi[x(t),sx(t)]+\sum_{i=1}^{n}\frac{\mathscr{R}_{im}(s)}{\mathscr{D}(s)}P_i(t),\end{aligned}\tag{6.29}$$

where

$$\mathscr{D}(s)=\begin{vmatrix}\mathscr{D}_{11}(s)&\cdots&\mathscr{D}_{1n}(s)\\\cdots&\cdots&\cdots\\\mathscr{D}_{n1}(s)&\cdots&\mathscr{D}_{nn}(s)\end{vmatrix}\tag{6.30}$$

is the characteristic determinant defining the linear part of the system; $\mathscr{R}_{pq}(s)$ are the algebraic adjunct of the $(pq)-th$ elements. Because of symmetry of the matrix, the coefficients of the quadratic form $\mathscr{R}_{pq}(s)\equiv\mathscr{R}_{qp}(s)$. Both here and later, we shall consider systems with *total dissipation* in which any motion is accompanied by energy dissipation. For this reason, we will not consider asymptotic time-damped processes caused by non-zero initial conditions.

Subtracting the first equation in (6.29) from the second we obtain for coordinate x, similar to (6.26)

$$\mathcal{D}(s)x(t)+\mathcal{R}(s)\Phi[x(t),sx(t)]=\sum_{i=1}^{n}\mathcal{S}_i(s)P_i(t), \tag{6.31}$$

where $\mathcal{R}(s)=\mathcal{R}_{kk}(s)+\mathcal{R}_{mm}(s)-2\mathcal{R}_{km}(s); \mathcal{S}_i(s)=\mathcal{R}_{im}(s)-\mathcal{R}_{ik}(s)$. The equations for $\mathcal{D}(s),\mathcal{R}(s),\mathcal{S}_i(s)$ are represented as polynomials of the differential operator s in which the degree of $\mathcal{D}(s)$ is $2n$, the degree of $\mathcal{R}(s)$ is $2n$ - 2 and the degree of $\mathcal{S}_i(s)$ does not exceed $2n$ - 2.

After defining the law of change of $x(t)$ from (6.31), the displacement of the impacting mass can be found from equation (6.29). For the remaining masses we obtain similarly

$$x_l(t)=\left[\frac{\mathcal{R}_{kl}(s)}{\mathcal{D}(s)}-\frac{\mathcal{R}_{ml}(s)}{\mathcal{D}(s)}\right]\Phi[x(t),sx(t)]+\sum_{i=1}^{n}\frac{\mathcal{R}_{il}(s)}{\mathcal{D}(s)}P_i(t) \tag{6.32}$$

$$(l=1,2,\ldots,n).$$

We notice that in equations (6.29), (6.32) terms of the form $\mathcal{R}_{pq}(s)/\mathcal{D}(s)=L_{pq}(s)$ represent *dynamic compliance operators* of the linear portion of the system connecting the displacement of the *p-th* mass with the force applied to the *q-th* mass

$$x_p(t)=L_{pq}(s)P_q(t) \tag{6.33}$$

while $L_{pq}(s)\equiv L_{qp}(s)$. Later the operator $L_{pq}(s)$ for the transmission of excitation from one coordinate to another will be called the *transfer* dynamic compliance operator to distinguish it from the *local* dynamic compliance operator obtained from the transfer one when $p\equiv q$.

Because (6.33) is linear, the expression for the operator $L_{pq}(s)$ coincides with the relationship for the Laplace transforms of the functions transformed by operator $L_{pq}(s)$ [79]

$$L_{pq}(s)=\frac{\mathcal{L}\{x_p(t)\}}{\mathcal{L}\{P_q(t)\}}=\frac{\int_0^\infty x_p(t)\exp(-st)dt}{\int_0^\infty P_q(t)\exp(-st)dt}.$$

Here, $\mathcal{L}$ is the Laplace transform operator; s is a complex value. For this reason, when using the operator description (6.33) later, we will widely use the conditions of identity between the operator L_{pq}(s) and the Laplace transform. In particular, it follows that

$$x_p(t)=\int_0^t h_{pq}(t-\tau)P_q(\tau)d\tau, \tag{6.34}$$

where the impulse response $h_{pq}(t)$ is the inverse Laplace integral from the dynamic compliance operator [78]

$$h_{pq}(t) = \mathscr{L}^{-1}\{L_{pq}(s)\} = \frac{1}{2\pi j}\int_{\alpha - j\infty}^{\alpha + j\infty} L_{pq}(s)\exp(st)ds \quad (t > 0). \tag{6.35}$$

Here, $\mathscr{L}^{-1}$ is the inverse Laplace transform operator; $j^2 = -1$; α is the abscissa of convergence chosen to satisfy the condition

$$\int_0^{\infty} \left| h_{pq}(t) \right| \exp(-\alpha t)dt < \infty .$$

For a system with total dissipation $\alpha = 0$. In this case, when $s = j\omega$ the Laplace transform becomes the Fourier transform.

Dynamic compliance operators can be constructed not only for a system with a finite number of degrees of freedom, but for also for continuous systems and for systems consisting of a mixture of structures including elements with lumped and distributed parameters. In general, these operators can directly form equations like (6.29) for elastic systems of sufficiently general form.

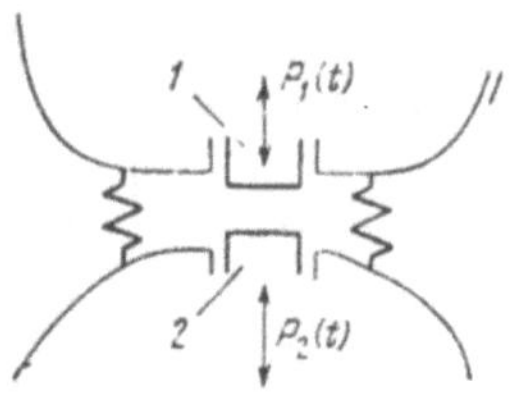

Fig. 6.4

Let us examine initially a one-dimensional vibro-impact interaction, (Fig. 6.4) between two elements *1*, *2* (particles, plane sections, etc.) of a linear stationary system of general structure under the action of external forces $P_1(t)$ and $P_2(t)$ applied to the contacting elements. Calling their absolute displacements $x_1(t)$ and $x_2(t)$ and introducing the characteristic of impact pair $\Phi(x, sx)$, where $x(t) = x_2(t) - x_1(t)$, we will write the corresponding displacement through dynamic compliance operators for the elements

$$\begin{aligned} x_1(t) &= L_{11}(s)\{\Phi[x(t), sx(t)] + P_1(t)\} + L_{21}(s)\{-\Phi[x(t), sx(t)] + P_2(t)\}, \\ x_2(t) &= L_{12}(s)\{\Phi[x(t), sx(t)] + P_1(t)\} + L_{22}(s)\{-\Phi[x(t), sx(t)] + P_2(t)\}. \end{aligned} \tag{6.36}$$

Thus, to describe the interaction of impact pairs, it is necessary to know the matrix of dynamic compliance operators for impact pair $\Lambda(s)$

$$\Lambda(s) = \begin{Vmatrix} L_{11}(s) & L_{21}(s) \\ L_{12}(s) & L_{22}(s) \end{Vmatrix}. \tag{6.37}$$

as well as the force characteristics, $\Phi(x, sx)$. This matrix is symmetrical since $L_{12}(s) \equiv L_{21}(s)$. In the event when the elements of the impact pairs belong to a disconnected system, $L_{12}(s) \equiv 0$ and matrix (6.37) takes the diagonal form.

Introducing the nomenclature $L_{11}(s)P_1(t) + L_{21}(s)P_2(t) = x_{10}(t)$; $L_{12}(s)P_1(t) + L_{22}(s)P_2(t) = x_{20}(t)$ and rewriting the system of equations (6.36) using this nomenclature, we obtain

$$\begin{aligned} x_1(t) &= x_{10}(t) + [L_{11}(s) - L_{21}(s)]\Phi[x(t), sx(t)], \\ x_2(t) &= x_{20}(t) + [L_{12}(s) - L_{22}(s)]\Phi[x(t), sx(t)]. \end{aligned} \tag{6.38}$$

The expressions for $x_{10}(t)$ and $x_{20}(t)$ define the vibrations of the elements of the impact pair in the absence of impact interactions between them and completely characterise the external excitation applied to the impact pairs. In many cases, this relationship and the elements of the dynamic compliance operator matrix can be defined from the results of direct measurement on the object under study. For example, it is easy to apply when investigating vibro-impact interactions of two independent mechanical systems since in this case, $L_{12}(s) \equiv 0$ and the characteristics specified can be defined separately for each system. Applying some standard of force excitation in the form of a unit function $\eta(t)$ or an impulse function $\delta(t)$ to the elements of the impacting pair under zero initial conditions and taking the reactions as $g(t)$ or $h(t)$ we can define the dynamic compliance of $i-th$ element $(i = 1, 2)$ by the formula for the Laplace transform

$$L_{ii}(s) = s\int_0^\infty g(\tau)\exp(-s\tau)d\tau, \qquad L_{ii}(s) = \int_0^\infty h(\tau)\exp(-s\tau)d\tau.$$

Since, when $s = j\omega$, the corresponding operator coincides with the *dynamic compliance* of the systems, they can be measured directly as the response on the standard sweep sine test. As a result, we find

$$L_{ii}(j\omega) = |L_{ii}(j\omega)|\exp[j\varphi_{ii}(\omega)],$$

where $|L_{ii}(j\omega)|$ is the gain coefficient of the harmonic signal at frequency ω; $\varphi_{ii}(\omega)$ is the value of its phase shift. Currently, special apparatus is available to measure dynamic compliance [75, 199]. The value of x_{i0} can also be directly recorded by applying an external force $P_i(t)$.

Similar methods to describe the characteristics can usually be used even in the presence of linear coupling between the elements of the impact pairs, i.e., when $L_{12}(s) \not\equiv 0$. In order to exclude vibro-impact interaction in this case, it is necessary, with great care, to normalise the standard signal. To record the value of $x_{i0}(t)$ we can alternately switch on the sources of excitation with subsequent summation of the resulting measurements.

Let us return to the examination of the system of equations (6.38). Subtracting the first equation from the second we obtain the expression for the relative coordinate x:

$$x(t) + L(s)\Phi[x(t), sx(t)] = \sum_{i=1}^{2} B_i(s)P_i(t). \tag{6.39}$$

Here, $L(s) = L_{11}(s) + L_{22}(s) - 2L_{12}(s)$; $B_1(s) = L_{12}(s) - L_{11}(s)$; $B_2(s) = L_{22}(s) - L_{21}(s)$. Writing the operator equations (6.39) in the form $L(s) = \mathcal{R}(s)/\mathcal{D}(s)$; $B_i(s) = \mathcal{S}_i(s)/\mathcal{D}(s)$, we come once again to equations of type (6.31). However, in the presence of elements with distributed parameters in the system, the corresponding operator will not be rational but transcendental functions for the argument s.

Proceeding from the description of the operators as transfer functions to the time domain similar to (6.34) we obtain from (6.39) nonlinear integral equations which completely describe the process of the relative change in coordinate $x(t)$, either in transient or stationary regimes

$$x(t)+\int_0^t h(t-\tau)\Phi[x(\tau),\dot{x}(\tau)]d\tau=\sum_{i=1}^{2}\int_0^t h_i(t-\tau)P_i(\tau)d\tau\,. \tag{6.40}$$

Here, $h(t)=\mathcal{L}^{-1}\{L(s)\};\ h_i(t)=\mathcal{L}^{-1}\{B_i(s)\}$. The right-hand side of equation (6.40) is a known time function, characterising the relative motion of the impact pairs, in the absence of impact interactions. Calling this $\vartheta(t)$ we obtain

$$x(t)+\int_0^t h(t-\tau)\Phi[x(\tau),\dot{x}(\tau)]d\tau=\vartheta(t)\,. \tag{6.41}$$

Equation (6.41) is similar to the integral equation for contact interaction (1.44) considered in §1 and is generalised in cases of *vibro-impact* interaction, characterised by the systematic change from contacting to non-contacting motion of the elements of the impact pair. This describes a process starting from zero initial conditions. Taking into account the extinction of transient motion in the systems described, integral equations for the stationary process can be written in the following form, which do not depend on the initial conditions

$$x(t)+\int_{-\infty}^t h(t-\tau)\Phi[x(\tau),\dot{x}(\tau)]d\tau=\vartheta(t)\,. \tag{6.42}$$

3. The transformation of the variables in equation (6.41) can be represented by the schematic diagram shown in Fig. 6.5. Here, *1* is a nonlinear element transforming the dynamic process $x(t)$ studied into the force impulses $F(t)=\Phi[x(t),\dot{x}(t)]$ corresponding to the characteristics of the impact pair, *2* is a linear dynamic block describing the elastic vibration $\tilde{\chi}(t)$

$$\tilde{\chi}(t)=\int_0^t h(t-\tau)F(\tau)d\tau \tag{6.43}$$

under the effect of the force impulse, the shaded sector of the comparator *3* denotes subtraction. A similar structural scheme has traditionally been used for a wide range of nonlinear systems: relay operated [212], with variable structure [83] etc. The specific features of each class of system reflects to a large extent the character of the transformation produced by the nonlinear elements.

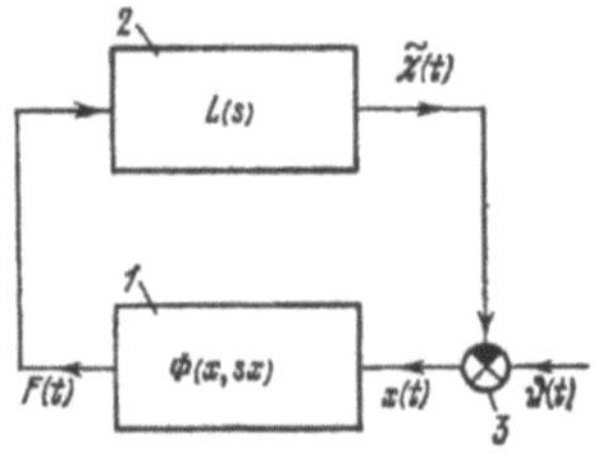

Fig. 6.5

As already mentioned, the characteristic of vibro-impact systems appears more strongly in the equations when using the hypotheses of stereomechanical theory for impact,. hence we will consider transformations which, for example, consist of nonlinear elements with the characteristics given in (6.19). We will assume that collisions occur at a certain instant of time t_ν with relative velocity $\dot{x}_{\nu-}$ $(\nu=1,2,\ldots)$. We will define the impulse J_ν of the force (6.19) for both phases of impact at the instant t_ν:

$$J_\nu = \int_{t_\nu - 0}^{t_\nu + 0} \frac{M\dot{x}_{\nu-}^2}{2}\delta_+[x(t)-\Delta]dt = \frac{M\dot{x}_{\nu-}^2}{2}\left[\int_{\Delta}^{\Delta+\varepsilon}\frac{\delta_+(x-\Delta)dx}{\dot{x}_{\rm I}} + \int_{\Delta+\varepsilon}^{\Delta}\frac{\delta_+(x-\Delta)dx}{\dot{x}_{\rm II}}\right],$$

where $\dot{x}_{\rm I}, \dot{x}_{\rm II}$ are the average velocities in the I and II phases of impact. Since $\dot{x}_{\rm I} = \dot{x}_{\nu-}/2, \dot{x}_{\rm II} = -\dot{x}_{\nu-}/2$ in the case considered using (6.15) we find $J_\nu = 2M\dot{x}_{\nu-}$. This result, of course, can be obtained straight away from impulse-momentum equation. Thus, as a result of transformation of process $x(t)$ by the nonlinear element, the outcome will be a series of force impulses, e.g.

$$F(t) = \Phi_E[x(t), \dot{x}(t)] = 2M\sum_\nu \dot{x}_{\nu-}\delta(t-t_\nu)$$

where t_ν is the instant when $x(t)$ reaches the value $x = \Delta$. For a periodic process with one collision per period T, taking into account the energy loss during impact, we obtain, similarly

$$F_1(t) = M(1+R)\dot{x}_- \sum_{\nu=-\infty}^{\infty}\delta(t-\nu T). \tag{6.44}$$

For symmetrical vibro-impact process, caused by double-sided impact pairs with alternate impacts on the stops

$$F_2(t) = M(1+R)\dot{x}_-\left[\sum_{\nu=-\infty}^{\infty}\delta(t-\nu T) - \sum_{\nu=-\infty}^{\infty}\delta(t-\nu T-\frac{T}{2})\right] = F_1(t) - F_1\left(t-\frac{T}{2}\right). \tag{6.45}$$

Consequently, the nonlinear elements in the schematic structure of vibro-

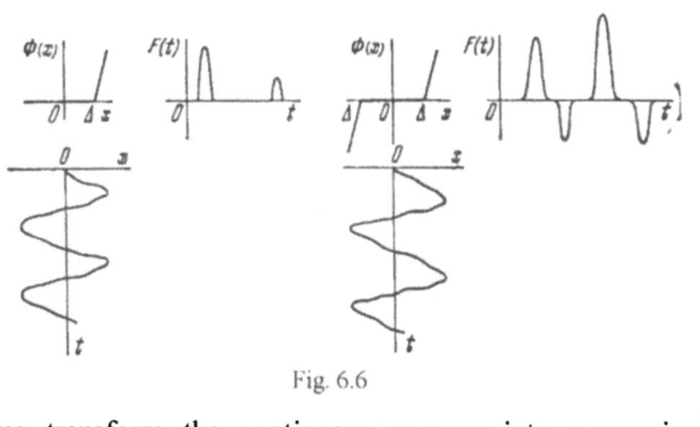

Fig. 6.6

impact systems transform the continuous process into successive impulses, modulated by the velocity of the input process at the instant when the process reaches the threshold value. Such nonlinear components of the structural scheme we will call *impact elements*. The impact elements have *positional-impulse* effects which specifically combine the effects of the relayed and impulse elements. When using descriptions of vibro-impact systems based on other impact theories, the actual form of transformation applied to the impacting elements changes and is reflected by the form of the generating impulse, but the general character of the effects of impact elements is preserved. Fig. 6.6 shows examples of transformations representing impact elements with characteristics (6.7), (6.8).

The force impulse generated by the impacting elements is absorbed by the linear block *2* (Fig, 6.5) for which, the *transfer function*, according to the results of section 2, takes the form of a dynamic compliance operator for the elastic system $L(s) = \mathcal{R}(s)/\mathcal{D}(s)$. We will pause here to consider the characteristics of dynamic compliance which has considerable importance for the study that follows.

For the linear part of the system (6.28), $L(s)$ takes the form of a rational fraction

$$L(s) = \frac{\mathcal{R}(s)}{\mathcal{D}(s)} = \frac{h_0 s^{2n-2} + h_1 s^{2n-3} + \ldots + h_{2n-2}}{d_0 s^{2n} + d_1 s^{2n-1} + \ldots + d_{2n}}. \tag{6.46}$$

This shows that the dynamic compliance $L(j\omega)$ has a *filtering effect*, i.e.,

$$\lim_{\omega \to \infty} |L(j\omega)| = 0. \tag{6.47}$$

Remembering that the impulse response $h(t)$ for the system with operators of the form (6.46) can be defined by the formula [97]

$$h(t) = \sum_{k=1}^{l} \sum_{m=1}^{\nu_k} \frac{H_{km}}{(\nu_k - m)!} t^{\nu_k - 1} \exp(s_k t) \qquad (t \geq 0), \tag{6.48}$$

where

$$H_{km} = \frac{1}{(m-1)!} \left[\frac{d^{m-1}}{ds^{m-1}} \frac{(s - s_k)^{\nu_k} \mathcal{R}(s)}{\mathcal{D}(s)} \right]_{s=s_k}.$$

Here, l is the number of different roots s_k of the characteristic equation for the system

$$\mathcal{D}(s) = 0; \tag{6.49}$$

where ν_k is a multiple for the $k-th$ root $(\nu_1 + \nu_2 + \ldots + \nu_l = 2n)$. In the following we will limit consideration to systems having only simple roots. In this case,

$$h(t) = \sum_{k=1}^{2n} \frac{\mathcal{R}(s_k)}{\mathcal{D}'(s_k)} \exp(s_k t) \qquad (t \geq 0). \tag{6.50}$$

where $\mathcal{D}'(s_k) = (d\mathcal{D}/ds)_{s=s_k}$.

We are interested mainly in the resonant response of vibro-impact systems. These are displayed if the system is weakly dissipative, i.e., if the coefficient n_{pq} of the quadratic form B which represents the dissipative effect of the linear part of (6.28) is small.

Let us rewrite the equations for the linear part of the system (6.28) with normal coordinates $z_1, \ldots, z_2$. Using [95]

$$s^2 z_p + \sum_{q=1}^{n} b_{pq} s z_q + \Omega_p^2 z_p = Q_p(t) \qquad (p = 1, \ldots, n), \tag{6.51}$$

whence

$$z_p = \sum_{q=1}^{n} A_{pq} x_q, \quad Q_p(t) = \sum_{q=1}^{n} A_{pq} P_q(t), \tag{6.52}$$

where Ω_p are the natural frequencies of the conservative system; b_{pq} are small quantities: A_{pq} is the normalised coefficient of the $p-th$ mode of vibration. According to (6.51) the characteristic equation (6.49) now takes the following form:

$$\mathscr{D}(s)=\begin{vmatrix} s^2+b_{11}+\Omega_1^2 & b_{12}s & \cdots & b_{1n}s \\ b_{21}s & s^2+b_{22}s+\Omega_2^2 & \cdots & b_{2n}s \\ \cdots\cdots & \cdots\cdots & \cdots & \cdots\cdots \\ b_{n1}s & b_{n2}s & \cdots & s^2+b_{nn}s+\Omega_n^2 \end{vmatrix}=0. \qquad (6.53)$$

Expanding the determinant and neglecting the products of small terms we obtain

$$\mathscr{D}(s)=\prod_{\nu=1}^{n}(s^2+2r_\nu\Omega_\nu s+\Omega_\nu^2)=0 \qquad (r_\nu=b_{\nu\nu}/(2\Omega_\nu)). \qquad (6.54)$$

As a result, the roots of the characteristic equation will be complex with the conjugate pairs

$$s_\nu\approx -r_\nu\Omega_\nu\pm j\Omega_\nu, \qquad (6.55)$$

On the basis of the assumptions made the solutions for the equations (6.51) can be written in the following operator form:

$$z_p(t)=\frac{Q_p(t)}{s^2+2r_p\Omega_p s+\Omega_p^2} \qquad (6.56)$$

Using (6.52), (6.56) we find the connection between the original coordinates and the excitation of the linear system

$$x_p(t)=\sum_{\nu=1}^{n}A_{\nu p}z_\nu(t)=\sum_{\nu=1}^{n}\frac{A_{\nu p}}{s^2+2r_\nu\Omega_\nu s+\Omega_\nu^2}\sum_{k=1}^{n}A_{\nu k}P_k(t). \qquad (6.57)$$

Assuming that $P_k(t)\equiv 0$ when $k\neq q$ in (6.56), $P_q(t)=1$, we obtain in agreement with (6.33) the equation for the dynamic compliance operator

$$L_{pq}(s)=\sum_{\nu=1}^{n}\frac{A_{\nu p}A_{\nu q}}{s^2+2r_\nu\Omega_\nu s+\Omega_\nu^2}. \qquad (6.58)$$

According to (6.58)

$$L_{pq}(s)\equiv L_{qp}(s) \qquad (6.59)$$

When $s=j\omega$, ignoring small terms of second order we find

$$L_{pq}(j\omega)=\sum_{\nu=1}^{n}\frac{A_{\nu p}A_{\nu q}}{(\Omega_\nu^2-\omega^2)^2+4r_\nu^2\Omega_\nu^2\omega^2}(\Omega_\nu^2-\omega^2-j2r_\nu\Omega_\nu\omega). \qquad (6.60)$$

Note that when the energy dissipation is neglected, then the imaginary part of the dynamic compliance reduces to zero. The modulus of dynamic compliance is found to be

$$\left|L_{pq}(j\omega)\right|=\sum_{\nu=1}^{n}\frac{A_{\nu p}A_{\nu q}}{\sqrt{(\Omega_\nu^2-\omega^2)^2+4r_\nu^2\Omega_\nu^2\omega^2}}. \qquad (6.61)$$

Hence, the dynamic compliance operator for the linear system studied with *n* degrees of freedom, can be represented in the form of a sum of *n* resonant filters, with natural frequencies of the conservative system. When $\omega=\Omega_\nu$, a peak occurs in the dynamic compliance due to the appearance of the small quantity $2r_\nu\Omega_\nu^2$ in the denominator of the *ν-th* term. The width of the peak defines the band-width of the filter, and is proportional to the value $r_\nu\Omega_\nu$. As the number of modes of vibration v increases the "amplitude" of the peak decreases rapidly. Fig 6.7 shows examples of the expression (6.61) for the case $p = q$. A detailed analysis of frequency characteristics of dynamic compliance as described here can be found in [123].

In a similar way, the equations for the dynamic compliance operators of a series of systems with distributed parameters can be presented, for example, straight elastic rods undergoing axial, torsional or transverse vibrations [24].* The number of vibration modes for this increases infinitely, and the coefficients for each mode remain functions of the continuous coordinate x, which represents the position of the section under consideration. Denoting these coefficients as $A_v(x)$ we find when transferring the action to section $x = p$ from the concentrated force applied at the section $x = q$,

$$L_{pq}(s) = \sum_{v=1}^{\infty} \frac{A_v(p)A_v(q)}{s^2 + 2r_v\Omega_v s + \Omega_v^2}. \tag{6.62}$$

During the treatment of such systems, the mechanism of energy dissipation is frequently represented in the form of internal friction in the material expressed in terms of the absorbtion coefficient $\varkappa$ (see §1 section 2). In this case, $r_v = \varkappa\Omega_v/(4\pi\omega)$ [176] and the equations for the dynamic compliance take the form

$$L_{pq}(j\omega) = \sum_{v=1}^{\infty} \frac{A_v(p)A_v(q)}{(\Omega_v^2 - \omega^2)^2 + \varkappa^2\Omega_v^4/(4\pi^2)} \left(\Omega_v^2 - \omega^2 - j\frac{\varkappa\Omega_v^2}{2\pi}\right), \tag{6.63}$$

whence

$$\left|L_{pq}(j\omega)\right| = \sum_{v=1}^{\infty} \frac{A_v(p)A_v(q)}{\sqrt{(\Omega_v^2 - \omega^2)^2 + \varkappa^2\Omega_v^2/(4\pi^2)}}. \tag{6.64}$$

As the mode number v increases, the second term under the radical grows rapidly, therefore the practical value can only take a few initial modes.

Frequently during the study of the mathematical model of real linear systems, the equations for dynamic compliance can be determined directly by searching for a steady state solution due to the effect of harmonic forces with unit amplitude. A convenient procedure will be explained later (see §15 section 1, §16, section 2).

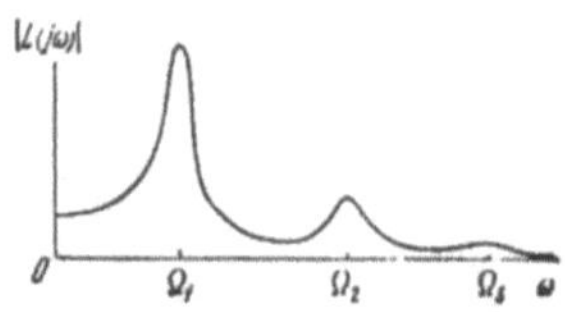

Fig. 6.7

4. To understand the character of the transformation of the variables occurring in the description of vibro-impact systems of the type which can be reduced to the structural scheme of Fig. 6.5, let us consider a process undergoing a *T-periodic* sequence of impulses of the form (6.44) or (6.45), through a linear dynamic block with characteristic $L(s)$, represented as the sum of dynamic compliance operators of the form (6.58). Impulses are produced by impacting elements when the system performs periodic motion with one impact against each stop during a period.

* In [24] instead of dynamic compliance, the name *harmonic influence coefficient* is used.

Functions (6.44), (6.45) define the following Fourier series [191]:

$$F_1(t) = \frac{J}{T}\sum_{m=-\infty}^{\infty}\exp(mj\omega t) \qquad (j^2 = -1), \tag{6.65}$$

$$F_2(t) = \frac{2J}{T}\sum_{m=-\infty}^{\infty}\exp(2m+1)j\omega t \tag{6.66}$$

where J is the value of the impact impulse; $\omega = 2\pi / T$.

These series are divergent, however, they can be considered as *generalised* periodic functions, which allows regular operations to be performed with them, such as addition and multiplication, differentiation, integration etc. [191, 196].

Applying the principle of superposition as the excitation (6.65) or (6.66) pass through a linear stationary system with frequency characteristic $L(j\omega)$ the resulting periodic process will take the following form:

$$\chi_{1J}(t) = \frac{J}{T}\sum_{m=-\infty}^{\infty}L(mj\omega)\exp(mj\omega t) \tag{6.67}$$

in the non-symmetrical case, and

$$\chi_{2J}(t) = \frac{2J}{T}\sum_{m=-\infty}^{\infty}L[(2m-1)j\omega]\exp(2m-1)j\omega t \tag{6.68}$$

in the symmetrical. For some functions $L(j\omega)$, series (6.67), (6.68) can converge. Since only these cases conform to actual process in vibro-impact systems, we can find the general criterion for convergence of the series (6.67), (6.68). Assuming an integration of quadratic function of $\chi_{iJ}(t)$ and using Parseval's equality [90] we find for the non-symmetrical case

$$\int_0^T \chi_{1J}^2(t)dt = \frac{J^2}{T}\sum_{m=-\infty}^{\infty}L(mj\omega)L(-mj\omega) = \frac{J^2}{T}|L(0)|^2 + \frac{2J^2}{T}\sum_{m=1}^{\infty}|L(mj\omega)|^2 . \tag{6.69}$$

A similar equality can be found for the series (6.68).

The filtering effect of the linear part of the system, (6.47), is a necessary condition for the convergence of series (6.69). Since for sufficiently large m, each term in (6.61) is of the same order as the quantity $(m\omega)^{-2}$, series (6.69) also demonstrates sufficient conditions for convergence [90]. For this it is necessary that the purely imaginary poles and poles with positive real parts are absent from the operator $L(s)$. Since the poles of the operator $L(s)$ are roots of the characteristic equation (6.53), according to (6.65) the conditions indicated are satisfied for the elastic system considered, because of total dissipation. The existence of zero poles for the linear part of the system is also permissible. This corresponds to free motion as a rigid body.

Thus, the real qualities of a mechanical system ensure that the harmonic series (6.68), (6.69) converge to the periodic functions $\chi_{1J}(t)$ or $\chi_{2J}(t)$ with period T. These functions, which are the stationary reactions of the linear system to the periodic δ-function, have been given the

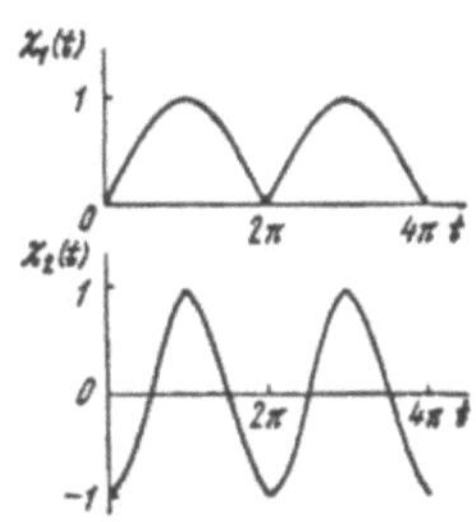

Fig. 6.8

name *impulse-frequency characteristics of the first and second types* [191]. Below, we shall call them *periodic Green's functions*, following mechanical traditions. It can be shown that for linear operators of type (6.46), with simple poles s_k, the periodic Green's functions $\chi_1(t)$ or $\chi_2(t)$ (the absence of the index J implies that the functions are considered when J = 1) can be defined by the formulas

$$\chi_1(t) = \sum_{k=1}^{2n} \frac{\mathcal{R}(s_k)}{\mathcal{D}'(s_k)} \frac{\exp(s_k t)}{1-\exp(s_k t)} \quad (t \in (0,T)),$$

$$\chi_2(t) = \sum_{k=1}^{2n} \frac{\mathcal{R}(s_k)}{\mathcal{D}'(s_k)} \frac{\exp(s_k t)}{1+\exp(s_k T/2)} \quad \left(t \in \left(0,\frac{T}{2}\right)\right), \tag{6.70}$$

and if, among the roots s_k there are purely imaginary ones, then frequency $\omega = 2\pi/T$ must not be equal to or a multiple of them. From now on, for brevity in such cases, when the specific characteristics $\chi_1(t)$ or $\chi_2(t)$ have no role we will omit the numerical index.

The periodic Green's functions are linearly related to the value of the impulse i.e.,

$$\chi_J(t) = J\chi(t), \tag{6.71}$$

We note that the degree of the polynomial in the denominator which is two orders greater than that of the numerator in the operator $L(s)$ ensures that the function $\chi(t)$ is continuous along the whole time axis. However, the derivative, $\dot{\chi}_1(t)$ takes a jump when $t = mT$, and $\dot{\chi}_2(t)$ when $t = mT/2$ $(m = 0, \pm 1, \ldots)$ [191].

Suppose, for example,

$$L(s) = (s^2 + bs + \Omega^2)^{-1}, \tag{6.72}$$

i.e., the linear part of the system has represents an oscillator with viscous damping. Then from formulae (6.67), (6.68) we find

$$\chi_1(t) = \frac{1}{T} \sum_{m=-\infty}^{\infty} \frac{\exp(mj\omega t)}{bmj\omega + \Omega^2 - m^2\omega^2}, \tag{6.73}$$

$$\chi_2(t) = \frac{2}{T} \sum_{m=-\infty}^{\infty} \frac{\exp[(2m-1)j\omega t]}{b(2m-1)j\omega + \Omega^2 - (2m-1)^2\omega^2} \tag{6.74}$$

Assuming that $\Omega^2 > b^2/4$ we find from formula (6.70)

$$\chi_1(t) = \frac{\exp(-bt/2)}{\Omega_1} \frac{\sin\Omega_1 t + \exp(-bT/2)\sin\Omega_1(T-t)}{1+\exp(-bT) - 2\exp(-bT/2)\cos\Omega_1 T} \quad (t \in [0,T]), \tag{6.75}$$

$$\chi_2(t) = \frac{\exp(-bt/2)}{\Omega_1} \frac{\sin\Omega_1 t + \exp(-bT/4)\sin\Omega_1(t-T/2)}{1+\exp(-bT) + 2\exp(-bT/4)\cos\Omega_1 T/2} \quad (t \in [0,T/2]), \tag{6.76}$$

where $\Omega_1 = \sqrt{\Omega^2 - b^2/4}$ is the natural frequency of the oscillator. Fig. 6.8 shows examples obtained for this relationship constructed when $T = 2\pi$ s, $b = 0$,

$\Omega = 0.51/s$. Note that periodic Green's functions have the dimensions of [*s/kg*].

5. With the help of periodic Green's functions, the periodic single impact solution $x(t)$ of the equation of a vibro-impact system for relative motion of the elements of the impact pair can be written in the form (see Fig. 6.5)

$$x(t) = \vartheta(t) - \chi_J(t), \tag{6.77}$$

where $\vartheta(t)$ is the periodic relative motion of the elements obtained from the solution of the corresponding linear equations with prescribed excitation when impact interactions are neglected. Note that in the same form (6.77) we have previously obtained the final expression for the law of motion when solving the problem of forced vibration of an oscillator impacting against stops (see (2.14), (3.13)). In those cases, $\vartheta(t)$ was written in the form

$$\vartheta(t) = a\cos(\omega t + \varphi), \tag{6.78}$$

where a is a value found from the solution of the linear problem; φ is the unknown phase of the motion relative to the instant of collision which is considered as the origin for time measurement. The solution $\vartheta(t)$ can be written in the form (6.78) for systems of arbitrary dimension excited by harmonic external force.

The circumstances specified allow us to develop the following compact form for exact periodic solutions. Using (6.71), (6.78), the periodic single impact solution (6.77) takes the form

$$x(t) = a\cos(\omega t + \varphi) - J\chi(t), \tag{6.79}$$

where the unknown phase φ and impulse J are defined by the conditions of collisions

$$x(0) = \Delta, \qquad J = (1+R)M\dot{x}_-(0), \tag{6.80}$$

Taking account of the coordinate axis direction we must also have $J \geq 0$.

Applying the conditions (6.80) to (6.79), we find

$$a\cos\varphi = \Delta + J\chi(0), \qquad a\sin\varphi = -\frac{J}{\omega}\left[\dot{\chi}_-(0) + \frac{1}{(1+R)M}\right]. \tag{6.81}$$

Here, $\dot{\chi}_-(0)$ is the left value of the derivative of the periodic Green's functions at the point of discontinuity. Since the periodic Green's function is usually assigned in the sector [0, *T*] in the non-symmetrical case or [0,*T*/2] in the symmetrical, then the value of $\dot{\chi}_-(0)$ is defined in the first case from the equation $\dot{\chi}_-(0) = \dot{\chi}_{1-}(T)$ and in the second case from the equation $\dot{\chi}_-(0) = -\dot{\chi}_{2-}(T/2)$.

Solving the system of equations (6.81) in terms of the unknowns we obtain finally

$$J = \frac{-\Delta\chi(0) \pm \sqrt{a^2\chi^2(0) - \dfrac{1}{\omega^2}\left[\dot{\chi}_-(0) + \dfrac{1}{(1+R)M}\right]^2 (\Delta^2 - a^2)}}{\chi^2(0) + \dfrac{1}{\omega^2}\left[\dot{\chi}_-(0) + \dfrac{1}{(1+R)M}\right]^2} \quad (J \geq 0), \tag{6.82}$$

$$\sin\varphi = -\frac{J}{\omega a}\left[\dot{\chi}_{-}(0) + \frac{1}{(1+R)M}\right], \quad \cos\varphi = \frac{\Delta + J\chi(0)}{a}, \tag{6.83}$$

For perodic multi-impact regimes with v consecutive impacts per period against each stop the solution can be found in the form

$$x(t) = \vartheta(t) - \sum_{i=1}^{v} J_i\chi(t - t_i), \tag{6.84}$$

The unknown value of the impact impulse J_i and the instant of collision t_i can be found from a system of transcendental equations obtained from (6.84) using collision conditions like (6.80):

$$x(t_i) = \Delta, \quad J_i = (1+R)M\dot{x}_{-}(t_i) \quad (J_i \geq 0)$$

The solutions (6.79), (6.84) must be tested in the absence of additional crossings: $x(t) \leq \Delta$ in the case of single-sided stops and $|x(t)| \leq \Delta$ for symmetrical double-sided stops. In §14 section 2, a method to study the stability of these solutions will be described.

Calculation schemes which represent exact methods to find periodic solutions, are to be preferred when dealing with large systems by comparison to the procedures applied in Chapter 1, since the use of periodic Green's functions in the form of series (6.67), (6.68) allows us to avoid time-consuming procedures to find the roots of the characteristic equations obtained when writing the general solution in the form (5.6).

Moreover, this method allows the study of systems with distributed parameters, and also of systems where the linear elements are described only in terms of the dynamic compliance characteristics of the contacting elements. An important detail is also that, by limiting series (6.67), (6.68) to finite terms using filtering effects of linear parts of the system, a continuous approximation to the exact solution can be obtained. In particular, by neglecting terms of the series corresponding to $m > 1$ we obtain a harmonic approximation, which has an important meaning for the understanding of the qualitative picture of the process, as will be shown later.

It is just as important that it is possible to expand the operator $L(j\omega)$, which appears in (6.67), (6.68) as a series in terms of the natural mode of vibration for the linear part of the system, and to reduce the basic mode, which in practice influences the regimes considered in the frequency range investigated. This procedure was introduced into the study of vibro-impact systems in the works of Zevin [224, 226].

Using periodic Green's functions allows us to convert equation (6.42) to Hammerstein's type of integral equation when we describe the periodic motion of a vibro-impact system in terms of any form of force characteristics for an impact pair $\Phi(x,\dot{x})$, allowing us to construct effective schemes for continuous approximation and to estimate the errors of the approximate solutions [191]. For a steady periodic solution $x(t)$ expanding the periodic function $F(t) = \Phi[x(t), \dot{x}(t)]$ as a Fourier series, we find

$$F(t) = \sum_{m=-\infty}^{\infty} F_m \exp(mj\omega t),$$

where the coefficients of the series are defined by the formula

$$F_m = \frac{1}{T}\int_0^T F(\tau)\exp(-mj\omega\tau)d\tau . \tag{6.85}$$

For the basic structural scheme of Fig. 6.5, we find, using superposition of the motion on the output from the linear block,

$$x(t) = \vartheta(t) - \sum_{m=-\infty}^{\infty} L(mj\omega)F_m \exp(mj\omega t) . \tag{6.86}$$

Substituting (6.85) into (6.86) and changing the order of integration and summation we obtain

$$x(t) = \vartheta(t) - \int_0^T \left[\frac{1}{T}\sum_{m=-\infty}^{\infty} L(mj\omega)\exp[mj\omega(t-\tau)]\right]F(\tau)d\tau$$

Using (6.67), (6.71) we can see that the term in the square brackets represents the periodic Green's function $\chi(t-\tau)$ for the linear part of the system. Using this we eventually obtain

$$x(t) = \vartheta(t) - \int_0^T \chi(t-\tau)\Phi[x(\tau),\dot{x}(\tau)]d\tau , \tag{6.87}$$

where $\vartheta(t+T) = \vartheta(t)$.

According to equation (6.87) the character of the periodic motion of the vibro-impact system depends weakly on the actual description of the force interaction during impact and consequently on the form of the function $\Phi(x,\dot{x})$. Obviously, examining the integral on the right-hand side

$$\int_0^T \chi(t-\tau)\Phi[x(\tau),\dot{x}(\tau)]d\tau = \int_0^\varepsilon \chi(t-\tau)\Phi[x(\tau),\dot{x}(\tau)]d\tau ,$$

where the small value ε limits the length of the interval during which $\Phi[x(\tau),\dot{x}(\tau)] \not\equiv 0$ we find using average theorem and assuming continuous periodic Green's functions *,

$$\begin{aligned}\int_0^\varepsilon \chi(t-\tau)\Phi[x(\tau),\dot{x}(\tau)]d\tau &= \chi[t-\varepsilon\theta(t)]\int_0^\varepsilon \Phi[x(\tau),\dot{x}(\tau)]d\tau = \\ &= J\chi[t-\varepsilon\theta(t)] \qquad (0<\theta(t)<1).\end{aligned} \tag{6.88}$$

Here

$$J = \int_0^\varepsilon \Phi[x(\tau),\dot{x}(\tau)]d\tau \tag{6.89}$$

is the impulse force interaction during impact.

Noting that $\varepsilon\theta(t) << T$ and consequently for continuous periodic Green's function $\chi[t-\varepsilon\theta(t)] \approx \chi(t)$, we conclude that the force characteristic $\Phi(x,\dot{x})$ in practice only influences the motion through its integral (6.89) in small areas of

---

* A sufficient condition for continuous periodic Green's function corresponds to the conditions for uniform convergence of the Fourier series of the form (6.67), (6.68) and consequently can be written in the form [185] $|L(mj\omega)| = O(1/m^{1+\lambda})$, where λ is an arbitrary positive number.

integration ε and consequently it can be approximated by any convenient expression leading to the identical value for the integral. For this reason, during the study of motion of vibro-impact systems, it is expedient to use the simplest method to describe the contact interaction.

From the transformation (6.88) the following important point follows, that whatever method is used to describe contact interaction for vibro-impact systems, when the linear part is represented by continuous periodic Green's function, then the single impact solution of (6.87) is close to the following:

$$x(t) \approx \vartheta(t) - J\chi(t) . \tag{6.90}$$

The general scheme for successive approximation for equation (6.87) takes the form

$$x_{i+1}(t) = \vartheta(t) - \int_0^T \chi(t-\tau)\Phi[x_i(\tau), \dot{x}_i(\tau)]d\tau , \tag{6.91}$$

where the correct choice for the initial approximation $x_0(t)$ has an important influence on the convergence of the iteration process. Calculation by continuous approximation does not require us to revert to the integral equation (6.87) but can in practice be done by introducing the general operator

$$x_{i+1}(t) = \vartheta(t) - L(j\omega)\Phi[x_i(t), \dot{x}_i(t)] . \tag{6.92}$$

A routine method to find the initial approximation for the periodic solution of a nonlinear system is the method of harmonic linearisation. A description of this method, adapted for systems with vibro-impact interactions and used for the analysis of the basic qualitative effects of solutions for various models of vibro-impact systems, will be dealt with at considerable length in the following two chapters of this book.

In conclusion, we direct the reader's attention to the possibility of directly constructing *transfer* processes for vibro-impact systems based on (6.42) using numerical procedures as described, for example, for integral equations in the monograph [169].

§7. Harmonic linearisation of vibro-impact systems

1. Let us now turn to a description of the idea and methods of harmonic linearisation of vibro-impact systems [34, 37]. The basis of the harmonic linearisation method lies in the principle of *harmonic balance*.

Let us examine the operator equation (see §6)

$$x(t) + L(s)\Phi[x(t), sx(t)] = \vartheta(t) , \tag{7.1}$$

where $\vartheta(t)$ is a given periodic function with period $T = 2\pi/\omega$. Assuming that (7.1) has a periodic solution of identical period T, we can write it in the form of a Fourier series

$$x(t) = \sum_{m=-\infty}^{\infty} x_m \exp(mj\omega t) , \tag{7.2}$$

$$\dot{x}(t) = j\omega \sum_{m=-\infty}^{\infty} x_m m \exp(mj\omega t) . \tag{7.3}$$

Substituting (7.2), (7.3) in the equation for the nonlinearity $\Phi(x, \dot{x})$, we obtain a similar T-periodic function of time $F(t) = \Phi[x(t), \dot{x}(t)]$, which can be represented by the following Fourier series:

$$F(t) = \sum_{m=-\infty}^{\infty} F_m \exp(mj\omega t), \tag{7.4}$$

where the coefficients F_m are defined by formula (6.85). Finally, the expansion of the external periodic excitation $\vartheta(t)$ can be written in the form

$$\vartheta(t) = \sum_{m=-\infty}^{\infty} \vartheta_m \exp(mj\omega t) \tag{7.5}$$

Substituting series (7.2) to (7.5) in equation (7.1) and equating the coefficients for identical exponents we find

$$x_m + L(mj\omega)F_m(x_0, x_1, x_{-1}, \ldots) = \vartheta_m \quad (m = 0, \pm 1, \pm 2, \ldots) \tag{7.6}$$

The solution of the transcendental equation (7.6) as an infinite series is not generally possible, but with several additional assumptions, this system can be truncated to finite values.

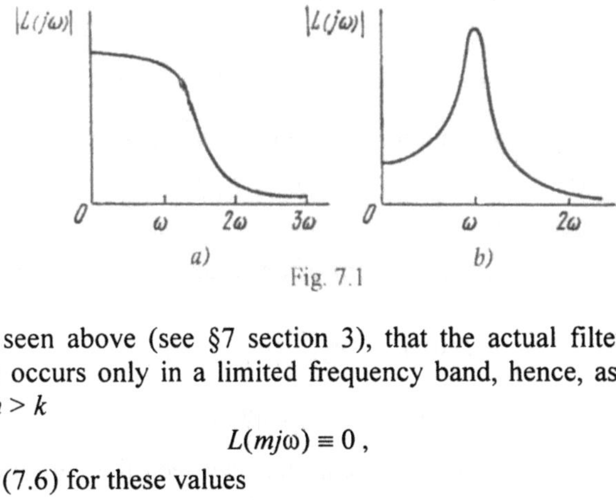

Fig. 7.1

We have seen above (see §7 section 3), that the actual filtering effect of a linear system occurs only in a limited frequency band, hence, assuming that for some value $m > k$

$$L(mj\omega) \equiv 0, \tag{7.7}$$

we find from (7.6) for these values

$$x_m = \vartheta_m \quad (|m| > k) \tag{7.8}$$

The linear system with frequency characteristics (7.7) is known as an *ideal filter*. In such cases, when the actual system can be idealised in the same way, then using (7.8) the system of equations (7.6) becomes as follows:

$$x_m + L(mj\omega)F_m(x_0, x_1, x_{-1}, \ldots, x_k, x_{-k}) = \vartheta_m \tag{7.9}$$
$$(m = 0, \pm 1, \ldots, \pm k)$$

and can be solved by either numerical or analytical procedures.

The simplest such system of finite equations arises when $k = 1$, i.e. in the case when the ideal filter does not pass harmonics higher than the fundamental.

The predominance of the fundamental over other harmonics in the vibration spectrum of the system is not only possible when limiting the bandwidth passed by the linear system (Fig. 7.1, *a*), but also when it demonstrates a strong resonance response (Fig. 7.1, *b*). Assuming that the vibration frequency coincides with resonance, we may say in this case that $L(j\omega) >> L(mj\omega)$ $(|m| > 1)$, and consequently idealisation (7.7) remains justifiable.

For the actual system, condition (7.7) is not exactly satisfied, and so we can only proceed to an *approximate* search for a periodic solution in the form of limited Fourier series when either *filtering* or *resonance* effects are present within the linear parts of the system [1]. Obviously, the closer the equation represents the effects described for the system, the better the degree of approximation. As was shown in §6 section 3, the actual characteristics of elastic systems ensure the effects described, hence it is completely justifiable to apply the approximations described in the study of vibro-impact systems.

When considering exact solutions to the problem of vibration of oscillators with stops, we saw that the basic qualitative information of the nonlinear behaviour of vibro-impact systems is completely reflected by the first harmonic, thus the search for and analysis of the first harmonic is expedient even in those cases when the filtering or resonance effects of the system do not appear strongly in the equations. The quickness and clearness of information about the basic nonlinear effects of the system obtained in this way allow us to form the initial concepts to construct rational numerical procedures and to estimate them qualitatively.

Suppose $\vartheta(t) = m_\vartheta + a\cos(\omega t + \varphi)$, then $\vartheta_0 = m_\vartheta$, $\vartheta_{\pm 1} = \frac{a}{2}\exp(\pm j\varphi)$, $\vartheta_m \equiv 0$ when $|m| > 1$. Searching for a solution $x(t)$ in the form

$$x(t) = m_x + a_x \cos\omega t + x_H(t) = x_h(t) + x_H(t) \tag{7.10}$$

where $x_H(t)$ contains higher harmonic components only, we find $x_0 = m_x$, $x_{\pm 1} = a_x/2$. From the system of equations (7.6) we obtain for the first two components

$$m_x + L(0)m_f = m_\vartheta \quad (m_F \equiv F_0), \tag{7.11}$$

$$\frac{a_x}{2} + L(j\omega)F_1 = \frac{a}{2}\exp(j\varphi). \tag{7.12}$$

Equation (7.11) determines the connection between the periodic components of the solution and the constant component m_x:

$$m_x = m_\vartheta - L(0)m_F. \tag{7.13}$$

Using (6.85) we can rewrite equation (7.12) in the form

$$a_x + L(j\omega)\frac{2}{T}\int_0^T F(\tau)\exp(-j\omega\tau)d\tau = a\exp(j\varphi). \tag{7.14}$$

We can represent the integral in equation (7.14) in the following form:

$$\frac{2}{T}\int_0^T F(\tau)\exp(-j\omega\tau)d\tau =$$
$$= \frac{2}{T}\int_0^T \Phi[x(\tau),\dot{x}(\tau)](\cos\omega\tau - j\sin\omega\tau)d\tau = a_x(k + j\omega\beta), \tag{7.15}$$

where

$$k = \frac{2}{Ta_x}\int_0^T \Phi[x(\tau), \dot{x}(\tau)]\cos\omega\tau d\tau,$$
$$\beta = \frac{2}{T\omega a_x}\int_0^T \Phi[x(\tau), \dot{x}(\tau)]\sin\omega\tau d\tau, \tag{7.16}$$

As a result, equation (7.14) can be written as:

$$a_x[1 + L(j\omega)(k + j\omega\beta)] = a\exp(j\varphi) \tag{7.17}$$

This is an exact expression for the parameters of the fundamental. For known values of k and β, separating (7.17) into real and imaginary parts, we obtain a pair of equations to define a_x and φ:

$$\sin\varphi = \frac{a_x}{a}[k\operatorname{Im}L(j\omega) + \omega\beta\operatorname{Re}L(j\omega)], \tag{7.18}$$

$$\cos\varphi = \frac{a_x}{a}[1 + k\operatorname{Re}L(j\omega) - \omega\beta\operatorname{Im}(j\omega)]. \tag{7.19}$$

From these equations we find

$$a_x = \frac{a}{\sqrt{[1 + k\operatorname{Re}L(j\omega) - \omega\beta\operatorname{Im}L(j\omega)]^2 + [k\operatorname{Im}L(j\omega) + \omega\beta\operatorname{Re}L(j\omega)]^2}}, \tag{7.20}$$

$$\tan\varphi = \frac{k\operatorname{Im}L(j\omega) + \omega\beta\operatorname{Re}L(j\omega)}{1 + k\operatorname{Re}L(j\omega) - \omega\beta\operatorname{Im}L(j\omega)}. \tag{7.21}$$

In fact, according to (6.85) and (7.16) the coefficients m_F, k and β in turn depend on the parameters of the solution for $x(t)$, hence expressions (7.13), (7.20), (7.21) can be used as equations to find these parameters.

Note that the system of defining equations (7.11), (7.17) which are appropriate in the search for components of the solution $x_h(t) = m_x + a_x\cos\omega t$, can be found directly from the original equation (7.1) by the substituting in it for the nonlinear function $\Phi(x, \dot{x})$, the linearised equation

$$\Phi(x, \dot{x}) = m_F + k(x_h - m_x) + \beta\dot{x}_h \tag{7.22}$$

The substitution indicated differs from the usual linearisation for a nonlinear function in the fact that the coefficients m_F, k and β depend on unknown parameters in the solution $x(t)$.

Finding the coefficients m_F, k and β is connected with the formation of several *a priori* concepts concerning the relative character of the periodic solution $x(t)$. In such cases, when the operator for the linear part of system $L(s)$ gives rise to vigorous qualities for the filtering or resonance effect, the value of $x_H(t)$ in equation (7.10) becomes small and from (6.85), (7.16), changing the integration variable, we obtain

$$m_F = \frac{1}{2\pi}\int_0^{2\pi}\Phi(m_x + a_x\cos\psi, \; -a_x\omega\sin\psi)d\psi \quad (\psi = \omega\tau), \tag{7.23}$$

$$k = \frac{1}{\pi a_x}\int_0^{2\pi}\Phi(m_x + a_x\cos\psi, \; -a_x\omega\sin\psi)\cos\psi d\psi, \tag{7.24}$$

$$\beta = -\frac{1}{\pi a_x \omega}\int_0^{2\pi}\Phi(m_x + a_x\cos\psi, -a_x\omega\sin\psi)\sin\psi d\psi . \qquad (7.25)$$

The selection of coefficients for equation (7.22) in the form (7.23) to (7.25) is called *harmonic linearisation* [181]. Here, the linearisation coefficients depend on m_x and a_x. Substituting (7.23) to (7.25) into (7.13), (7.20) we obtain a pair of coupled equations to find m_x and a_x. After finding these values the phase φ is found from (7.21).

As an example, we will look at the forced vibration of a linear oscillator with symmetric rigid stops (Fig. 3.1). Introducing the static characteristics for a double-sided impact pair $\Phi(x)$ (loss of energy during impact is not taken into account here), we will write, bearing in mind condition (3.8), the differential equation of the system in the following form:

$$M\ddot{x} + n\dot{x} + \Phi(x) = P\cos(\omega t + \varphi) , \qquad (7.26)$$

where, because of the symmetry of the system $\Phi(-x) = -\Phi(x)$ and, as a result $m_x \equiv 0$. Searching for an approximate solution $x(t)$ for equation (7.26) in the form

$$x(t) \approx a_x \cos\omega t , \qquad (7.27)$$

we expand the symmetrical periodic function $\Phi(a_x\cos\omega t)$ as a Fourier series, limiting the expansion to the first term

$$\Phi(a_x\cos\omega t) = k(a_x)a_x\cos\omega t + \ldots \approx k(a_x)x , \qquad (7.28)$$

where

$$k(a_x) = \frac{1}{\pi a_x}\int_0^{2\pi}\Phi(a_x\cos\psi)\cos\psi d\psi \quad (\psi = \omega t) . \qquad (7.29)$$

Substituting (7.28) in (7.26) we will rewrite the latter in operator form

$$[Ms^2 + ns + c + k(a_x)]x = P\cos(\omega t + \varphi) . \qquad (7.30)$$

For the solution of the linearised equation (7.29) in the form (7.27) we find

$$a_x = \frac{P}{|M(j\omega)^2 + nj\omega + c + k(a_x)|} = \frac{P}{\sqrt{[c + k(a_x) - M\omega^2]^2 + n^2\omega^2}} , \qquad (7.31)$$

$$\tan\varphi = n\omega/[c + k(a_x) - M\omega^2] . \qquad (7.32)$$

These equations are similar to (7.20), (7.21). The first, taking into account (7.29) defines the unknown amplitude of the fundamental a_x. Once that has been found the solution for φ is obtained from (7.32).

For the system considered (7.26) the amplitude-frequency characteristics can be constructed without solving equation (7.31) based on the following reasoning. The function $\omega_0(a_x) = \sqrt{c + k(a_x)}$ which fixes the relationship between the natural frequency of the linearised system and the amplitude of vibration, represents an approximate equation for the backbone curve. We will use it to replace the exact backbone curve for the nonlinear system considered, which takes the simple form *1* (Fig, 7.2).

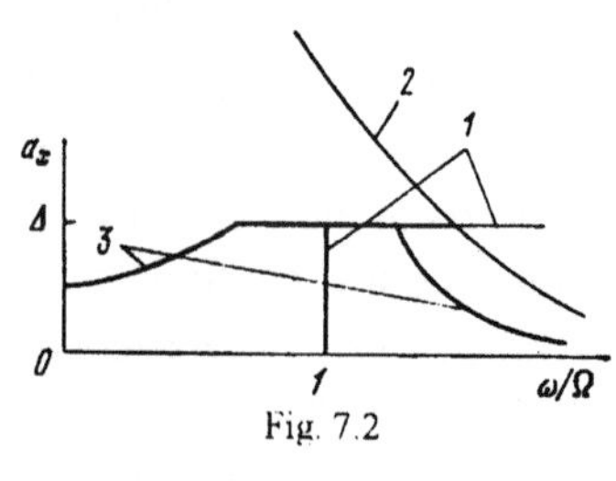

Fig. 7.2

Substituting $\omega^2 = c + k(a_x)$ in (7.31), we obtain the equation for the energy boundary of vibration

$$a_x = P/(n\omega) . \qquad (7.33)$$

According to (7.31), the intersection of curve *2* (7.33) with the backbone curve gives the maximum value of the amplitude characteristic. Adding the branches of the amplitude characteristics for the linear oscillator *3* to Fig 7.2, we can draw the full amplitude frequency characteristics for the system at the fundamental (compare to Fig. 3.9).

Hence, this approach gives a simple method to understand the basic qualitative principles of forced vibration in systems and to find expressions for them that are convenient to calculate. Later, we will make a detailed study of forced vibrations for the systems considered, using approximate methods and making comparisons with exact solutions (see §9).

2. The approximation of periodic functions as a finite sum of Fourier series ensures minimum mean square deviation [90]. On this basis, harmonic linearisation methods, where similar approximations to the real solution are limited to the basic harmonic components, can be given another interpretation allowing it to be related to other forms of equivalent linearisation and to obtain convenient formula for the calculation of the linearisation coefficients [124].

Let us substitute the nonlinear function $\Phi(x,\dot{x})$ by the linear one $m_F + k(x - m_x) + \beta\dot{x}$, choosing the coefficients m_F, k, and β from the conditions of minimum of mean square deviation in an infinite time interval. Minimising the value

$$\Theta(m_F, k, \beta) = \lim_{T\to\infty} \frac{1}{T}\int_0^T \{\Phi[x(t),\dot{x}(t)] - m_F - kx^0 - \beta\dot{x}^0\}^2\,dt,$$

where $x^0 = x - m_x$ we find

$$\frac{\partial\Theta}{\partial m_F} = -\lim_{T\to\infty}\frac{2}{T}\int_0^T [\Phi(x,\dot{x}) - m_F - kx^0 - \beta\dot{x}^0]dt = 0,$$

$$\frac{\partial\Theta}{\partial k} = -\lim_{T\to\infty}\frac{2}{T}\int_0^T [\Phi(x,\dot{x}) - m_F - kx^0 - \beta\dot{x}^0]x^0 dt = 0, \tag{7.34}$$

$$\frac{\partial\Theta}{\partial \beta} = -\lim_{T\to\infty}\frac{2}{T}\int_0^T [\Phi(x,\dot{x}) - m_F - kx^0 - \beta\dot{x}^0]\dot{x}^0 dt = 0.$$

Assuming

$$x(t) \approx m_x + a_x\cos\omega t, \quad \dot{x}(t) \approx -a_x\omega\sin\omega t, \tag{7.35}$$

we find

$$\lim_{T\to\infty}\frac{1}{T}\int_0^T x^0(t)\dot{x}^0(t)dt = 0, \quad \lim_{T\to\infty}\frac{1}{T}\int_0^T x^0(t)dt = 0,$$

$$\lim_{T\to\infty}\frac{1}{T}\int_0^T \dot{x}(t)dt = 0.$$

Using these relationships we find from (7.34)

$$m_F = \lim_{T\to\infty}\frac{1}{T}\int_0^T \Phi(x,\dot{x})dt, \tag{7.36}$$

$$k = \frac{1}{\sigma_x^2}\lim_{T\to\infty}\frac{1}{T}\int_0^T \Phi(x,\dot{x})x^0 dt, \tag{7.37}$$

$$\beta = \frac{1}{\sigma_{\dot{x}}^2}\lim_{T\to\infty}\frac{1}{T}\int_0^T \Phi(x,\dot{x})\dot{x}^0 dt, \tag{7.38}$$

where we make use of the following notation:

$$\sigma_x^2 = \lim_{T\to\infty}\frac{1}{T}\int_0^T [x^0(t)]^2 dt = \frac{a_x^2}{2}, \quad \sigma_{\dot{x}}^2 = \lim_{T\to\infty}\frac{1}{T}\int_0^T [\dot{x}^0(t)]^2 dt = \frac{a_x^2\omega^2}{2}. \tag{7.39}$$

In [124] it is shown that in a solution of the form (7.35) condition (7.34) is sufficient to find the minimum function $\Theta(m_F, k, \beta)$.

Introducing the mean square value (7.39) for the (7.35) allows the identification of formulae for harmonic and statistical linearisation. To complete this synthesis we introduce *distribution function* $W(u)$ for the deterministic process $x(t)$ as relative length of time, during the course of which the condition $x(t) \le u$ [128] is satisfied. Using Heaviside's function $\eta(x)$ we find

$$W(u) = \lim_{T\to\infty} \frac{1}{T} \int_0^T \eta[u - x(t)]dt \,. \tag{7.40}$$

In agreement with (7.40), we obtain for the *density function* $w = dW / dt$

$$w(u) = \lim_{T\to\infty} \frac{1}{T} \int_0^T \delta[u - x(t)]dt \,, \tag{7.41}$$

where $\delta(x) = d\eta(x)/dx$ is the Dirac function. By definition

$$\int_{x_1}^{x_2} w(u)du = 1 \,, \tag{7.42}$$

where x_1 and x_2 are the minimum and maximum values of $x(t)$.

For the process (7.35), for example, we find (Fig. 7.3)

$$W(u) = \frac{T-(t_2-t_1)}{T} = \frac{1}{T}\left[T - \left(\frac{\pi}{\omega} - \frac{1}{\omega}\arcsin\frac{u-m_x}{a_x}\right) + \frac{1}{\omega}\arcsin\frac{u-m_x}{a_x}\right] =$$

$$= \frac{1}{2} + \frac{1}{\pi}\arcsin\frac{u-m_x}{a_x}$$

whence

$$w(u) = \left[\pi\sqrt{a_x^2 - (u-m_x)^2}\right]^{-1} . \tag{7.43}$$

Similarly introducing a joint distribution function $W(u,v)$ for the processes $x(t)$ and $\dot{x}(t)$ as relative time durations satisfying the conditions $x(t) \le u, \dot{x}(t) \le v$, i.e.

$$W(u,v) = \lim_{T\to\infty} \frac{1}{T} \int_0^T \eta[u - x(t)]\eta[v - \dot{x}(t)]dt \,. \tag{7.44}$$

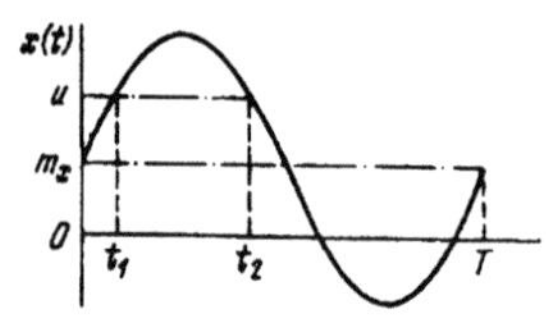

Fig. 7.3

To satisfy the density function $w(u,v)$ we obtain from (7.44)

$$w(u,v) = \frac{\partial^2 W(u,v)}{\partial u \partial v} = \lim_{T\to\infty} \int_0^T \delta[u - x(t)]\delta[v - \dot{x}(t)]dt \,, \tag{7.45}$$

whence

$$\int_{x_1}^{x_2}\int_{v_1}^{v_2} w(u,v)dudv = 1 \,, \tag{7.46}$$

where segment $[x_1, x_2], [v_1, v_2]$ includes the region of values of functions

$x(t)$ and $\dot{x}(t)$.

In particular, taking into account that between the coordinates and velocity in process (7.35) there exists a deterministic link

$$v = \pm\omega\sqrt{a_x^2 - (u - m_x)^2}\,, \tag{7.47}$$

we obtain using (7.43),

$$w(u,v) = \frac{1}{\pi\sqrt{a_x^2 - (u - m_x)^2}}\delta[v \pm \omega\sqrt{a_x^2 - (u - m_x)^2}\,]\,. \tag{7.48}$$

The $\pm$ sign in the argument of the δ – function is an abbreviation meaning the sum of two δ – functions, one of which takes the positive sign and the other the negative sign.

Introducing the representation (7.45), equations (7.36) to (7.38) for the calculation of the linearisation coefficients from the conditions of minimum mean square deviation, can be transformed to the following form:

$$\begin{aligned} m_F &= \lim_{T\to\infty}\frac{1}{T}\int_0^T \Phi(x,\dot{x})dt = \\ &= \lim_{T\to\infty}\frac{1}{T}\int_0^T dt\int_{x_1}^{x_2}\int_{v_1}^{v_2}\Phi(u,v)\delta[u - x(t)]\delta[v - \dot{x}(t)]du dv = \\ &= \int_{x_1}^{x_2}\int_{v_1}^{v_2}\Phi(u,v)du dv\lim_{T\to\infty}\frac{1}{T}\int_0^T \delta[u - x(t)]\delta[v - \dot{x}(t)]dt = \\ &= \int_{x_1}^{x_2}\int_{v_1}^{v_1}\Phi(u,v)w(u,v)du dv, \end{aligned} \tag{7.49}$$

$$\begin{aligned} k &= \frac{1}{\sigma_x^2}\lim_{T\to\infty}\int_0^T \Phi(x,\dot{x})(x - m_x)dt = \\ &= \frac{1}{\sigma_x^2}\int_{x_1}^{x_2}\int_{v_1}^{v_2}(u - m_x)\Phi(u,v)w(u,v)du dv, \end{aligned} \tag{7.50}$$

$$\begin{aligned} \beta &= \frac{1}{\sigma_{\dot{x}}^2}\lim_{T\to\infty}\int_0^T \Phi(x,\dot{x})\dot{x}^0(t)dt = \\ &= \frac{1}{\sigma_{\dot{x}}^2}\int_{x_1}^{x_2}\int_{v_1}^{v_2}v\Phi(u,v)w(u,v)du dv \end{aligned} \tag{7.51}$$

similar to the equations for the statistical linearisation coefficients (see §8). During the development of equations (7.49) to (7.51) we made use of the following quality for the δ – function :

$$\int_{x_1}^{x_2}\Phi(x)\delta(x - a)dx = \Phi(a) \quad (x_1 < a < x_2) \tag{7.52}$$

for any continuous function $\Phi(x)$.

The linearisation of the nonlinearities $\Phi(x,\dot{x})$ using (7.49) to (7.51) on the distribution $w(u,v)$ of a deterministic process $x(t)$ was called *distribution function linearisation* by the authors M Z Kolovsky and A A Pervosvansky, [128, 130]. They showed that this method could be used effectively to study polyharmonic processes. This is due to the fact that the definition of the linearisation coefficients is based on the initial moments of distribution and are only weakly dependent on higher moments of distribution. As a result, distribution functions can be approximated for a wide class of function which makes it possible to simplify their application considerably. On this basis, similar simple methods are proposed to calculate the harmonic linearisation coefficients [124].

3. We will now present the formulae for the harmonic linearisation coefficients for the basic concept of force characteristics of impact pairs. Methods for calculating the linearisation coefficients for a nonlinear function $\Phi(x)$ of type (6.7), (6.8) by the formula (7.23) to (7.25) are presented in detail in the monograph [181], hence we will immediately introduce the final equations. For the characteristics of a single-sided interaction (6.7) we find

$$m_F = -c_2 a_x \left\{ \frac{\Delta - m_x}{2a_x} - \frac{1}{\pi}\left[\frac{\Delta - m_x}{a_x} \arcsin\frac{\Delta - m_x}{a_x} + \sqrt{1 - \left(\frac{\Delta - m_x}{a_x}\right)^2} \right] \right\}, \tag{7.53}$$

$$k = c_2 \left\{ \frac{1}{2} - \frac{1}{\pi}\left[\arcsin\frac{\Delta - m_x}{a_x} + \frac{\Delta - m_x}{a_x}\sqrt{1 - \left(\frac{\Delta - m_x}{a_x}\right)^2} \right] \right\}, \tag{7.54}$$

where $a_x \geq \Delta - m_x$. The latter condition represents the presence of the impact interaction.

For the double-sided impact pair (7.8) when $a_x \geq \Delta + |m_x|$ we find

$$\begin{aligned} m_F = {} & \frac{c_2 a_x}{\pi}\left[\sqrt{1 - \left(\frac{\Delta - m_x}{a_x}\right)^2} - \sqrt{1 + \left(\frac{\Delta + m_x}{a_x}\right)^2} \right] + c_2 m_x + \\ & + \frac{c_2}{\pi}\left[\Delta\left(\arcsin\frac{\Delta - m_x}{a_x} - \arcsin\frac{\Delta + m_x}{a_x} \right) - \right. \\ & \left. - m_x\left(\arcsin\frac{\Delta - m_x}{a_x} - \arcsin\frac{\Delta + m_x}{a_x} \right) \right], \end{aligned} \tag{7.55}$$

$$\begin{aligned} k = {} & c_2 - \frac{c_2}{\pi}\left[\arcsin\frac{\Delta - m_x}{a_x} + \arcsin\frac{\Delta + m_x}{a_x} + \right. \\ & \left. + \frac{\Delta - m_x}{a_x}\sqrt{1 - \left(\frac{\Delta - m_x}{a_x}\right)^2} + \frac{\Delta + m_x}{a_x}\sqrt{1 - \left(\frac{\Delta + m_x}{a_x}\right)^2} \right]. \end{aligned} \tag{7.56}$$

When studying symmetrical processes, if the condition $x(t + T/2) = -x(t)$ is satisfied, the values $m_x \equiv 0, m_F \equiv 0$ and formula (7.56) is substantially simplified.

The characteristics (6.7) and (6.8) reflect only the effect of elastic interaction during impact. To calculate the loss of energy in this case arising from the introduction of the coefficient of restitution R, the following considerations can be used. Let us choose the linearisation coefficient β so that the equivalent dissipation due to the force of viscous friction $\beta\dot{x}$ in (7.22) gives rise to the same energy loss as there was during impact. The energy loss of mass M for one impact

$$E_{d2} = \frac{M\dot{x}_-^2}{2}(1 - R^2) \qquad \left(M = \frac{M_1 M_2}{M_1 + M_2} \right). \tag{7.57}$$

The value of the pre-impact velocity $\dot{x}_-$ is found approximately from the first harmonic of process $x(t)$. Bearing in mind, for example, the resonance characteristics of the vibro-impact regime of interest, we can assume that impact occurs at maximum velocity

$$\dot{x}_- \approx a_x\omega . \tag{7.58}$$

Substituting (7.58) into (7.57) we find

$$E_{d2} = \tfrac{1}{2} Ma_x^2\omega^2(1-R^2). \tag{7.59}$$

We define the work of the linear force $\beta\dot{x}$ during harmonic vibration (7.35):

$$E_{d1} = \int_0^{2\pi\omega} \beta\dot{x}^2 dt = \int_0^{2\pi\omega} \beta a_x^2\omega^2 \sin^2 \omega t dt = \pi\beta\omega a_x^2 . \tag{7.60}$$

Equating the energy values (7.59 and (7.60) we find an equation for the linearisation coefficient β :

$$\beta = M\omega(1-R^2)/2\pi \approx \rho M\omega \quad (0 \le r \le 0.16) . \tag{7.61}$$

For double-sided collisions, the value of the coefficient β must be double:

$$\beta = M\omega(1-R^2)/\pi \approx \rho M\omega \quad (0 \le r \le 0.32). \tag{7.62}$$

Note that in the formulae (7.53) and (7.54) we can define the transfer when $c_2 \to \infty$ to the situation for a.rigid stop*. Since a system with such limits satisfies the condition

$$m_x + a_x = \Delta, \tag{7.63}$$

as $c_2 \to \infty$ the value of $(\Delta - m_x)/a_x \to \alpha = 1$ To define the limiting transition we eliminate c_2 from (7.53) and (7.54), dividing the first relationship by the second. Applying the limit when $\alpha \to 1$ and expanding the indeterminacy, we find

$$m_F / k = a_x / 2. \tag{7.64}$$

Using condition (7.13) we can obtain formulae for the linearisation coefficients which depend only on one unknown parameter a_x of process x(t) from equations (7.63), (7.64):

$$m_F = (m_\vartheta - \Delta + a_x)L^{-1}(0), \tag{7.65}$$

$$k = \frac{2}{L(0)}\left(1 + \frac{m_\vartheta - \Delta}{a_x}\right). \tag{7.66}$$

Hence in the case of rigid stops, (7.53), (7.54) can be replaced by the simple relationships (7.55), (7.66). As well as significantly simplifying the formula, the convenience of such a substitution is that it is possible to eliminate the coefficient c_2 , whose actual value is complex to define.

Let us now move to the calculation of the harmonic linearisation coefficients for force characteristics of the type (6.13) to (6.16). Calculation with such functions can conveniently be carried out using the basic formula (7.48) to (7.50) linearised in terms of the distribution function.

From §6, section 3, it follows that when applying force characteristics impact pair containing the δ − function , when finding the change of impact force with time using transformation $F(t) = \Phi[x(t), \dot{x}(t)]$, the presence of the discontinuous function $\dot{x}(t)$ at the instant of impact formally causes the value of the coefficient to be doubled during transfer by comparison with the continuous case because of using the *average* velocity of the corresponding impact phase. During the approximation of such force characteristics for regular (linear) functions this information appears lost in the process of equivalent linearisation and the succeeding calculations for the solution of the linearised equations. To compensate for this defect when calculating the linearisation coefficients (both harmonic and statistical) we will double the force functions (6.13) to (6.16).

From (7.49) to (7.51), bearing in mind (6.18), and also taking account of the fact that for the force characteristic under consideration $\Phi_E(u,-v) = \Phi_E(u,v)$, $\Phi_D(u,-v) = \Phi_D(u,v)$ $(u \equiv x,\ v \equiv \dot{x})$, we obtain for the harmonic approximation (7.35) bearing in mind the symmetrical functions $w(u,v)$ relative to $v = 0$:

*The occurrence of this limiting transformation was brought to the author's attention by V K Astashev.

$$m_F = \int_{m_x-a_x}^{m_x+a_x}\int_{-a_x\omega}^{a_x\omega}\Phi_E(u,v)w(u,v)dudv\,, \tag{7.67}$$

$$k = \frac{1}{\sigma_x^2}\int_{m_x-a_x}^{m_x+a_x}\int_{-a_x\omega}^{a_x\omega}(u-m_x)\Phi_E(u,v)w(u,v)dudv\,, \tag{7.68}$$

$$\beta = \frac{1}{\sigma_{\dot{x}}^2}\int_{m_x-a_x}^{m_x+a_x}\int_{-a_x\omega}^{a_x\omega}v\Phi_D(u,v)w(u,v)dudv \tag{7.69}$$

Substituting in (7.67) to (7.69) the double value of the force characteristics (6.13) to (6.14) and the density function (7.48), (retaining only the velocity distribution before impact by using the sense of equations (6.13) to (6.14)) we find for single-sided impact pairs, using condition (7.52) and the terminology (7.39)

$$m_F = \int_{m_x-a_x}^{m_x+a_x}\int_{-a_x\omega}^{a_x\omega}\frac{Mv^2(1+R^2)\delta_+(u-\Delta)}{2\pi\sqrt{a_x^2-(u-m_x)^2}}\delta[v-\omega\sqrt{a_x^2-(u-m_x)^2}\,]dudv =$$

$$= \frac{M(1+R^2)}{2\pi}\int_{m_x-a_x}^{m_x+a_x}\frac{\delta_+(u-\Delta)du}{\sqrt{a_x^2-(u-m_x)^2}}\int_{-a_x\omega}^{a_x\omega}v^2\delta[v-\omega\sqrt{a_x^2-(u-m_x)^2}\,]dv =$$

$$= \frac{M\omega^2(1+R^2)}{2\pi}\int_{m_x-a}^{m_x+a}\sqrt{a_x^2-(u-m_x)^2}\,\delta_+(u-\Delta)du =$$

$$= \frac{M\omega^2(1+R^2)}{2\pi}\sqrt{a_x^2-(\Delta-m_x)^2}\,, \tag{7.70}$$

$$k = \frac{2}{a_x^2}\int_{m_x-a_x}^{m_x+a_x}\int_{-a_x\omega}^{a_x\omega}(u-m_x)\frac{Mv^2(1+R^2)\delta_+(u-\Delta)}{2\pi\sqrt{a_x^2-(u-m_x)^2}}\delta[v-\omega\sqrt{a_x^2-(u-m_x)^2}\,]dudv =$$

$$= \frac{M(1+R^2)}{\pi a_x^2}\int_{m_x-a_x}^{m_x+a_x}\frac{(u-m_x)\delta_+(u-\Delta)du}{\sqrt{a_x^2-(u-m_x)^2}}\int_{-a_x\omega}^{a_x\omega}v^2\delta[v-\omega\sqrt{a_x^2-(u-m_x)^2}\,]dv =$$

$$= \frac{M\omega^2(1+R^2)}{\pi a_x^2}\int_{m_x-a_x}^{m_x+a_x}(u-m_x)\sqrt{a_x^2-(u-m_x)^2}\,\delta_+(u-\Delta)du =$$

$$= \frac{M\omega^2(1+R^2)}{\pi a_x^2}(\Delta-m_x)\sqrt{a_x^2-(\Delta-m_x)^2}\,, \tag{7.71}$$

$$\beta = \frac{1}{a_x^2\omega^2}\int_{m_x-a_x}^{m_x+a_x}\int_{-a_x\omega}^{a_x\omega}\frac{Mv^3(1-R^2)\delta_+(u-\Delta)\operatorname{sign}v}{\pi\sqrt{a_x^2-(u-m_x)^2}}\delta[v-\omega\sqrt{a_x^2-(u-m_x)^2}\,]dudv =$$

$$= \frac{M(1-R^2)}{\pi a_x^2\omega^2}\int_{m_x-a_x}^{m_x+a_x}\frac{\delta_+(u-\Delta)du}{\sqrt{a_x^2-(u-m_x)^2}}\int_{-a_x\omega}^{a_x\omega}v^3\operatorname{sign}v\,\delta[v-\omega\sqrt{a_x^2-(u-m_x)^2}\,]dv =$$

$$= \frac{M\omega(1-R^2)}{\pi a_x^2}\int_{m_x-a_x}^{m_x+a_x}[a_x^2-(u-m_x)^2]\delta_+(u-\Delta)du = \frac{M\omega^2(1-R^2)}{\pi a_x^2}[a_x^2-(\Delta-m_x)^2]. \tag{7.72}$$

Naturally, the expressions shown apply when $a_x \geq \Delta - m_x$.

The similar calculation carried out for a double-sided impact pair using regulations (6.15) to (6.16) we find when $a_x \geq |m_x| + \Delta$

$$m_F = \frac{M\omega^2(1+R^2)}{2\pi}\left[\sqrt{a_x^2-(\Delta-m_x)^2}-\sqrt{a_x^2-(\Delta+m_x)^2}\,\right], \tag{7.73}$$

$$k=\frac{M\omega^2(1+R^2)}{\pi a_x^2}\left[(\Delta-m_x)\sqrt{a_x^2-(\Delta-m_x)^2}+(\Delta+m_x)\sqrt{a_x^2-(\Delta+m_x)^2}\right] \tag{7.74}$$

$$\beta=\frac{2M\omega(1-R^2)}{\pi a_x^2}(a_x^2-\Delta^2-m_x^2). \tag{7.75}$$

It must be mentioned that formula (7.70) to (7.75) can only be used over a comparatively narrow area of change in the argument a_x/Δ, i.e., the change of input parameters within small limits corresponds to every possible form of vibro-impact regime. This implies that in the form presented, the harmonic linearisation method is too crude. Small errors in the parameters of the process $x(t)$ can significantly influence the value of the instantaneous force when higher harmonics are ignored.

The principal method of improving the accuracy of the values of the harmonic linearisation coefficients is to by improve the initial information relative to the character of the solution sought. It is possible for example, to calculate the coefficients using more harmonics [124, 209, 214], or to improve the accuracy of the distribution density, $w(u,v)$ for process $x(t)$ by using limited regions for change in $x(t)$, i.e., limiting $w(u,v)$ by the condition $w(u,v)=0$ when $x>\Delta$ in the non-symmetrical case and $|x|>\Delta$ in the symmetrical (see §8, section 4). However, the complexity of the corresponding calculation procedure is expediently matched with the possibility of direct numerical solution by successive approximations of the type (6.91), (6.92). It is often possible to refine the coefficients of linearisation arising from the specific solution of the problem. Thus in proper cases we will continue with this problem.

4. Let us now look at another modification to method of harmonic linearisation of vibro-impact systems, which is distinguished by the fact that the linearisation coefficients are not found from the parameters of the process $x(t)$, but as parameters of the force $F(t)=\Phi[x(t),\dot{x}(t)]$, which is developed at the output of the impacting elements (Fig. 6.5) [37]. In the following, to distinguish these two linearisation methods, we shall refer to them as *motion linearisation* and *force linearisation.*

It is convenient to use force linearisation when the force elastic characteristics of impact pairs are described by the relationship $F=\Phi(x)$, shown in Fig. 6.2. Direct linearisation of these functions is not possible since their values are ambiguous and unlimited during impact, hence we will linearise them by reverse function $x=X(F)$, which characterises the relationship of relative motion of the elements of the impact pair to the impact forces occurring between them. For the relationships shown in Fig. 6.2, *b* - *d*, the function $x=X(F)$ can be represented by the following equations (Fig. 7.4):

$$x=\begin{cases}+\Delta & \text{when} \quad F>0,\\ -\Delta & \text{when} \quad F<0,\end{cases} \tag{7.76}$$

$$x=\frac{1}{c}[F-(F-c\Delta)\eta(F-c\Delta)], \tag{7.77}$$

$$x=-\Delta+\frac{1}{c}[(F+c\Delta)\eta(F+c\Delta)-(F-c\Delta)\eta(F-c\Delta)]. \tag{7.78}$$

In the case of the characteristic $\Phi(x)$, shown in Fig. 7.2, *a,* the reverse

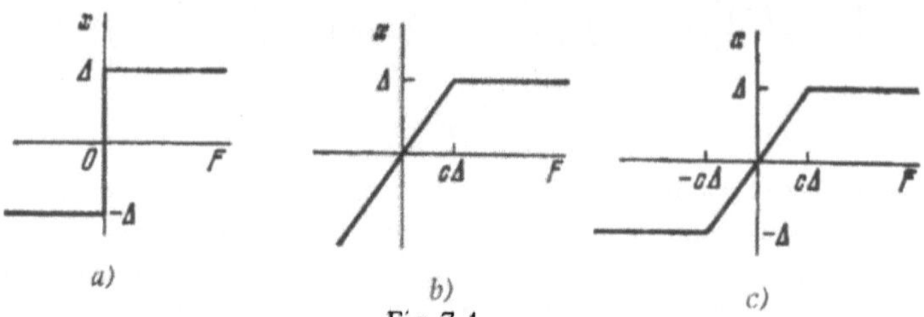

Fig. 7.4

relationship does not exist, hence we will not use this approach in this case.

To linearise the function $x = X(F)$, we will assume

$$x \approx m_x + q(F - m_F) \quad (q = 1/k) \tag{7.79}$$

The linearisation coefficients m_x and q are found from the condition for the minimum mean square error (see section 3) given by the formula

$$m_x = \int_{F_1}^{F_2} X(F) w(F) dF, \tag{7.80}$$

$$q = \frac{1}{\sigma_x^2} \int_{F_1}^{F_2} (F - m_F) X(F) w(F) dF. \tag{7.81}$$

Here F_1, F_2 are minimum and maximum values of process $F(t)$, σ_F^2 is the mean square value of the centred process $F^0(t) = F(t) - m_F$; $w(F)$ is the density function of $F(t)$.

Using the low sensitivity to the input process for nonlinearities of types (7.76) to (7.78), on the type of output action [214], when they are linearise, a fairly crude assumption can be made for the relative form of $F(t)$. Thus, limiting our choice to the simplest, we assume

$$F(t) \approx m_F + a_F \cos \omega t. \tag{7.82}$$

Bearing in mind that the density function of $F(t)$ in this case takes the form

$$w(F) = [\pi \sqrt{a_F^2 - (F - m_F)^2}\,]^{-1}, \tag{7.83}$$

$\sigma_F^2 = a_F^2 / 2$, and from (7.80), (7.81) we find the following values of the coefficients of harmonic linearisation.

For the characteristic (7.76) (Fig. 7.4, *a*) when $a_F \geq |m_F|$

$$m_x = \frac{2\Delta}{\pi} \arcsin \frac{m_F}{a_F}, \tag{7.84}$$

$$q = \frac{4\Delta}{\pi a_F} \sqrt{1 - \left(\frac{m_F}{a_F}\right)^2}. \tag{7.85}$$

For the characteristic (7.77) (Fig. 7.4, *b*) when $a_F \geq c\Delta - m_F$

$$m_x = \frac{a_F}{c} \left\{ \frac{c\Delta + m_F}{2a_F} - \frac{1}{\pi} \left[\frac{c\Delta - m_F}{a_F} \arcsin \frac{c\Delta - m_F}{a_F} + \sqrt{1 - \left(\frac{c\Delta - m_F}{a_F}\right)^2} \right] \right\}, \tag{7.86}$$

$$q=\frac{1}{c}\left\{\frac{1}{2}+\frac{1}{\pi}\left[\arcsin\frac{c\Delta-m_F}{a_F}+\frac{c\Delta-m_F}{a_F}\sqrt{1-\left(\frac{c\Delta-m_F}{a_F}\right)^2}\right]\right\} \tag{7.87}$$

For characteristic (7.78) (Fig. 7.4, c) when $a_F \geq c\Delta + |m_F|$

$$m_x=\frac{1}{\pi c}\left[\sqrt{1-\left(\frac{c\Delta+m_F}{a_F}\right)^2}-\sqrt{1-\left(\frac{c\Delta-m_F}{a_F}\right)^2}+\right. \\ \left.+(c\Delta+m_F)\arcsin\frac{c\Delta+m_F}{a_F}-(c\Delta-m_F)\arcsin\frac{c\Delta-m_F}{a_F}\right], \tag{7.88}$$

$$q=\frac{1}{\pi c}\left[\arcsin\frac{c\Delta-m_F}{a_F}+\arcsin\frac{c\Delta+m_F}{a_F}+\right. \\ \left.+\frac{c\Delta-m_F}{a_F}\sqrt{1-\left(\frac{c\Delta-m_F}{a_F}\right)^2}+\frac{c\Delta+m_F}{a_F}\sqrt{1-\left(\frac{c\Delta+m_F}{a_F}\right)^2}\right]. \tag{7.89}$$

To study the latter case for double-sided impact pairs with elastic coupling the linearisation coefficients can be found in another way using the following argument. Let us represent the static elastic characteristics of this impact pair $F=\Phi(x)$ (Fig. 6.2, *c*) as the sum of two characteristics, one linear (Fig. 7.5, *a*) and one nonlinear (Fig. 7.5, *b*). For this we assume conditionally that when the maximum force developed by the linear spring is constant and equal to $c\Delta$ during the occurrence of a vibro-impact regime, then the maximum force in nonlinear "elasticity" will be $a_F-c\Delta$. Linearising these characteristics in relation to this force with the help of the examples described above and using (7.84), (7.85) we obtain when $a_F \geq c\Delta + |m_F|$

$$m_x=\frac{2\Delta}{\pi}\arcsin\frac{m_F}{a_F-c\Delta} \tag{7.90}$$

$$q=\left[c+\frac{\pi(a_F-c\Delta)}{4\Delta\sqrt{1-[m_F/(a_F-c\Delta)]}}\right]^{-1} \tag{7.91}$$

Comparing the numerical value of the linearisation coefficients in the forms (7.90), (7.91) to the corresponding values (7.88) , (7.89) shows close agreement. This detail can be used by simplifying formula (7.87) for the case when $m_F=0$, which will be studied frequently later. The simplification shown is using the following analogy.

According to (7.87, (7.89) the expression for q when $m_F\equiv 0$ takes the form

$$q_1=\frac{2}{\pi c}\left[\arcsin\frac{c\Delta}{a_F}+\frac{c\Delta}{a_F}\sqrt{1-\left(\frac{c\Delta}{a_F}\right)^2}\right] \tag{7.92}$$

for a double-sided impact pair and

$$q_2=\frac{1}{c}\left\{\frac{1}{2}+\frac{1}{\pi}\left[\arcsin\frac{c\Delta}{a_F}+\frac{c\Delta}{a_F}\sqrt{1-\left(\frac{c\Delta}{a_F}\right)^2}\right]\right\} \tag{7.93}$$

for a single-sided impact pair. Note that the relationship in the square brackets in (7.92), (7.93) differ only because for the first case the value of Δ from the physical sense is essentially positive, but in the second case, changes sign to negative when gap changes to interference. At the same time, even in this case only a positive value of Δ can be used, changing the sign before the radical when gap changes to interference. Taking this into account, we can eliminate the general expression in square brackets from (7.92), (7.93). As a result we find

$$q_2=(1\pm cq_1)/(2c) \tag{7.94}$$

Substituting in (7.94) for q_1, the equivalent value (7.91), taken when $m_{k} \equiv 0$ we eventually find

$$q_2 = \frac{1}{2c}\left(1 \pm \frac{1}{1-\pi/4+\pi a_{k}/(4c\Delta)}\right) \tag{7.95}$$

Here the positive sign indicates a gap, and the minus sign interference, the quantity Δ is equal to the absolute value of the gap or interference.

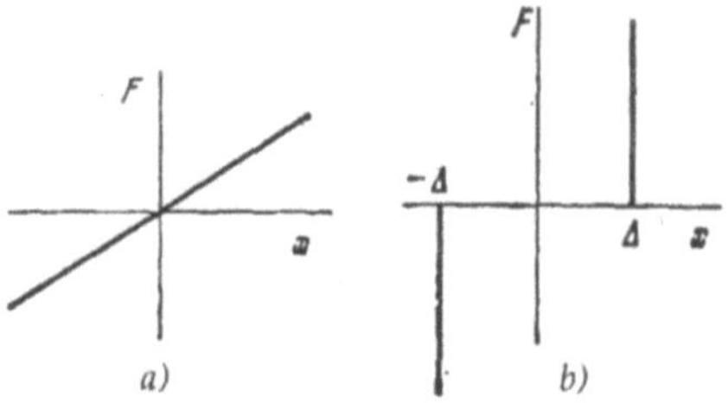

Fig. 7.5

When it is necessary to calculate energy dissipation during impact when using harmonic linearisation by force the equivalent coefficients of viscous friction β can be taken in the form (7.61), (7.62).

In conclusion, note that the coefficient of harmonic linearisation can be calculated in principle for even more complicated contact interactions, for example, in the form of characteristics (6.5), (6.6) although in this case integrals are developed which it is not possible to evaluate through elementary functions. This difficulty can be overcome by simplifying the approximate expressions for calculating the linearisation coefficients [124] or using special graphical methods to construct them [211]. However, taking account of the insensitivity of periodic solutions to the specific character terms of contact interactions as defined in §7, such complicated procedures for finding the first approximation are hardly justified. Characteristics of type (6.5), (6.6) remain useful in schemes for successive approximation (6.91) when, for example, the contact forces are to be further refined, however, even in this case it is convenient to construct the initial approximation from whatever simplification possible, resulting, say, from harmonic linearisation of the equations with characteristics of the type (6.7), (6.8). After finding the highest approximation, the maximum contact forces are defined using (7.5), (7.6) from one of the formula

$$\max|\Phi(x)| = \Phi(x_{\max}) \quad \text{or} \quad \max|\Phi(x)| = \Phi(x_{\min}) \tag{7.96}$$

§8. Statistical linearisation of vibro-impact systems

1. The description of the equations of motion for vibro-impact systems in §6 gives a convenient basis for studying the statistical dynamics of such systems. Problems in statistical dynamics for vibro-impact systems occur when studying the vibration of machines and structures with regard to colliding elements of kinematic pairs or couplings, when designing systems of vibration isolation, during the study of interference effects in systems for automatic control, or in similar devices consisting of mechanical elements with clearance, etc. In the majority of the cases mentioned, random vibration is caused by "external" sources of wide band excitation: jet engines, turbulent media in which components move, irregularity in the profile of the path of transport devices, noise in electronic instruments in control circuits, etc. However, even in the case of periodic external excitation the presence in the structure of a large number of couplings, creates "internal" sources of random excitation. These irregular impact interactions, caused by additional mobility in gaps, leading to dynamic instability of the structure, or caused by specific impact regimes in conditions of poor synchronisation, are characterised by the appearance of *irregular clatter* *.

* The study of vibration accompanying repeated attenuated collisions, was carried out in the work of R F Nagaev [165, 167] and M I Feigin [84, 87].

The methods of statistical dynamics allow us to study a combination of various real processes which arise under identical conditions, and to combine them by common probability descriptions. The basis for this description lies in the introduction of *random functions*, hence we need to deal with some background knowledge from the theory of random functions, before further investigation [184, 204].**

The random process or random function $x(t)$ with deterministic argument t (time) is a function lying in the range $-\infty < t < \infty$, which is defined to have a random variable $x(t_1)$ at each fixed instant of time t_1. The full probability characteristic of random variable $x(t_1)$ is known as the *distribution function* $W_1(u_1,t_1)$, which characterises the probability that the variable $X(t_1)$ of the random variable $x(t_1)$ satisfies the condition $X(t_1) < u_1$. Hence

$$W_1(u_1,t_1) = \mathcal{P}\{X(t_1) \le u_1\}, \tag{8.1}$$

where $\mathcal{P}$ is the probability occurrence $(0 \le \mathcal{P} \le 1)$. For any argument t_1, the value $W(u_1,t_1)$ characterises the probability of the condition of process $x(t_1)$ in each fixed instant of time and is known as its *one-dimensional distribution function*. In order to establish the connection between the state of process $x(t)$ in consecutive instants of time, the *multi-dimensional distribution function* is introduced as

$$W_n(u_1,t_1,\ldots,u_n,t_n) = \mathcal{P}\{X(t_1) \le u_1,\ldots,X(t_n) \le u_n\}. \tag{8.2}$$

Alongside the distribution function for describing random processes, the *n-dimensional density function* of a random process is given by

$$w_n(u_1,t_1,\ldots,u_n,t_n) = \frac{\partial^n W_n}{\partial u_1 \ldots \partial u_2}. \tag{8.3}$$

When analysing random processes an important quantity are their *moments*, which are functions defined generally as follows

$$m_{i_1\ldots i_n}(t_1,\ldots,t_n) = \int_{-\infty}^{\infty}\ldots\int_{-\infty}^{\infty} u_1^{i_1}\ldots u_n^{i_n} w_n(u_1,t_1,\ldots,u_n,t_n)du_1\ldots du_n. \tag{8.4}$$

Function (8.4) is called the n-dimensional moment of the *q-th* order $(q = i_1 + \ldots + i_n)$. Moment characteristics are symmetrical relative to all their arguments. Apart from moments of form (8.4), called *initial*, we also introduce *central* moments defined by

$$\mu_{i_1\ldots i_n}(t_1,\ldots,t_n) = \int_{-\infty}^{\infty}\ldots\int_{-\infty}^{\infty}[u_1 - m_1(t_1)]^{i_1}\ldots[u_n - m_n(t_n)]^{i_n} \times \\ \times w(u_1,t_1,\ldots u_n,t_n)du_1\ldots du_n \tag{8.5}$$

where $m_1(t_1)$, ..., $m_1(t_n)$ are expected values characterising the statistical average value.

**Translator's Note. For texts in English covering similar material, the reader can consult A B Clarke, R L Disney "Probability and random processes for engineers," (Wiley) or M O'Flynn "Probabilities, random variables and random processes," (Harper & Row).

A typical feature of moments is that as their order increases, the information in them about the random processes principally reflects the part played by the occasional large amplitude peaks. Thus the practical study of random processes, is limited, as a rule, to several fundamental moments. The most important is the first order moment

$$m_1(t) = m_x(t) = \int_{-\infty}^{\infty} u w_1(u,t) du\,, \tag{8.6}$$

known as the *expected*, *mean* or *average value* of random process $x(t)$, and the central second order moment

$$\mu_{11}(t_1,t_2) = R_x(t_1,t_2) = \int_{-\infty}^{\infty} [u_1 - m_x(t_1)][u_2 - m_x(t_2)] w(u_1,t_1,u_2,t_2) du_1 du_2 \tag{8.7}$$

known as the *covariance* of the process. Substituting $t_1 = t_2 = t$ in (8.7) we obtain the *mean square value* of the random process as

$$\sigma_x^2(t) = R_x(t,t) = \int_{-\infty}^{\infty} [u - m_x(t)]^2 w_1(u,t) du\,. \tag{8.8}$$

The function $\sigma_x(t)$ is called the standard deviation of random process $x(t)$.
Using moments we can reduce the density function of random processes.

In applications, the *normal* or *Gaussian* distribution plays an important role. Clarifying this, the sum of the largest quantity of statistically independent random processes is always close to the normal for the process, provided none of the components dominates the others. The probability characteristics of a normal process are completely defined by its expected value and covariance. For example, the one-dimensional density function takes the form

$$w_1(u,t) = \frac{1}{\sqrt{2\pi}\sigma_x(t)} \exp\left\{ -\frac{[u - m_x(t)]^2}{2\sigma_x^2(t)} \right\}. \tag{8.9}$$

We shall study only *stationary* random processes. A random process is called *stationary in the narrow sense* when its distribution function does not change when the time origin changes

$$W_n(u_1, t_1 + \tau, \dots, u_n, t_n + \tau) \equiv W_n(u_1, t_1, \dots, u_n, t_n) \tag{8.10}$$

for all n and τ. As a result for such processes

$$W_1 = W_1(u_1); \quad W_2 = W_2(u_1, u_2, \tau)\,, \tag{8.11}$$

i.e. a one-dimensional distribution function is independent of time, but a two-dimensional depends only on the difference $\tau = t_2 - t_1$. Similar features are possessed by the density function of stationary processes. As a result, from (8.6) to (8.8) we find for the stationary case

$$\begin{gathered} m_x(t) = m_x = \text{const}, \quad R_x(t_1,t_2) = R_x(t_2 - t_1) = R_x(\tau), \\ \sigma_x^2(t) = \sigma_x^2 = R_x(0) = \text{const}. \end{gathered} \tag{8.12}$$

A random process satisfying conditions (8.12) belongs to the class of process known as *stationary in the wide sense* or a *strictly stationary* random process. Of course, conditions (8.12) is, in general, insufficient to satisfy the identity (8.10), however, for a normal process the meanings of stationary in wide and narrow senses coincide.

An important feature of stationary processes is the possibility of representing them spectrally. Applying a Fourier transformation to the covariance, we find

$$S_x(\omega) = \int_{-\infty}^{\infty} R_x(\tau)\exp(-j\omega\tau)d\tau . \tag{8.13}$$

Function $S_x(\omega)$ is called the *spectral density* of the stationary random process. The integral of this term

$$\int_{\omega_1}^{\omega_2} S_x(\omega)d\omega$$

defines a value proportional to the mean power of the process in the frequency range $[\omega_1, \omega_2]$. From the expression for the inverse Fourier transform we find from (8.13)

$$R_x(\tau) = \frac{1}{2\pi}\int_{-\infty}^{\infty} S_x(\omega)\exp(j\omega\tau)d\omega . \tag{8.14}$$

Relationships (8.13) and (8.14) are generally known as the Wiener-Khinchine relations.

Assuming $t = 0$ in (8.14), we find

$$R_x(0) = \sigma_x^2 = \frac{1}{2\pi}\int_{-\infty}^{\infty} S_x(\omega)d\omega \tag{8.15}$$

As both are even functions of their arguments, the convolution and spectral density are connected to those characteristic conditions so that the larger the *time correlation* of random process t_0 ensuring the condition $|R_x(\tau)| \approx 0$ when $t > t_0$, then the narrower the spectrum occurring in the frequency process $S_x(\omega)$.

A stationary process, whose spectral density is constant throughout the frequency range from $-\infty$ to $+\infty$ is called *white noise*. According to (8.15) white noise has an infinitely large mean square value, hence it can be considered only as the idealisation of an actual wideband random process. Taking $S_x(\omega) = S_0 = \text{const}$ in (8.14) we obtain for white noise

$$R_x(\tau) = \frac{S_0}{2\pi}\int_{-\infty}^{\infty}\exp(j\omega\tau)d\omega = S_0\delta(\tau) , \tag{8.16}$$

where $\delta(\tau)$ is the Dirac function.

A stationary random process proceeds uniformly with time, i.e. its statistical characteristics do not depend on the chosen initial instant of observation. Hence, one actual waveform of a stationary process continuing for a sufficiently long time can include all data and conditions of the random process. Stationary random processes, all forms of which can be represented in this way, are called *ergodic**. A sufficient condition for a stationary normal process to be ergodic takes the form [204]

---

* Strictly speaking, in an ergodic process the existence of separate unrepresentative waveforms is also accepted (producing a *set of zero dimensions*).

$$\int_{-\infty}^{\infty} | R_x(\tau) | \, d\tau < \infty . \tag{8.17}$$

For ergodic processes

$$W_1(u_1) = \lim_{T\to\infty} \frac{1}{2T} \int_{-T}^{T} \eta[u_1 - x(t)]dt \tag{8.18}$$

where $\eta(u)$ is the unit step function; $x(t)$ is the waveform of the ergodic process,

$$m_x = \lim_{T\to\infty} \frac{1}{2T} \int_{-T}^{T} x(t)dt , \tag{8.19}$$

$$R_x(\tau) = \lim_{T\to\infty} \frac{1}{2T} \int_{-T}^{T} x(t)x(t+\tau)dt . \tag{8.20}$$

These relationships provide the basis for calculating the statistical characteristics of stationary ergodic processes by experimentally recording their waveform.

To describe the statistical connection between two stationary processes $x(t)$ and $y(t)$ we introduce *cross-correlation functions* as

$$\begin{aligned} R_{xy}(t_1, t_2) &= \langle [x(t_1) - m_x][y(t_2) - m_y] \rangle, \\ R_{yx}(t_1, t_2) &= \langle [y(t_1) - m_y][x(t_2) - m_x] \rangle. \end{aligned} \tag{8.21}$$

Both here and later, square brackets mean the operation of statistical averaging of an ensemble of actual waveforms. If function (8.21) depends only on the difference $\tau - t_2 - t_1$, then processes $x(t)$ and $y(t)$ are said to be *stationary coupled.* For such processes

$$R_{xy}(\tau) = R_{yx}(-\tau) . \tag{8.22}$$

Using the cross-correlation function the density function for a normally distributed random vector [x, y] is defined as

$$\begin{aligned} w(u,v) &= \frac{\partial^2}{\partial u \partial v} \mathscr{P}\{X \le u, Y \le v\} = \frac{1}{2\pi\sqrt{\sigma_x^2\sigma_y^2 - R_{xy}^2}} \times \\ &\times \exp\left\{ - \frac{\sigma_y^2(u-m_x)^2 - 2R_{xy}(u-m_x)(v-m_y) + \sigma_x^2(v-m_y)^2}{2(\sigma_x^2\sigma_y^2 - R_{xy}^2)} \right\} \end{aligned} \tag{8.23}$$

In the absence of a correlation between processes $x(t)$ and $y(t)$ the value $R_{xy} \equiv 0$ and expression (8.23) is substantially simplified

$$w(u,v) = \frac{1}{2\pi\sigma_x\sigma_y} \exp\left[-\frac{(u-m_x)^2}{2\sigma_x^2} - \frac{(v-m_y)^2}{2\sigma_y^2} \right]. \tag{8.24}$$

Comparing (8.24) and (8.9) we can see that the joint density function in this case is equal to the product of the density functions produced for these processes. Of course this fact follows directly from the nature of the statistical independence of the occurrences.

Random process $x(t)$ is considered to be differentiable at point t, if for any number ε convergent to zero, there is a limit for a random sequence of values*

$$[x(t+\varepsilon)-x(t)]/\varepsilon .$$

A random value appearing as the limit of a random sequence is called the *derivative of the random process* (function) $x(t)$ at point t and will be designated as $\dot{x}(t)$. A sufficient condition for differentiability of a stationary random function is when the second derivative of the correlation function exists at $\tau = 0$ [204]:

$$d^2 R_x(\tau)/d\tau^2 \,|_{\tau=0} .$$

The knowledge of the joint probability density and its derivative $w(u,v)$ allows the calculation of the average value f of the events when the process $x(t)$ crosses level Δ in unit time. This value is defined by the formula [204]

$$f = \int_0^{\infty} v w(\Delta, v) dv . \tag{8.25}$$

The inverse value $t_0 = 1/f$ gives the average time between crossings. It can be shown that for a differentiable stationary process the ordinate of the random function and its derivative at the same instant are noncorrelated, and for a normal process and independent random variables. From the normality of process $x(t)$, comes normality $\dot{x}(t)$ [184]. For this reason, during calculation from (8.25) the normal law is used in form (8.24) where $y(t)=\dot{x}(t)$.

Under the influence of a stationary differentiable random process $x(t)$ on a linear system with constant parameters described by the operator $L(s)$, the mathematical expectation and the spectral density reaction of the system $z(t)$ are expressed through similar characteristics of excitation in the following form [184]:

$$m_z = L(0) m_x, \tag{8.26}$$

$$S_z(\omega) = | L(j\omega) |^2 \, S_x(\omega) . \tag{8.27}$$

During this, within the limits of the actual frequency range, function $L(j\omega)$ must not take an infinite value. It is important to note that the specified transformation does not violate the normality of the process.

2. We will now turn to the application of ideas and methods of statistical linearisation to vibro-impact systems [37, 38]. Suppose that for the system described by equation (6.39), a stationary random normal excitation $P_i(t)$ acts. Taking into account that during linear stationary transformation the random process remains stationary and normal, we can write the resulting equations in the form

$$x(t) + L(s)\Phi[x(t), sx(t)] = \vartheta(t) , \tag{8.28}$$

where $\vartheta(t)$ is the given stationary normal random process.

Following the *statistical linearisation method* [113], we will substitute the nonlinear function $\Phi(x,\dot{x})$ by the linear one

$$\Phi(x,\dot{x}) = m_F + k(x - m_x) + \beta\dot{x} , \tag{8.29}$$

selecting the linearisation coefficients m_F, k and β so as to minimise the

* The limit of a sequence of random variables is a random value, the root-mean-square variation of which, from the elements of the sequence, tends to zero when $\varepsilon \to 0$.

statistical mean square value of the residual

$$\langle\Theta(m_F,k,\beta)\rangle = \int_{-\infty}^{\infty}\int_{-\infty}^{\infty}[\Phi(u,v)-m_F-k(u-m_x)-\beta\dot{v}]^2 w(u,v)du dv\,. \quad (8.30)$$

Here $w(u,v)$ is the joint density function of processes $x(t)$ and $y(t)$. From the necessary conditions of extremum existence $\partial\langle\Theta\rangle/\partial m_F = \partial\langle\Theta\rangle/\partial k = \partial\langle\Theta\rangle/\partial\beta = 0$ we obtain equations to determine the coefficients of statistical linearisation

$$m_F = \int_{-\infty}^{\infty}\int_{-\infty}^{\infty}\Phi(u,v)w(u,v)du dv\,, \quad (8.31)$$

$$k = \frac{1}{\sigma_x^2}\int_{-\infty}^{\infty}\int_{-\infty}^{\infty}(u-m_x)\Phi(u,v)w(u,v)du dv\,, \quad (8.32)$$

$$\beta = \frac{1}{\sigma_y^2}\int_{-\infty}^{\infty}\int_{-\infty}^{\infty}v\Phi(u,v)w(u,v)du dv\,. \quad (8.33)$$

Carrying out the substitution (8.29) in (8.28) and assuming that $\vartheta(t) = m_\vartheta + \vartheta^0(t)$ where $m_\vartheta = \text{const}$, $\vartheta^0(t)$ is the centred component of the process: $\vartheta^0(t) = \vartheta(t) - m_\vartheta$,

$$x(t) + L(s)\{m_F + k[x(t)-m_x] + \beta s x(t)\} = m_\vartheta + \vartheta^0(t)\,. \quad (8.34)$$

Searching for the solution of (8.34) in the form

$$x(t) = m_x + x^0(t)\,, \quad (8.35)$$

where $m_x = \text{const}$, $x^0(t) = x(t) - m_x$, we find from (8.34) for constant and centred components

$$m_x + L(0)m_F = m_\vartheta\,, \quad (8.36)$$

$$x^0(t) + L(s)[kx^0(t) + \beta s x^0(t)] = \vartheta^0(t)\,. \quad (8.37)$$

From (8.37) we find

$$x^0(t) = \frac{1}{1+L(s)(k+\beta s)}\vartheta^0(t)\,. \quad (8.38)$$

Making use of the formula for the linear transformation of the spectral density (8.27) we obtain from (8.38):

$$S_x(\omega) = \frac{S_\vartheta(\omega)}{|1+L(j\omega)(k+\beta j\omega)|^2}\,. \quad (8.39)$$

The mean square value of $x(t)$ can be calculated from (8.15) by using equation (8.39) as

$$\sigma_x^2 = \frac{1}{2\pi}\int_{-\infty}^{\infty}\frac{S_\vartheta(\omega)d\omega}{|1+L(j\omega)(k+\beta j\omega)|^2} \quad (8.40)$$

and noting that according (8.27) $S_y(\omega)=\omega^2 S_x(\omega)$, the mean square value of velocity of process $x(t)$ as:

$$\sigma_y^2 = \frac{1}{2\pi}\int_{-\infty}^{\infty}\frac{S_\vartheta(\omega)\omega^2 d\omega}{|1+L(j\omega)(k+\beta j\omega)|^2}. \tag{8.41}$$

For integrals of type (8.40), (8.41)

$$I_n = \frac{1}{2\pi}\int_{-\infty}^{\infty}\frac{g_n(j\omega)d\omega}{|h_n(j\omega)|^2},$$

where $h_n(s) = a_0 s^n + a_1 s^{n-1} + \ldots + a_n$; $g_n(s) = c_0 s^{2n-2} + c_1 s^{2n-4} + \ldots + c_{n-1}$, having the general formula, presented, for example in [184].

In what follows, we shall be interested in general in the specified statistical characteristics of process $x(t)$, the calculation of which will consist of the nature of *spectral theory* for stationary random processes. In addition, note that the values of m_x, σ_x, σ_v in equations (8.36), (8.40). (8.41) have been shown to depend on the linearisation coefficients, which , in their turn, depend on the unknown statistical characteristics of process $x(t)$.

According to (8.31) to (8.33), in order to find the final relationship it is necessary, as in the harmonic linearisation method, to express some initial considerations relative to the character of the process examined, particularly its density function $w(u,v)$. In order to achieve that, we shall return to the structural scheme of the system examined (Fig. 6.5), taking into account the effects the random excitation $\vartheta(t)$ has on it.

During the transformation of a random process in nonlinear elements $\Phi(x,sx)$ the normality of the process is violated. That is why the input signal $F(t)$ of the linear filter $L(s)$ can be substantially different from the normal. We shall examine the transformation of $F(t)$ by a linear filter [177]. Using (6.43) we find, changing the variable $t-\tau=\tau'$ and extending the upper limit to infinity, bearing in mind that $h(t-\tau)\equiv 0$ when $\tau > t$,

$$\tilde{\chi}(t) = \int_0^{\infty} h(\tau')F(t-\tau')d\tau'. \tag{8.42}$$

Designating a variable further as τ, we will divide the interval of its variation as $(0,\tau_1)$, (τ_1,τ_2), ..., (τ_n,τ_{n+1}), ..., transforming the integral into an infinite sum, using the averaging theorem

$$\tilde{\chi}(t) = \sum_{m=0}^{\infty} h_m F(t-\tau_m^*) = \sum_{m=0}^{\infty}\xi_m. \tag{8.43}$$

Here

$$h_m = \int_{\tau_m}^{\tau_{m+1}} h(\tau)d\tau,\quad \xi_m = h_m F(t-\tau_m^*),\quad \tau_m \le \tau_m^* \le \tau_{m+1}.$$

We will choose the sub-intervals in such a way that the random value $F(t-\tau_m^*)$ and consequently ξ_m are statistically independent. Obviously, the more strongly process $F(t)$ is correlated then the longer the sub interval $T_m = \tau_{m+1} - \tau_m$, from which one can say that $R_F(t)\approx 0$ when $t > T_m$. Here

$R_F(t)$ is the correlation function of process $F(t)$.

As we already pointed out in section 1, the distribution sum of large statistically independent random processes, and consequently their values, tends to normality, if the summed components are statistically non dominant. In the case considered, this condition is satisfied if function $h(\tau)$ changes sufficiently slowly, i.e. if its value does not change significantly over a large number of sub-intervals. The maximum rate of damping $h(\tau)$ according to (6.50) is defined the magnitude of the greatest value of the real part $r_k\Omega_k$ of root s_k (see 6.55) of the characteristic equation of the system (6.54). Denoting the inverse value of $r_k\Omega_k$, as

$$T_0 = \min_k[1/(r_k\Omega_k)], \tag{8.44}$$

we can define the condition for the normality of a random process when passing through a linear filter in the following manner:

$$T_m / T_0 << 1, \tag{8.45}$$

i.e. the correlation time for an external excitation T_m must be considerably less than the characteristic time of damping for transient components of a linear system. Since the time correlation defines the width of the frequency band of a random process, and the value T_0 according to (6.61) the pass band of the filter, condition (8.45) can be thought of in the following physical terms: to normalise a random process transformed by a linear filter, the pass band of the frequency characteristics of the latter $|L(j\omega)|$ must be considerably narrower than the range of signuificance of the frequency of the spectral density $S_F(\omega)$ of the external excitation $F(t)$. Above all it is important that $T_m << T$, where $T = 2\pi/\Omega_k$. In the strongest sense, by normalising a process, it is understood that the signal at the output from a linear filter is closer to normality than at input.

To satisfy the requirements described above it is possible to assume approximately that the incoming nonlinear element $\Phi(x, sx)$ (Fig. 6.5) of a random excitation $x(t)$, existing as the solution of equation (8.28), has a normal distribution, which allows us to use coefficients (8.31) to (8.33) with the density function (8.24) during calculation, taking $\dot{x}(t) = y(t)$.

As a result the specified coefficients are dependent on m_x, σ_x, σ_y and the system of finite equations (8.36), (8,.40), (8.41) allows us finally to define these values.

In such cases, when the nonlinear characteristics take the form $\Phi(x)$, i.e. they reflect only the effect of elastic interaction, $\beta \equiv 0$ and the formula for calculating the linearisation coefficients is simplified considerably:

$$m_F = \int_{-\infty}^{\infty} \Phi(u) w(u) du, \tag{8.46}$$

$$k = \frac{1}{\sigma_x^2} \int_{-\infty}^{\infty} (u - m_x) \Phi(u) w(u) du. \tag{8.47}$$

Here $w(u)$ is the single-dimensional density function of process $x(t)$, given by expression (8.9) when $m_x = \text{const}, \sigma_x = \text{const}$.

It is interesting to note that, as shown in [177] for a normal process $x(t)$, approximating the nonlinear function $\Phi(x)$ by the linear relationship $m_F + k(x - m_x)$ is the best possible in terms of the minimum root-mean-square divergence of all possible linear approximations. It is also shown that calculation using the method described of the linearised value of the correlation function of process $x(t)$ and consequently the standard deviation, always show less than their exact values. There exists the possibility of statistical linearisation leading to the known overestimation of the values [113], however calculation of the linearisation coefficients using this method is considerably more difficult.

The practical calculation frequently uses values, existing as the half-sum of the coefficients, obtained from both methods. We will limit the first method to the calculation of linearisation coefficients, giving an adequate idea of the basic qualitative features of statistical dynamics of vibro-impact systems.

3. Let us turn to the coefficients for the statistical linearisation for the basic force characteristics of an impact pair. For nonlinearities of type (6.7), (6.8) the calculation according to (8.46) to (8.47) and using (8.9) is performed in monograph [113], hence we arrive directly at the final expressions. For the characteristics of single-sided interaction (7.7) we find

$$m_F = -c_2\sigma_x\left\{\frac{\Delta - m_x}{\sigma_x}\left[\frac{1}{2} + \mathcal{F}\left(\frac{\Delta - m_x}{\sigma_x}\right)\right] - \frac{1}{\sqrt{2\pi}}\exp\left[-\frac{(\Delta - m_x)^2}{2\sigma_x^2}\right]\right\}, \tag{8.48}$$

$$k = c_2\left[\frac{1}{2} + \mathcal{F}\left(\frac{\Delta - m_x}{\sigma_x}\right)\right], \tag{8.49}$$

where $\mathcal{F}(u) = \frac{1}{\sqrt{2\pi}}\int_0^u \exp\left(-\frac{\lambda^2}{2}\right)d\lambda$ is the probability integral, the value of which is tabulated in [184]. For double-sided impact pairs (6.8)

$$\begin{aligned} m_F = c_2\Bigg\{ & m_x - (\Delta + m_x)\mathcal{F}\left(\frac{\Delta + m_x}{\sigma_x}\right) + (\Delta - m_x)\mathcal{F}\left(\frac{\Delta - m_x}{\sigma_x}\right) + \\ & + \frac{\sigma_x}{\sqrt{2\pi}}\left[\exp\left(-\frac{(\Delta - m_x)^2}{2\sigma_x^2}\right) - \exp\left(-\frac{(\Delta + m_x)^2}{2\sigma_x^2}\right)\right]\Bigg\}, \end{aligned} \tag{8.50}$$

$$k = c_2\left[1 - \mathcal{F}\left(\frac{\Delta + m_x}{\sigma_x}\right) - \mathcal{F}\left(\frac{\Delta - m_x}{\sigma_x}\right)\right]. \tag{8.51}$$

When it is necessary to take energy dissipation during collisions into account in cases when using the statistical characteristic $\Phi(x)$ it is possible to make use of reasoning, similar to that which was used during harmonic linearisation.

Let us define the mean power of the energy dissipated during impact as N. Due to the suggested density function of process $x(t)$ (8.24), the standard deviation of its velocity does not depend on the state u and is equal to σ_y. As a result, the standard deviation of the collision velocity, when $u = \Delta$, is also equal to σ_y. As a result, using (7.57), we find during impact onto one stop

$$N_{d2} = \frac{1 - R^2}{2} Mf\sigma_y^2. \tag{8.52}$$

Here f is the average number of collisions in unit time against one stop.

The average power dispersed by the force of viscous friction $\beta\dot{x}$ under normal distribution of process $\dot{x}(t)=y(t)$ equals

$$N_{d1} = \int_{-\infty}^{\infty}\beta v^2 w(v)dv = \beta\sigma_y^2 . \tag{8.53}$$

Equating the values (8.52) and (8.53) we find

$$\beta = \frac{1-R^2}{2}Mf . \tag{8.54}$$

The value of f is found from (8.25) as the average number of intersections when $y > 0$ of process $x(t)$ at level $u = \Delta$. Substituting (8.24) in (8.25) we find

$$f = \frac{\sigma_y}{2\pi\sigma_x}\exp\left[-\frac{(\Delta - m_x)^2}{2\sigma_x^2}\right]. \tag{8.55}$$

Substituting (8.55) into (8.54) we finally obtain

$$\beta = \frac{(1-R^2)M\sigma_y}{4\pi\sigma_x}\exp\left[-\frac{(\Delta - m_x)^2}{2\sigma_x^2}\right]. \tag{8.56}$$

Naturally, for regimes with double-sided impacts the value of f and β must be doubled.

From time to time to simplify the calculation of the value of f, the problem can be simplified as before on the basis of a spectral analysis of the external excitation and natural characteristics of the system. For example, in (8.54) we can put

$$f = \omega_0 /(2\pi) , \tag{8.57}$$

where ω_0 is the dominant frequency in the spectrum of the process.

4. When calculating the coefficients of the statistically linearised force characteristics of the impact pair $\Phi(x,\dot{x})$ of type (6.13) to (6.16) from equation (8.31) to (8.33) the accuracy of calculation can be increased by refining the law of distribution $w(u,v)$ making use of additional information reflecting the specific nonlinearity examined. During calculation for systems with rigid stops it is convenient to take into account the limiting region of measurement of $x(t)$, i.e. to apply to $w(u,v)$ the condition

$$w(u,v) \equiv 0 \quad \text{when} \quad u > \Delta \tag{8.58}$$

for single-sided impacts pairs and

$$w(u,v) \equiv 0 \quad \text{when} \quad |u| > \Delta \tag{8.59}$$

for double-sided. Frequently, when disturbing systems with normal random processes it is expedient to choose as the value of $w(u,v)$ the truncated normal law [38] (see also §13, section 3)

$$w(u,v) = C_1\exp\left[-\frac{(u-m_x)^2}{2\sigma_x^2} - \frac{v^2}{2\sigma_y^2}\right][1-\eta(u-\Delta)] \tag{8.60}$$

for the single-sided case and

$$w(u,v) = C_2\exp\left[-\frac{(u-m_x)^2}{2\sigma_x^2} - \frac{v^2}{2\sigma_y^2}\right][1-\eta(|u|-\Delta)] \tag{8.61}$$

for the double-sided. Here $\eta(u)$ is the unit step function; C1, C2 are constants defined from the normalisation conditions

$$\int_{-\infty}^{\Delta}\int_{-\infty}^{\infty} w(u,v)dudv = 1 \tag{8.62}$$

for the single-sided case and

$$\int_{-\Delta}^{\Delta}\int_{-\infty}^{\infty} w(u,v)dudv = 1 \tag{8.63}$$

for the double-sided.

Substituting (8.60) into (8.62), we find

$$\int_{-\infty}^{\Delta} du \int_{-\infty}^{\infty} C_1 \exp\left[-\frac{(u-m_x)^2}{2\sigma_x^2} - \frac{v^2}{2\sigma_y^2}\right][1-\eta(u-\Delta)]dv =$$

$$= C_1 \int_{-\infty}^{\Delta} \exp\left[-\frac{(u-m_x)^2}{2\sigma_x^2}\right] du \int_{-\infty}^{\infty} \exp\left(-\frac{v^2}{2\sigma_y^2}\right) dv =$$

$$= 2\pi\sigma_x\sigma_y C_1\left[\frac{1}{2} + \mathcal{F}\left(\frac{\Delta - m_x}{\sigma_x}\right)\right].$$

Eventually we find

$$C_1 = \left\{2\pi\sigma_x\sigma_y\left[\frac{1}{2} + \mathcal{F}\left(\frac{\Delta - m_x}{\sigma_x}\right)\right]\right\}^{-1}. \tag{8.64}$$

Substituting (8.61) into (8.63) and following a similar calculation, we find for the double-sided impact pairs

$$C_2 = \left\{2\pi\sigma_x\sigma_y\left[\mathcal{F}\left(\frac{\Delta - m_x}{\sigma_x}\right) + \mathcal{F}\left(\frac{\Delta + m_x}{\sigma_x}\right)\right]\right\}^{-1} \tag{8.65}$$

Noting that for the characteristics (6.13) to (6.16) $\Phi_E(u,-v) = \Phi(u,v)$, $\Phi_D(u,-v) = \Phi_D(u,v)$ and also, according to (8.60), (8.61) $w(u,-v) = w(u,v)$ from the formulae (8.31) to (8.33) we find for the single-sided impact pair, taking (6.18) into account

$$m_F = \int_{-\infty}^{\Delta}\int_{-\infty}^{\infty} \Phi_E(u,v)w(u,v)du dv\,, \tag{8.66}$$

$$k = \frac{1}{\sigma_x^2}\int_{-\infty}^{\Delta}\int_{-\infty}^{\infty} (u-m_x)\Phi_E(u,v)w(u,v)du dv\,, \tag{8.67}$$

$$\beta = \frac{1}{\sigma_y^2}\int_{-\infty}^{\Delta}\int_{-\infty}^{\infty} v\Phi_E(u,v)w(u,v)du dv\,. \tag{8.68}$$

We shall substitute into (8.66) to (8.68) the double value of force characteristics (6.13) to (6.14) (see the note on p103) and also the formula for the law of distribution (8.60). During this, note that, since the value of the velocity before impact appears in equations (6.13) to (6.14), during calculation of the coefficients of linearised density distribution then the velocity in (8.60) and (8.61) must be halved. Using (8.64) we find

$$m_F = \int_{-\infty}^{\Delta}\int_{-\infty}^{\infty} \frac{Mv^2(1+R^2)\delta_+(u-\Delta)}{8\pi\sigma_x\sigma_y\left[\frac{1}{2} + \mathcal{F}\left(\frac{\Delta - m_x}{\sigma_x}\right)\right]} \exp\left[-\frac{(u-m_x)^2}{2\sigma_x^2} - \frac{v^2}{2\sigma_y^2}\right][1-\eta(u-\Delta)]du dv =$$

$$= \frac{M(1+R^2)}{8\pi\sigma_x\sigma_y\left[\frac{1}{2} + \mathcal{F}\left(\frac{\Delta - m_x}{\sigma_x}\right)\right]} \int_{-\infty}^{\Delta} \delta_+(u-\Delta)\exp\left[-\frac{(u-m_x)^2}{2\sigma_x^2}\right] du \int_{-\infty}^{\infty} v^2 \exp\left(-\frac{v^2}{2\sigma_y^2}\right) dv = \tag{8.69}$$

$$= \frac{M(1+R^2)}{4\sqrt{2\pi}} \frac{\exp\left[-\frac{(\Delta - m_x)^2}{2\sigma_x^2}\right]}{\frac{1}{2} + \mathcal{F}\left(\frac{\Delta - m_x}{\sigma_x}\right)} \frac{\sigma_y^2}{\sigma_x^2},$$

$$k=\frac{1}{\sigma_x^2}\int_{-\infty}^{\Delta}\int_{-\infty}^{\infty}(u-m_x)\frac{Mv^2(1+R^2)\delta_+(u-\Delta)}{8\pi\sigma_x\sigma_y\left[\frac{1}{2}+\mathscr{F}\left(\frac{\Delta-m_x}{\sigma_x}\right)\right]}\exp\left[-\frac{(u-m_x)^2}{2\sigma_x^2}-\frac{v^2}{2\sigma_y^2}\right][1-\eta(u-\Delta)]dudv=$$

$$=\frac{M(1+R^2)}{8\pi\sigma_x\sigma_y\left[\frac{1}{2}+\mathscr{F}\left(\frac{\Delta-m_x}{\sigma_x}\right)\right]}\int_{-\infty}^{\Delta}(u-m_x)\delta_+(u-\Delta)\exp\left[-\frac{(u-m_x)^2}{2\sigma_x^2}\right]du\int_{-\infty}^{\infty}v^2\exp\left(-\frac{v^2}{2\sigma_y^2}\right)dv=$$

$$=\frac{M(1+R^2)}{4\sqrt{2\pi}}\frac{(\Delta-m_x)\exp\left[-\frac{(\Delta-m_x)^2}{2\sigma_x^2}\right]}{1+\mathscr{F}\left(\frac{\Delta-m_x}{\sigma_x}\right)}\frac{\sigma_y^2}{\sigma_x^3},\tag{8.70}$$

$$\beta=\frac{1}{\sigma_x^2}\int_{-\infty}^{\Delta}\int_{-\infty}^{\infty}\frac{Mv^3(1-R^2)\delta_+(u-\Delta)\mathrm{sign}v}{8\pi\sigma_x\sigma_y\left[\frac{1}{2}+\mathscr{F}\left(\frac{\Delta-m_x}{\sigma_x}\right)\right]}\exp\left[-\frac{(u-m_x)^2}{2\sigma_x^2}-\frac{v^2}{2\sigma_y^2}\right][1-\eta(u-\Delta)]dudv=$$

$$=\frac{M(1-R^2)}{8\pi\sigma_x\sigma_y^3\left[\frac{1}{2}+\mathscr{F}\left(\frac{\Delta-m_x}{\sigma_x}\right)\right]}\int_{-\infty}^{\Delta}\delta_+(u-\Delta)\exp\left[-\frac{(u-m_x)^2}{2\sigma_x^2}\right]du\int_{-\infty}^{\infty}v^3\mathrm{sign}v\exp\left(-\frac{v^2}{2\sigma_y^2}\right)dv=$$

$$=\frac{M(1-R^2)\exp\left[-\frac{(\Delta-m_x)^2}{2\sigma_x^2}\right]}{8\pi\sigma_x\sigma_y^3\left[\frac{1}{2}+\mathscr{F}\left(\frac{\Delta-m_x}{\sigma_x}\right)\right]}\int_{-\infty}^{\infty}v^3\mathrm{sign}v\exp\left(-\frac{v^2}{2\sigma_y^2}\right)dv=$$

$$=\frac{M(1-R^2)}{2\pi}\frac{\exp\left[-\frac{(\Delta-m_x)^2}{2\sigma_x^2}\right]}{\frac{1}{2}+\mathscr{F}\left(\frac{\Delta-m_x}{\sigma_x}\right)}\frac{\sigma_y}{\sigma_x}.\tag{8.71}$$

For the double-sided impact pair equations (8.31) to (8.33) take the following form:

$$m_F=\int_{-\Delta}^{\Delta}\int_{-\infty}^{\infty}\Phi_E(u,v)w(u,v)dudv\,,\tag{8.72}$$

$$k=\frac{1}{\sigma_x^2}\int_{-\Delta}^{\Delta}\int_{-\infty}^{\infty}(u-m_x)\Phi_E(u,v)w(u,v)dudv\,,\tag{8.73}$$

$$\beta=\int_{-\Delta}^{\Delta}\int_{-\infty}^{\infty}v\Phi_D(u,v)w(u,v)dudv\,,\tag{8.74}$$

where Φ_E and Φ_D are defined by doubling equations (6.15) to (6.16). Substituting for them in (8.72) to (8.74) and carrying out the calculation similar to that above, we find using (8.61),(8.65)

$$m_F=\frac{M(1+R^2)}{4\sqrt{2\pi}}\frac{\exp\left[-\frac{(\Delta-m_x)^2}{2\sigma_x^2}\right]-\exp\left[-\frac{(\Delta+m_x)^2}{2\sigma_x^2}\right]}{\mathscr{F}\left(\frac{\Delta-m_x}{\sigma_x}\right)+\mathscr{F}\left(\frac{\Delta+m_x}{\sigma_x}\right)}\frac{\sigma_y^2}{\sigma_x^2}\tag{8.75}$$

$$k = \frac{M(1+R^2)}{4\sqrt{2\pi}} \frac{(\Delta - m_x)\exp\left[-\frac{(\Delta - m_x)^2}{2\sigma_x^2}\right] + (\Delta + m_x)\exp\left[-\frac{(\Delta + m_x)^2}{2\sigma_x^2}\right]}{\mathcal{F}\left(\frac{\Delta - m_x}{\sigma_x}\right) + \mathcal{F}\left(\frac{\Delta + m_x}{\sigma_x}\right)} \frac{\sigma_y^2}{\sigma_x^3}, \tag{8.76}$$

$$\beta = \frac{M(1-R^2)}{2\pi} \frac{\exp\left[-\frac{(\Delta - m_x)^2}{2\sigma_x^2}\right] + \exp\left[-\frac{(\Delta + m_x)^2}{2\sigma_x^2}\right]}{\mathcal{F}\left(\frac{\Delta - m_x}{\sigma_x}\right) + \mathcal{F}\left(\frac{\Delta + m_x}{\sigma_x}\right)} \frac{\sigma_y}{\sigma_x}. \tag{8.77}$$

5. The coefficients of linearisation obtained above depend on the statistical characteristics $m_x, \sigma_x^2, \sigma_y^2$ for process $x(t)$ appears in the input signal for the nonlinearity examined describing the impact. Making use of the terminology described earlier (see § 7) the described method to find the coefficients we will call *statistical motion linearisation*. By analogy with harmonic linearisation we can construct a method of statistical *force* linearisation, during which the coefficients of linearisation are expressed in terms of the statistical characteristics of process $F(t)$, acting as the output of the impacting elements. As for the harmonic case, such a method allows the use of elastic statistical characteristics of the impact pair $F = \Phi(x)$, given by the form shown in Fig. 7.2, *c-d*, leading to a simple calculation relationship.

We shall replace the approximate inverse function $x = X(F)$ (7.76) to (7.78) (Fig. 7.4) by the linear relationship

$$x \approx m_x + q(F - m_F) \quad (q = 1/k) \tag{8.78}$$

and defining the linearisation coefficients m_x and q by the formulae

$$m_x = \int_{-\infty}^{\infty} X(F) w(F) dF, \tag{8.79}$$

$$q = \frac{1}{\sigma_F^2} \int_{-\infty}^{\infty} (F - m_F) X(F) w(F) dF \tag{8.80}$$

minimising the mean square value of the error. Here m_F, σ_F^2 and $w(F)$ are mathematical expectation, mean square value and one-dimensional density function of process $F(t)$.

The relationship for $w(F)$ is chosen as the result of the following considerations. The spectral characteristics of process $F(t)$ during transformation $F(t) = \Phi[x(t)]$ are defined corresponding to the characteristics of the input process $x(t)$ and additional distortions arising from the nonlinearity of the transformation. The use of inertia-free linearisation relationships during statistical linearisation does not allow the us to catch these distortions. Thus in the approximate solution when applying normalisation to process $x(t)$ the distribution function of process $F(t)$ also shows normality. In other words, the approximate solution reflects only the characteristics of the main harmonic components in the transmission band of the linearised system. However, as mentioned in §7 the weak sensitivity of the inverse transformation $x(t) = X[F(t)]$ of types (7.76) to (7.78) to the particular form $F(t)$, the presence in this process of additional components by comparison with the basic components cannot substantially influence the character of the transformed main harmonics. Hence, even in the case of statistical linearisation it is possible to use quite crude approximations for process $F(t)$. From this discussion, we can take as an approximation for the normal law of distribution of $F(t)$:

$$w(F) = \frac{1}{\sqrt{2\pi}\sigma_F} \exp\left[-\frac{(F - m_F)^2}{2\sigma_F^2}\right]. \tag{8.81}$$

As a result, for functions of type (7.76), (7.78) (Fig. 7.4 *a*, *b*), we can make use of the coefficients of statistical linearisation calculated from the formulae similar to (8.79) to (8.81) in monograph [113], because they are treated there as typical nonlinearities of automatic control systems: ideal relay characteristics and linear characteristics with saturation. We will quote the resulting formula using our designations here.

For function (7.76), we have

$$m_x = 2\Delta \mathcal{F}(m_F/\sigma_F), \tag{8.82}$$

$$q = \sqrt{\frac{2}{\pi}} \frac{\Delta}{\sigma_F} \exp\left(-\frac{m_F^2}{2\sigma_F^2}\right), \tag{8.83}$$

for function (7.78)

$$m_x = \frac{1}{c}\left\{(c\Delta + m_F)\mathcal{F}\left(\frac{c\Delta + m_F}{\sigma_F}\right) - (c\Delta - m_F)\mathcal{F}\left(\frac{c\Delta - m_F}{\sigma_F}\right) + \right.$$
$$\left. + \frac{\sigma_F}{\sqrt{2\pi}}\left[\exp\left(-\frac{(c\Delta + m_F)^2}{2\sigma_F^2}\right) - \exp\left(-\frac{(c\Delta - m_F)^2}{2\sigma_F^2}\right)\right]\right\} \tag{8.84}$$

$$q = \frac{1}{c}\left[\mathcal{F}\left(\frac{c\Delta + m_F}{\sigma_F}\right) + \mathcal{F}\left(\frac{c\Delta - m_F}{\sigma_F}\right)\right] \tag{8.85}$$

For function (7.77) from formulae (8.79), (8.80) we calculate directly using (8.81)

$$m_x = \int_{-\infty}^{\infty} \frac{1}{c}[F - (F - c\Delta)\eta(F - \Delta)] \frac{1}{\sqrt{2\pi}\sigma_F} \exp\left[-\frac{(F - m_F)^2}{2\sigma_F^2}\right] dF =$$
$$= \frac{1}{\sqrt{2\pi}\sigma_F c} \int_{-\infty}^{c\Delta} F \exp\left[-\frac{(F - m_F)^2}{2\sigma_F^2}\right] dF + \frac{\Delta}{\sqrt{2\pi}\sigma_F} \int_{c\Delta}^{\infty} \exp\left[-\frac{(F - m_F)^2}{2\sigma_F^2}\right] dF. \tag{8.86}$$

The first integral is obtained with the help of the substitution $(F - m_F)/\sigma_F = \lambda$. As a result we find, using the abbreviation $(c\Delta - m_F)/\sigma_F \equiv v$,

$$\frac{1}{\sqrt{2\pi}\sigma_F c} \int_{-\infty}^{c\Delta} F \exp\left[-\frac{(F - m_F)^2}{2\sigma_F^2}\right] dF = \frac{1}{\sqrt{2\pi} c} \int_{-\infty}^{v} (m_F + \sigma_F \lambda) \exp\left(-\frac{\lambda^2}{2}\right) d\lambda =$$
$$= \frac{m_F}{\sqrt{2\pi} c} \int_{-\infty}^{v} \exp\left(-\frac{\lambda^2}{2}\right) d\lambda - \frac{\sigma_F}{\sqrt{2\pi} c} \int_{-\infty}^{v} -\lambda \exp\left(-\frac{\lambda^2}{2}\right) d\lambda =$$
$$= \frac{m_F}{c}\left[\frac{1}{2} + \mathcal{F}\left(\frac{c\Delta - m_F}{\sigma_F}\right)\right] - \frac{\sigma_F}{\sqrt{2\pi} c} \exp\left[-\frac{(c\Delta - m_F)^2}{2\sigma_F^2}\right]$$

Noting that the second integral in the right-hand part of (8.86) can be calculated using the probability integral we find eventually

$$m_x = \frac{\sigma_F}{c}\left\{\frac{m_F}{\sigma_F}\left[\frac{1}{2} + \mathcal{F}\left(\frac{c\Delta - m_F}{\sigma_F}\right)\right] - \frac{1}{\sqrt{2\pi}} \exp\left[-\frac{(c\Delta - m_F)^2}{2\sigma_F^2}\right]\right\} + \Delta\left[\frac{1}{2} - \mathcal{F}\left(\frac{c\Delta - m_F}{\sigma_F}\right)\right]. \tag{8.87}$$

Further

$$q = \frac{1}{\sigma_F^2} \int_{-\infty}^{\infty} (F - m_F) \frac{1}{c}[F - (F - c\Delta)\eta(F - c\Delta)] \frac{1}{\sqrt{2\pi}\sigma_F} \exp\left[-\frac{(F - m_F)^2}{2\sigma_F^2}\right] dF =$$
$$= \frac{1}{\sigma_F^2}\left\{\frac{1}{\sqrt{2\pi}\sigma_F c} \int_{-\infty}^{c\Delta} F(F - m_F) \exp\left[-\frac{(F - m_F)^2}{2\sigma_F^2}\right] dF + \right.$$
$$\left. + \frac{\Delta}{\sqrt{2\pi}\sigma_F} \int_{c\Delta}^{\infty} (F - m_F) \exp\left[-\frac{(F - m_F)^2}{2\sigma_F^2}\right] dF\right\}$$

Carrying out in both integrals the change of variable $(F - m_F)/\sigma_F \equiv \lambda$, we find

$$q = \frac{1}{\sqrt{2\pi}\sigma_F c} \int_{-\infty}^{v} (m_F + \sigma_F \lambda) \lambda \exp\left(-\frac{\lambda^2}{2}\right) d\lambda + \frac{\Delta}{\sqrt{2\pi}\sigma_F} \int_{v}^{\infty} \lambda \exp\left(-\frac{\lambda^2}{2}\right) d\lambda =$$

$$= \frac{m_F}{\sqrt{2\pi}\sigma_F c}\int_{-\infty}^{v}\lambda\exp\left(-\frac{\lambda^2}{2}\right)d\lambda + \frac{1}{\sqrt{2\pi}}\int_{-\infty}^{v}\lambda^2\exp\left(-\frac{\lambda^2}{2}\right)d\lambda + \frac{\Delta}{\sqrt{2\pi}\sigma_F}\int_{v}^{\infty}\lambda\exp\left(-\frac{\lambda^2}{2}\right)d\lambda .$$

Integrating the second term by parts, we obtain finally

$$q = \frac{-m_F}{\sqrt{2\pi}\sigma_F c}\exp\left[-\frac{(c\Delta - m_F)^2}{2\sigma_F^2}\right] - \frac{1}{\sqrt{2\pi}c}\lambda\exp\left(-\frac{\lambda^2}{2}\right)\Bigg|_{-\infty}^{v} +$$

$$+\frac{1}{\sqrt{2\pi}c}\int_{-\infty}^{v}\exp\left(-\frac{\lambda^2}{2}\right)d\lambda + \frac{\Delta}{\sqrt{2\pi}\sigma_F}\exp\left[-\frac{(c\Delta - m_F)^2}{2\sigma_F^2}\right] =$$

$$= \frac{1}{c}\left[\frac{1}{2} + \mathscr{F}\left(\frac{c\Delta - m_F}{\sigma_F}\right)\right]. \tag{8.88}$$

The dissipation of energy in the system equivalent to the loss during collisions, can be calculated during the given method from formulae of type (8.54), (8.57). Later we will use the values obtained for the coefficients of statistical linearisation to examine the statistical dynamics of some characteristic vibro-impact systems. We will also compare approximate and exact solutions (see §13).

Chapter 3. Vibro-Impact processes in systems with single degree of freedom

I owe a lot to my engineering training because it taught me to tolerate approximations. Previously to that I thought ... one should just concentrate on exact equations all the time. Then I got the idea that in the actual world all our equations are only approximate.

Paul Dirac.

§9. Forced vibration

1. Using the methods of harmonic linearisation described in §7, we will continue by analysing forced vibration of a linear oscillator with rigid stops, which will occasionally be referred to as *impact oscillators* later on. We will analyse the basis of using the method of harmonic force and motion linearisation, and compare the results with the exact solutions obtained in §2 to §3. We will begin by studying the non-symmetrical system (Fig. 2.1, *a*), assuming as before, that the stop is arranged with gap (interference) Δ relative to the equilibrium condition of the system without stop.

A characteristic of unsymmetrical vibration, as mentioned in §2, consists of a constant component m_x in the periodic solution $x(t)$, arising because of the basic non-symmetry of the structure or external force acting on it. In addition, it is important to emphasise that the constant component of vibration in a nonlinear system can occur even during symmetrical excitation (see §2). Moreover, when a constant external load m_P is applied to a nonlinear system with a symmetrical structure, a constant displacement (drift) m_x can occur during vibration, which is not proportional to m_P.

A suitable case for the study of systems with both factors mentioned above influencing the appearance of drift, can be found in the following general form of the differential equation of the system:

$$(s^2+bs+\Omega^2)x+\Phi_E(x,sx)+\Phi_D(x,sx)=P(t) \tag{9.1}$$

Here $\Phi_E(x,sx)$ and $\Phi_D(x,sx)$ are the elastic and dissipative forces of the characteristic impact pair, while $\Phi_E(-x,sx)\neq\Phi_E(x,sx)$, $P(t)=m_P+a_P\cos(\omega t+\varphi)$, where in the general case $a_p=a_p(\omega)$.

Searching for an approximate periodic solution for equation (9.1) in the form

$$x(t)\approx m_x+a_x\cos\omega t,\qquad sx(t)\approx -a_x\omega\sin\omega t, \tag{9.2}$$

we can transform the periodic applied external excitation $P(t)$ using (9.2):

$$a_P(\cos\omega t\cos\varphi-\sin\omega t\sin\varphi)=\frac{a_P}{a_x}\left(\cos\varphi+s\frac{\sin\varphi}{\omega}\right)x^0, \tag{9.3}$$

where $x^0 = x - m_x$.

Substituting (9 .3) in (9 .1), we obtain

$$\left[s^2 + \left(b - \frac{a_P \sin\varphi}{a_x \omega}\right)s + \Omega^2 - \frac{a_P \cos\varphi}{a_x}\right]x^0 + \\ + \Omega^2 m_x + \Phi_E(x, sx) + \Phi_D(x, sx) = m_P . \tag{9.4}$$

Thus, using the character of the assumed solution the initial non-autonomous system can be converted to an "equivalent" autonomous one. Such a concept allows us to consider various types of vibration of nonlinear systems such as forced, parametric or self-sustained vibration in a similar manner, and this approach will be widely used later.

For the nonlinear functions $\Phi_E(x, sx), \Phi_D(x, sx)$, using their linearised expressions (see §7):

$$\Phi_E(x, sx) \approx m_F + k(x - m_x), \quad \Phi_D(x, sx) \approx \beta sx , \tag{9.5}$$

where the coefficients of harmonic linearisation m_F, k and β are found using the unknown parameters of the solution $x(t)$.

Substituting (9.5) into (9.4) and separating into similar terms, we obtain

$$\left[s^2 + \left(b + \beta - \frac{a_P \sin\varphi}{a_x \omega}\right)s + \left(\Omega^2 + k - \frac{a_P \cos\varphi}{a_x}\right)\right]x^0 + m_F + \Omega^2 m_x = m_P . \tag{9.6}$$

Separating the constant and centred components in (9.6), we find

$$m_F + \Omega^2 m_x = m_P , \tag{9.7}$$

$$\left[s^2 + \left(b + \beta - \frac{a_P \sin\varphi}{a_x \omega}\right)s + \Omega^2 + k - \frac{a_P \cos\varphi}{a_x}\right]x^0 = 0 . \tag{9.8}$$

Equation (9.7) represents the final relationship, defining the connection

$$m_x = m_x(a_x) . \tag{9.9}$$

This expression gives the relationship between the value of the drift to vibration of the system. In particular, using the method shown, relationship (9.9) can be found in practice without solving the differential equation for the system. Expression (9.9) allows us to exclude the value of m_x from the equation for coefficients k and β, after which equation (9.8) can be studied independently.

The linearised form of equation (9.8) has the following characteristic equation:

$$\mathscr{D}(s) = s^2 + \left(b + \beta - \frac{a_P \sin\varphi}{a_x \omega}\right)s + \Omega^2 + k - \frac{a_P \cos\varphi}{a_x} = 0 . \tag{9.10}$$

The conditions for the existence of periodic solution (9.2) for the linearised system (9.8) is the occurrence of a pair of imaginary roots $s = \pm j\omega$ to its characteristic equation (9.10). Substituting $s = j\omega$ into (9.10) and setting the real and imaginary parts to zero, we obtain a system of equations of the form

$$\sin\varphi = (b + \beta)\omega a_x / a_P , \tag{9.11}$$

$$\cos\varphi = (\Omega^2 + k - \omega^2) a_x / a_P . \tag{9.12}$$

These represent a particular case of expressions (7.18), (7.19).

Squaring the equations (9.11) and (9.12) and adding the left and right parts, we obtain the vibration amplitude in terms of the system parameters and the linearisation coefficients

$$a_x = a_P[(\Omega^2 + k - \omega^2)^2 + (b+\beta)^2\omega^2]^{-1/2}. \tag{9.13}$$

Using the relationship for the linearisation coefficients and equation (9.7), equation (9.13) is the equation for amplitude a_x, which can be found using either graphically or by calculation. Hence, from formulae (9.11), (9.12) the value of the phase φ can be found.

Let us examine equation (9.13). The relationship

$$\omega_0(a_x) = \sqrt{\Omega^2 + k(a_x)}, \tag{9.14}$$

as already discussed in §7, section 1, represents an approximation to the equation of the backbone curve on the basis of the fundamental for the system. Substituting $\omega = \omega_0$ in (9.13), we obtain the equation for the energy boundary of vibration

$$a_x = a_P/[(b+\beta)\omega], \tag{9.15}$$

which defines the maximum obtainable amplitude, since any violation of the equality $\omega = \omega_0$ leads to a reduction in the value a_x in (9.13) by comparison with (9.15). Note that equation (9.15) takes the same physical meaning as that obtained in § 3 for the exact boundary (3.17), however the structure of equation (9.15) is considerably simpler. In section 3 we will compare these relationships.

The intersections of curve *1* for the relationship (9.15) (Fig. 9.1) with the backbone curve *2* represents a solution to equation (9.13), defining an approximation for the maximum point of the amplitude-frequency characteristics of the system, the resonance branches of which are situated along the backbone curve. The remaining points of the amplitude curve can be found conveniently using of the following graphical example, fully described in [124]. Examining the natural frequency of the system using ω_0 as argument, we construct the graphical function *3* (Fig. 9.1)

$$a_x(\omega_0) = a_P[(\omega_0^2 - \omega^2)^2 + (b+\beta)^2\omega^2]^{-1/2}$$

for various fixed values ω with the remaining parameters of the system fixed. In this figure we draw the backbone curve *2* (9.14), for the identical parameter values. The intersection points of both curves determine the solution required. From Fig. 9.1 it can be seen that due to the character of the non-isochronous system several periodic solutions can occur for the same values of the constant parameters.

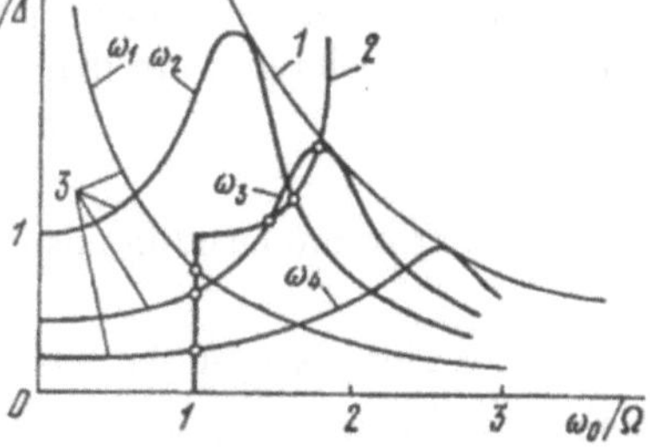

Fig. 9.1

For systems with single-sided impact pairs acceptable accuracy is obtained using the linearisation coefficients (7.65) to (7.66) and estimating the energy dissipation during impact from (7.61). In this case the procedure for an analytic solution can be considerably simplified, since the equation for the constant and periodic terms of the solution can be immediately separated: expression (9.9) takes the simple form (7.63), while equation (9.8) can be solved independently, since k is a function of only one unknown parameter a_x. We will now find this solution.

Noting that in the case examined the operator $L(s)$ takes the form (6.72) we find for formula (7.66)

$$L(0) = 1/\Omega^2 .$$

The value m_ϑ in (7.66) is expressed in terms of the parameters of the system considered in the following form:

$$m_\vartheta = m_P/\Omega^2 . \tag{9.16}$$

As a result, in this case the linearisation coefficients (7.65), (7.66) take the form

$$m_F = m_P + (a_x - \Delta)\Omega^2 , \tag{9.17}$$

$$k = 2\Omega^2\left(1 - \frac{\Delta - m_P/\Omega^2}{a_x}\right). \tag{9.18}$$

Substituting (9.18) in (9.14) we find for the natural frequency ω_0 of the nonlinear system

$$\left(\frac{\omega_0}{\Omega}\right)^2 = 3 - \frac{2(\Delta - m_P/\Omega^2)}{a_x} . \tag{9.19}$$

Expression (9.19) gives an obvious representation of the relative configuration of the backbone curves of the system. By satisfying the condition

$$m_P/\Omega^2 = \Delta , \tag{9.20}$$

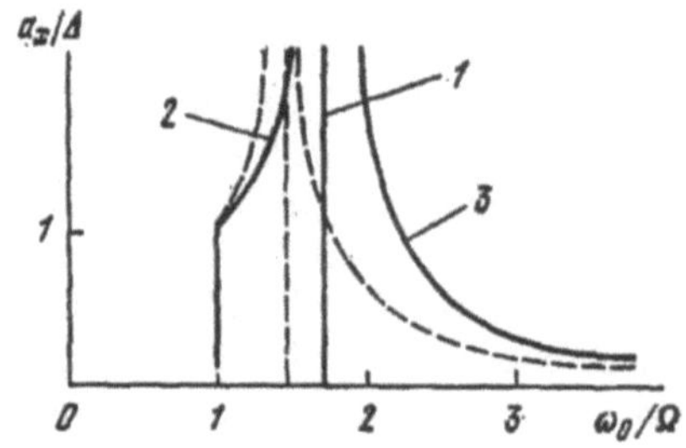

Fig. 9.2

i.e. when the static force m_P completely takes up the arranged gap or interference Δ, the system remains isochronous at the natural frequency

$$\omega_0 = \sqrt{3}\Omega \approx 1.73\Omega \tag{9.21}$$

and its backbone curve takes the form of a vertical line *1* (Fig. 9.2). When $m_P \equiv 0$ we find from (9.20) that the isochronous state in this case appears when $\Delta = 0$. The fact that the system considered is isochronous when $\Delta = 0$ was described in § 2 when considering exact solutions of the equations for free vibration. In that case the exact value of the frequency obtained was

$$\omega_0 = 2\Omega . \tag{9.22}$$

Comparing (9.21) with (9.22) shows the size of the error produced in this case by using the approximate approach.

The value $\Delta_0 = \Delta - (m_P/\Omega^2)$ determines the resulting set of the system. When $\Delta_0 > 0$ in the system implies a gap; when $\Delta_0 < 0$ implies interference. Fig. 9.2 shows the branches of the backbone curve for systems *2*, *3* using formula (9.19) with full lines for the cases $\Delta_0 > 0$ and $\Delta_0 < 0$. Comparing Fig. 9.2 with the graphical exact solution (Fig. 2.3) shows the very high quality correlation obtained by the approximation.

The amplitude-frequency characteristics were drawn for the system according to formula (9.13) using (9.18), (7.61). Fig. 9.3 shows several typical configurations for the amplitude-frequency characteristics of systems considered. The backbone lines and energy boundaries of vibration are shown dashed. For simplification the latter were drawn for $R = 1$. Naturally, when $\Delta_0 \le 0$ systems can only develop vibro-impact regimes of motion and the development of such regimes takes a different character. In the case when $a_P > \Omega^2\Delta_0$, the system

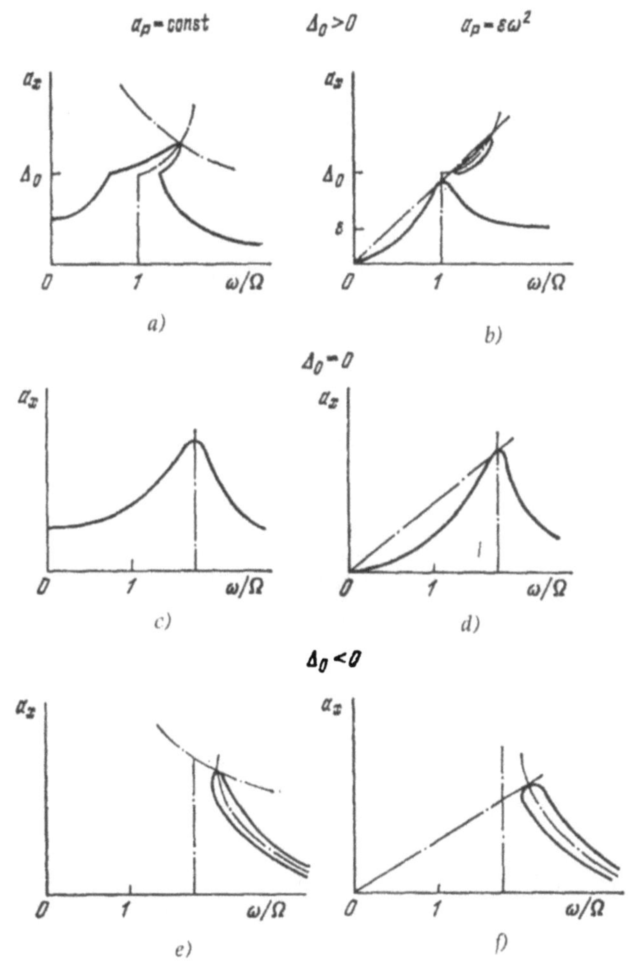

Fig. 9.3

excitation has a soft character, i.e. vibration exists whatever the initial conditions. However, vibro-impact regimes can exist in systems even when $a_P < \Omega^2\Delta_0$ (Fig. 9.3, *e*, *f*). In these cases the change to a vibro-impact regime can occur only as the result of hard excitation (an additional starting impulse). The value of interference does not influence the character of the backbone curves, so that increasing the contact force can lead to an increase in the value of the starting impulse needed but does not eliminate the possibility of establishing vibro-impact regimes. It follows from Fig. 9.3, *e*, *f* that such regimes can lead to particularly dangerous oscillation when slowing down ($\omega \to 0$).

With a gap ($\Delta > 0$) vibro-impact regimes can exist only when $a_x > \Delta_0$. In the case a_p = const, it is possible for regimes to develop into impact types smoothly as the frequency of excitation increases and there is a specific pulling typical for nonlinearities of hardening types (Fig. 9.3, *a*). When $a_p = \varepsilon\omega^2$ a situation can arise (Fig. 9.3, *b*) when the change from non-impact regimes to vibro-impact regimes requires hard excitation, hence the latter can be undetected when testing systems on vibration devices.

As can be seen from Fig. 9.3, *b* (see (9.15)), an effective method of combating vibro-impact regimes of the types considered is to increase the dissipation of the system.

2. We shall continue the basic search for non-symmetrical solutions to the equation of a linear oscillator with rigid stop using the method of harmonic *force* linearisation. In this case the elastic effects of a linear spring must be included in the basic nonlinear elastic characteristics of the system $\Phi(x)$, and the energy loss leads to the introduction of an equivalent dissipative force $\beta\dot{x}$ with coefficients given by (7.61).

The differential equation which results from applying the method studied can be written as:

$$[s^2 + (b+\beta)s]x + \Phi(x) = P(t) \tag{9.23}$$

or following the method described above for reducing the system using the character of the excitation and solution of the type (9.2), to equivalent autonomous differential equation (see (9.3)) of the following form:

$$\left[s^2 + \left(b + \beta - \frac{a_p \cos\varphi}{a_x}\right)s - \frac{a_p \cos\varphi}{a_x}\right]x^0 + \Phi(x) = m_p\,. \tag{9.24}$$

To linearise equation (9.24), we put

$$\Phi(x) \approx m_F + k(x - m_x)\,. \tag{9.25}$$

As a result from (9.24), (9.25) we find for the constant terms

$$m_F = m_p \tag{9.26}$$

Condition (9.26) can be accounted for immediately in the expression for $k = 1/q$ (see §7, section 4). Hence the value of k depends only on a_F and the linearised equation for the periodic component, as in the previous case, can be solved independently. It will take the following form:

$$\left[s^2 + \left(b + \beta - \frac{a_p \sin\varphi}{a_x\omega}\right)s + k - \frac{a_p \cos\varphi}{a_x}\right]x^0 = 0\,. \tag{9.27}$$

Substituting $s = j\omega$ in its characteristic equation and setting the real and imaginary parts to zero, we obtain a system of equations similar to (9.11), (9.12):

$$\sin\varphi = (b+\beta)\omega a_x / a_P, \tag{9.28}$$

$$\cos\varphi = (k-\omega^2)a_x / a_P. \tag{9.29}$$

Solving these equations for a_x, we find

$$a_x = a_P[(k^2-\omega^2)+(b+\beta)^2\omega^2]^{-1/2}. \tag{9.30}$$

Multiplying both sides of (9.30) by $\sqrt{k^2+\omega^2\beta^2}$ and using the relationship $a_F = a_x\sqrt{k^2+\omega^2\beta^2}$, we obtain eventually

$$a_F = a_P(k^2+\omega^2\beta^2)^{1/2}[(k^2-\omega^2)^2+(b+\beta)^2\omega^2]^{-1/2}. \tag{9.31}$$

Equation (9.31) makes it possible to find a_F, from which all the remaining parameters of the regime considered can be determined.

The value

$$\omega_0 = \sqrt{k(a_F)} \tag{9.32}$$

represents the natural frequency of the linearised system, which depends on the force amplitude a_F realised by the nonlinear element $\Phi(x)$, while

$$a_F = k(a_F)a_x. \tag{9.33}$$

We shall use the linearisation coefficients k and β in their forms (7.95), (7.61) in our analysis. Substituting (7.95) into (9.32) using here $c = \Omega^2$, we find the following equation for the natural frequency of the system:

$$\left(\frac{\omega_0}{\Omega}\right) = \left[2\left(1 \pm \frac{1}{1-\pi/4+\pi a_F/(4\Omega^2\Delta)}\right)\right]^{-1} \tag{9.34}$$

Expressing a_F in terms of a_x in (9.34) using (9.33), (7.95), we obtain the equation for the backbone curve of the system

$$\frac{a_x}{\Delta} = \frac{4}{\pi}\left[\pm\frac{1}{2-(\omega_0/\Omega)^2} - \frac{1}{(\omega_0/\Omega)^2}\left(1-\frac{\pi}{4}\right)\right]. \tag{9.35}$$

Here the positive sign before the first term implies a gap, while the negative sign implies interference. Equation (9.35) is shown by a dashed line in Fig. 9.2. It can be seen that in this case, the configuration of the backbone curves coincide closely with the exact results (Fig. 2.3): for an initial gap, ($\Delta > 0$) these are of hardening type, with interference ($\Delta < 0$) they are of softening type, while when $\Delta = 0$ the system remains isochronous. In addition, the isochronous frequency is found with a somewhat larger error and according to (9.35) takes the value

$$\omega_0 = \sqrt{2}\Omega \approx 1.41\Omega \tag{9.36}$$

The construction of the amplitude-frequency characteristics in the case considered according to formula (9.31) requires the values of coefficients k and β. For this the graphical method described in section 1 can be used (see Fig. 9.1). As a result we obtain the configuration of amplitude curves similar to those shown in Fig. 9.3. The unknown phase φ is found from (9.28), (9.29), using the known value of a_F.

Because the vibro-impact regimes studied are resonant, the accuracy of their resonance curves depends essentially on the accuracy of the relationship $\omega_0(a_x)$,

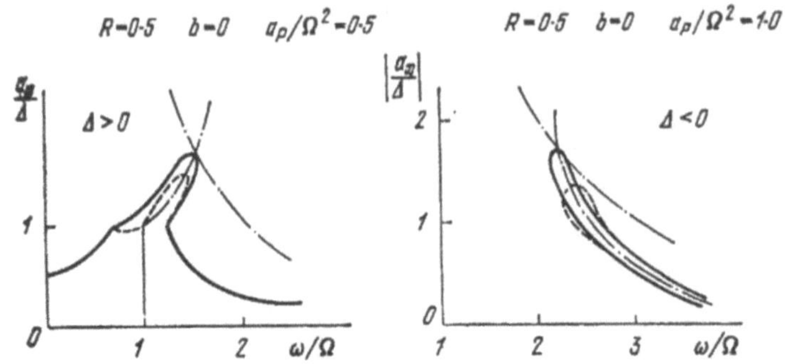

Fig. 9.4

expressed in terms of the linearisation coefficient k. For this relationship can use the exact backbone curves for systems similar to that which we already used in the example of §7, section 1, or the exact equations for the backbone curves for the fundamental. It is convenient to make such a substitution for systems with one degree of freedom, for which the relationship shown is easily found by the "stitching" method or its analogy. Nevertheless, for systems with higher degrees of freedom this refinement is not worthwhile in practice, since the application their stronger filtering effects automatically increases the accuracy of investigation using harmonic linearisation.

For the unsymmetrical system considered we will use the exact equations for their backbone curves (2.11), found from a study of the phase portrait for free vibration. Applying an external force m_P causes the value of Δ to change with increment (decrement) m_P/Δ^2, hence in the general case we find

$$\omega_0 = \frac{\Omega}{\arccos\left[1 - \dfrac{2a_x\Omega^2}{\Omega^2\Delta - m_P}\right]^{-1}}. \tag{9.37}$$

Fig. 9.4 shows examples of the amplitude-frequency characteristics for the system according to (9.13) using (9.14), (9.37), (7.61). For comparison the dotted line shows the exact relationship obtained in §2 (Fig. 2.9).

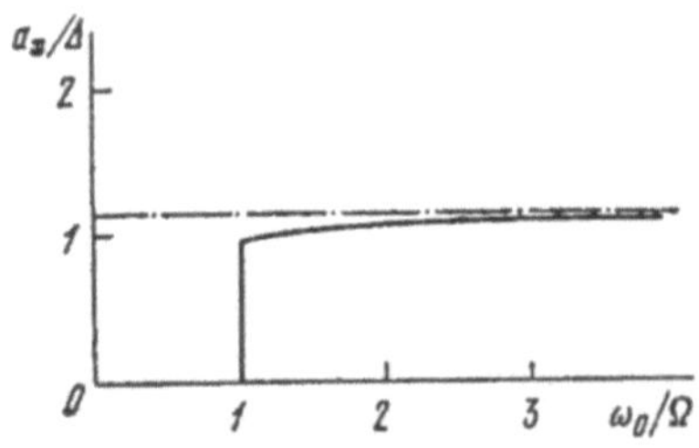

Fig. 9.5

3. Let us now consider the problem of forced vibration of a linear oscillator with symmetrical stops, excited by a force $P(t) = a_P(\omega)\cos(\omega t + \varphi)$ (Fig. 3.1). Using the method of harmonic force linearisation in this case gives a considerably simpler calculation procedure.

The equations for the system as before take the form (9.23) where $\Phi(x)$ is now the characteristic of a symmetrical impact pair (Fig. 6.2, b). As a result we can make use of the solution of equation (9.23) found in section 2, putting $m_F \equiv 0$ to account for the symmetry of the system and using the linearisation coefficient k and β in their forms (8.91), (8.92). In (7.91) we must have $m_F = 0$, $c = \Omega^2$, and so we find

$$k = \frac{1}{q} = \Omega^2\left(1 - \frac{\pi}{4} + \frac{\pi a_F}{4\Omega^2 \Delta}\right) \tag{9.38}$$

Eliminating a_F and k from equations (9.32), (9.33), (9.38), we find the equation for the backbone curves

$$\frac{a_x}{\Delta} = \frac{4}{\pi} - \frac{\Omega^2}{\omega_0^2}\left(\frac{4}{\pi} - 1\right) \qquad (\omega_0 \geq \Omega)\,. \tag{9.39}$$

According to (9.39) $1 \leq a_x/\Delta \leq 4/\pi$. Fig. 9.5 shows the backbone curves for the fundamental of the system, drawn using (9.39) (compare with the exact values in Fig. 3.4).

Substituting $\omega = \omega_0 = \sqrt{k}$ in (9.30) we obtain the equation for the energy boundary of vibration in the form (9.15). When $R = 1$, $\beta = 0$ and equation (9.15) takes the form

$$a_x = a_p/(b\omega)\,, \tag{9.40}$$

i.e. it coincides with the boundary of the existence domains for harmonic vibration of a linear oscillator with viscous friction. When a_p = const this is a hyperbola, hence its intersection with the hard backbone curve can occur at one point only and the amplitude-frequency characteristics will show pulling in domains of high frequency excitation (Fig. 9.6, *a*). Of course, impact in the system corresponds to the branches of the curves satisfying the condition $a_x/\Delta > 1$.

When $a_p = \varepsilon\omega^2$, i.e. during excitation, caused for example by an external unbalanced mass, equation (9.40) becomes linear at a relative frequency

$$a_x = \varepsilon\omega/b \tag{9.41}$$

and its intersection with the backbone curves leads to the configuration of the amplitude-frequency characteristics shown in Fig. 9.6, *b*, *c*. Hence depending on the ratio ε/b it is possible to have either smooth transformation to regimes with impacts due to the acceleration of the unbalanced mass (Fig. 9.6, *b*), or to proceed to such regimes as the result of hard excitation in the frequency range satisfying the condition (Fig.

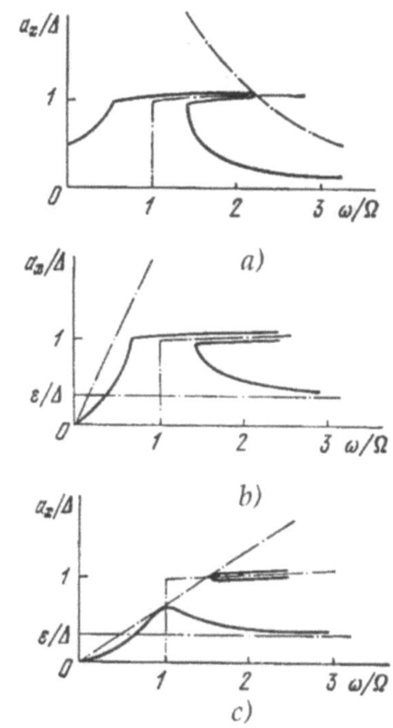

Fig. 9.6

9.6, *c*)

$$\omega > a_x b/\varepsilon \approx \Delta b/\varepsilon . \qquad (9.42)$$

In real systems the upper boundary for the indicated range is limited, because the finite power of the applied excitation at higher frequencies, prevents the law of excitation considered from being achieved (see §12).

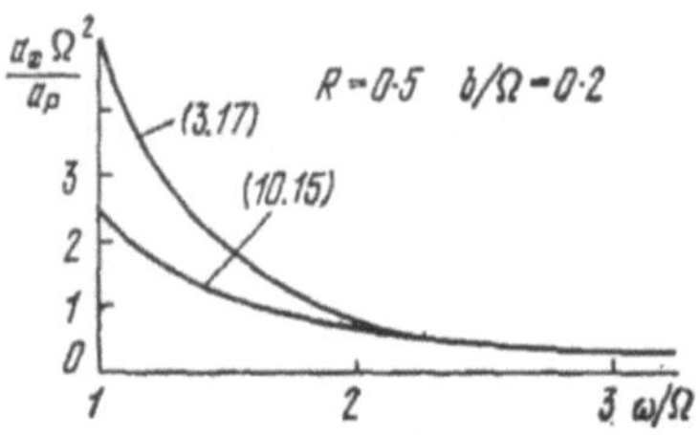

Fig. 9.7

Let us estimate the influence of both types of damping. Substituting (7.62) in (9.15), we find

$$a_x = a_p /(b\omega + \rho\omega^2) . \qquad (9.43)$$

The concept presented agrees with the result of analysis of the exact solution (see §3, section 5) on the dominant influence of the impact damping in the high-frequency range.

Fig. 9.7 shows compares the boundaries of existence domains for the regimes considered (energy boundaries of vibration), found using exact methods and the approximations from (9.15), (7.62). In the high-frequency range they virtually coincide. The discrepancy between these values close to the natural frequency of the linear oscillator is due to the fact that in this frequency range impacts have a weak character (see, for example, Fig. 3.10), since the value of the damping coefficient used (7.62), calculated from the assumption of a resonant phase regime, is over-estimated.

When $a_p = \varepsilon\omega^2$ from (9.43) we find

$$a_x = \frac{\varepsilon}{(b/\omega) + \rho} . \qquad (9.44)$$

From this we can obtain the upper limits for the lower boundary of the frequency band, in which the vibro-impact regime considered exists. Solving (9.44) for ω, we obtain the existence condition in the form

$$\omega \geq \frac{b}{(\varepsilon/a_x) - \rho} \approx \frac{b}{(\varepsilon/\Delta) - \rho} . \qquad (9.45)$$

From this its follows, for example, that when

$$\Delta \leq \varepsilon/\rho \qquad (9.46)$$

vibro-impact regimes exist in all post-resonance frequency bands in which the excitation is of the type considered.

As an example of use of the formula we will study the influence of energy dissipation on the clearance in a vibration isolation system, during the design of which the dynamic model considered is widely used. When the isolator has either natural or special stops, motion can arise due to the application of random jolts which violate the conditions for isolation, resulting in the establishment of vibro-impact regimes of the type considered ("clatter on the stops").

Suppose that an isolation system with a natural frequency of 10 *Hz* (Ω = 62.81/*s*) is affected by vibration excitation at a frequency of 100 *Hz* (Ω = 628.1/*s*) and magnitude $\varepsilon = 10^{-4} m$. The damping coefficient is b = 12.51/*s*. It is necessary to find the value of the free movement Δ, which will ensure the absence of clatter against the stops.

The formulae of the linear theory of vibration, neglecting the possibility of vibro-impact regimes due to hard excitation of the system gives

$$\Delta > a_x = \varepsilon\omega^2[(\Omega^2 - \omega^2)^2 + b^2\omega^2]^{-1/2} \approx 1.01 \cdot 10^{-4} m .$$

From equation (9.25), for example, when $R = 0.5$, we obtain using (7.62), $\Delta > a_x = 3.24\ 10^{-4}\ m$, finally from equation (9.42), neglecting the loss of energy during impact, we find $\Delta > a_x = \varepsilon\omega/b = 52\ 10^{-4} m$. Hence the gap taken into account must be about 4 times larger than the calculated amplitude of deformation of the suspension. If we do not take account of the energy losses during impact, then in the example considered the clearance in the isolator must be increased by a factor not less than 13. Note that the theoretical study of the phenomenon of deterioration of the isolator and the emergence in post-resonance domains of excitation frequencies in high-frequency regimes with impact against stops was made by Y I Iorish [109] and is investigated in detail in the monograph by M Z Kolovsky [124].

Later, as we study the quantitative features of various types of vibration of vibro-impact systems, for the case of single-sided impact pairs, we will use as a basis the harmonic motion linearisation, taking the coefficients k and β in the form (7.66), (7.61), while for double-sided pairs we shall use the method of harmonic force linearisation with coefficients (7.91), (7.62), since this will lead to considerably simpler calculation of the relationships and to satisfactory qualitative answers.

4. Harmonic linearisation of vibro-impact systems, while it gives a clear representation of the character of the fundamental during periodic vibration, does not provide full information about the form of the periodic motion. Of particular importance here is the question of whether or not sub-periodic regimes exist, since in these regimes the addition of energy to the system occurs at higher frequencies.

In practice, suppose that $x(t)$ represents a particular periodic motion for a vibro-impact system with one degree of freedom, having frequency l times less than the excitation frequency ω. This can be expanded in the form of a Fourier series

$$x(t) = \frac{x_0}{2} + \sum_{m=1}^{\infty} x_m \cos\left(\frac{m\omega}{l}t - \varphi_m\right). \tag{9.47}$$

Let us calculate the work E_e of the external force $P(t) = a_p\cos(\omega t + \varphi)$ on the motion considered (9.47). Using the conditions of orthogonality of the harmonic functions we find for the period of vibration

$$E_e = \int_0^{2\pi l/\omega} P(t)\dot{x}(t)dt = \pi a_P x_l l \sin\varphi_0, \tag{9.48}$$

where $\varphi_0 = \varphi + \varphi_l$. Hence, sub-periodic regimes can occur only when harmonics are present which have the excitation frequency ($x_l \not\equiv 0$) in the spectrum of motion. Correspondingly, we must identify approximate sub-periodic solutions from the equation of motion, taking account of this harmonic.

Effective methods to approximate multi-harmonic solutions exists by the distribution function linearisation discussed in §7 [124]. Other modifications to the calculation schemes for multi-harmonic linearisation are known [107]. All these examples can be applied to the study of sub-periodic motion of vibro-impact systems, but, since such applications have no special features compared to the traditionally used methods described, we will not continue with this here.

The specific character of vibro-impact systems makes it possible to propose a method of finding existence conditions for sub-periodic regimes using a much more complete representation of the solution sought. Let us find the work done by the dissipative force on the motion studied (9.47) for one period of vibration. For the system with viscous damping $b\dot{x}$ we find

$$E_{d1} = \int_0^{2\pi l/\omega} b\dot{x}^2 dt = \frac{\pi b\omega}{l}\sum_{m=1}^{\infty} x_m^2 m^2 . \tag{9.49}$$

To estimate the loss of energy during impact using the coefficient of restitution R, we shall substitute the equation for the pre-impact velocity $\dot{x}_-$ through the impulse J, reduced to a single mass in (7.57),

$$\dot{x}_- = J/(1+R) \tag{9.50}$$

and we find for the period in the case of single-sided collision

$$E_{d2} = \frac{M\dot{x}_-^2}{2}(1-R^2) = \frac{(1-R)J^2}{2(1+R)} .. \tag{9.51}$$

For double-sided collisions this value must be doubled.

Using (9.48), (9.49), (9.51) we can write the equation for the balance of work of the dissipative and excitation forces in the periodic motion considered (9.47)

$$\pi a_p x_l l \sin\varphi_0 = \frac{\pi b\omega}{l}\sum_{m=1}^{\infty} x_m^2 m^2 + \frac{(1-R)J}{2(1+R)} . \tag{9.52}$$

Expression (9.52) represents the energy conditions for the existence of motion (9.47). From this it can be seen that sub-periodic regimes occur in the system only for relatively small value of the coefficients used for defining energy dissipation. In addition, for the given value of a_P the defined levels of excitation can be reached only with significant amplitude of vibration, since due to the filtering effects of the linear part of the system, as l increases the value of x_l rapidly decreases. For large amplitudes of vibration larger inertia and elastic forces arise. The situation here is similar to resonance in a linear system, when vibration with considerable amplitude occurs in comparison to the relatively small value of the disturbing force, compensating for energy dissipation. For this reason, motion of the type described is known as *sub-periodic resonance* of order l or *fractional resonance* of order $1/l$. Resonance, corresponding to $l = 1$, is known as *fundamental*. Of course, resonant motion which appears at phase φ_0 in the region of $\pi/2$, ensures the best excitation.

It is possible to maintain vibration at high amplitude with a small excitation force only when the vibration is close to the natural vibration of the system, hence the latter is a good representation for the synthesis of resonant motion. Since the natural frequency ω_0 of the vibro-impact system considered satisfies the condition $\omega_0 \geq \Omega$ (see (9.14)), the frequency range in which sub-periodic resonance can exist is limited by the following condition:

$$\omega / l = \omega_0 \geq \Omega , \tag{9.53}$$

which corresponds, for example, with that found above by the exact method in expression (3.20). By finding the law of motion for vibration of the system by the stitching method and presenting this in the form (9.47), we can obtain from (9.52) the conditions for the existence of sub-periodic resonance of various multiples.

This procedure was developed by M Z Kolovsky and is explained in his monograph [124], to which the interested reader is referred.

At this stage, we will describe another representation of vibration during sub-periodic resonances, based on the application of periodic Green's functions [36]. Reverting to the basic description of single-impact periodic solutions for vibro-impact systems in the form (6.79), we note that during vibro-impact resonance the second term on the right side of equation (8.79) is dominant. This occurs because the first term describes forced vibration of the impact element pairs with negligible impact interaction, but the resonant frequency of such motion differs from the frequencies of the vibro-impact resonances. As a result vibration during vibro-impact resonances of various multiplicities is well described by the simplified expression

$$x(t) \approx -J\chi(t)\,. \tag{9.54}$$

As an example we will look at the exact solution (3.32) found in §3 for the periodic vibro-impact motion of an oscillator arranged between stops, writing this in the form:

$$x(t) = \mp a\cos\omega t + \sum_{m=1}^{\infty}\frac{4\zeta\Delta\cot[\pi l/(2\zeta)]}{\pi l[(m\zeta/l)^2-1}\left(1\pm\frac{a}{\Delta}\right)\cos\frac{m\omega}{l}t\,. \tag{9.55}$$

Here a is found from formula (3.10). At sub-periodic resonance frequencies $\zeta > l$, the value a becomes negligibly small and the exact solution (9.55) is well approximated by the term under the summation sign. Comparing (9.55) with (7.79), it can be seen that the term mentioned represents the periodic Green's functions of the system $\chi_J(t) = J\chi(t)$. Comparing this to the equation for free vibration (3.6), it can be seen that when $\zeta/l = \zeta_0$ they differ only by terms proportional to the value a/Δ. The latter is less than one, since according to the results of analysis for the exact solution (see §3, section 3) vibro-impact resonance as described only exists when $a < \Delta$.. Hence, just as the law of natural vibration, the periodic Green's functions represent a good approximation to the resonance solution for the equations of vibro-impact systems.

Using the representation of periodic Green's functions in the form of series (6.67), we can write the approximate solution sought for $x(t)$ of period T ($T = 2\pi l/\omega$) in accordance with (9.47), (9.54) in the following form:

$$x(t) = -\frac{J}{T}L(0) - \frac{2J}{T}\sum_{m=1}^{\infty}\left|L\left(\frac{mj\omega}{l}\right)\right|\cos\left[\frac{m\omega}{l}t + \arg L\left(\frac{mj\omega}{l}\right)\right]. \tag{9.56}$$

As a result the amplitude x_m of the corresponding harmonics takes the form

$$x_m = \frac{2J}{T}\left|L\left(\frac{mj\omega}{l}\right)\right|. \tag{9.57}$$

The value J is found from the condition of impact, that when $t = 0$, $x = \Delta$. Substituting in this condition the equation (9.54), we find

$$J = -\Delta/\chi(0)\,. \tag{9.58}$$

Since J has a non-negative value, in agreement with the coordinate axis direction assumed in this book, then the sign of the value $\chi(0)$ is opposite to the sign of Δ.

Using (9.57), (9.58) the equation for energy balance (9.52) takes the form

$$a_p\omega|L(j\omega)|\sin\varphi_0 = -\frac{\Delta}{\chi_1(0)}\left\{\frac{b\omega^3}{\pi l^3}\sum_{m=1}^{\infty}\left|L\left(\frac{mj\omega}{l}\right)\right|^2 m^2 + \frac{1-R}{2(1+R)}\right\}, \tag{9.59}$$

in which $\varphi_0 = \varphi - \arg L(j\omega)$. From (9.59) we find two possible values for the phase

$$\varphi_0 = \arcsin\frac{a_p\omega\,|\,L(j\omega)\,|}{-\dfrac{\Delta}{\chi_1(0)}\left\{\dfrac{b\omega^3}{\pi l^3}\displaystyle\sum_{m=1}^{\infty}\left|L\left(\frac{mj\omega}{l}\right)\right|^2 m^2 + \dfrac{1-R}{2(1+R)}\right\}},$$
$$\varphi_0 = \pi - \arcsin\frac{a_p\omega\,|\,L(j\omega)\,|}{-\dfrac{\Delta}{\chi_1(0)}\left\{\dfrac{b\omega^3}{\pi l^3}\displaystyle\sum_{m=1}^{\infty}\left|L\left(\frac{mj\omega}{l}\right)\right|^2 m^2 + \dfrac{1-R}{2(1+R)}\right\}}. \tag{9.60}$$

Since $|\sin\varphi_0| \le 1$,the existence conditions for resonant regimes can be written as:

$$a_p\omega|L(j\omega)| \ge -\frac{\Delta}{\chi_1(0)}\left\{\frac{b\omega^3}{\pi l^3}\sum_{m=1}^{\infty}\left|L\left(\frac{mj\omega}{l}\right)\right|^2 m^2 + \frac{1-R}{2(1+R)}\right\}. \tag{9.61}$$

Similar conditions for the case of double-sided collisions using (6.68) will be as follows:

$$a_p\omega|L(j\omega)| \ge -\frac{\Delta}{\chi_2(0)}\left\{\frac{2b\omega^3}{\pi l^3}\sum_{m=1}^{\infty}\left|L\left[\left(\frac{2m-1}{l}\right)j\omega\right]\right|^2 (2m-1)^2 + \frac{1-R}{2(1+R)}\right\}. \tag{9.62}$$

Using solutions in the form (9.54) makes it possible to obtain the law of motion in the form (using (9.58))

$$x(t) \approx \frac{\Delta}{\chi(0)}\chi(t). \tag{9.63}$$

From this all the remaining characteristics of motion can be found. Taking into consideration, for example, that energy dissipation in the system has been assumed to be small, and consequently, the periodic Green's functions are only slightly non-symmetric relative to $t = T/2$, the value of the amplitude of vibration a_x during single-sided collisions can be obtained from the approximate relationship

$$a_x \approx \frac{1}{2}\left[\Delta - x\left(\frac{T}{2}\right)\right] \approx \frac{\Delta}{2}\left[1 - \frac{1}{\chi_1(0)}\chi_1\left(\frac{T}{2}\right)\right]. \tag{9.64}$$

The value of the pre-impact velocity $\dot{x}_-$ is found from (9.50), (9.58):

$$\dot{x}_- = -\frac{\Delta}{\chi(0)(1+R)}. \tag{9.65}$$

Obviously when b is small the damping in the equation for $\chi(t)$ can be neglected, since away from the frequency of linear resonance ($\omega/l \ne \Omega$) the damping has minimal influence.

The permissibility of using the relationship obtained is proved since the condition

$$a_x > a\,, \tag{9.66}$$

is satisfied, where a as before is calculated from (3.10). When $\Delta \neq 0$ we can use a simpler and *sufficient* condition

$$|\Delta| > a\,, \tag{9.67}$$

which can be checked before the construction of approximate solutions. Let us express the conditions (9.61), (9.62) with the help of the parameters of the linear part of the vibro-impact system with one degree of freedom. On the basis of (6.72) we find

$$\left|L\left(\frac{mj\omega}{l}\right)\right| = \frac{1}{\left|\Omega^2 - \left(\frac{mj\omega}{l}\right)^2 + jb\frac{m\omega}{l}\right|} = \frac{1}{\sqrt{\left[\Omega^2 - \left(\frac{m\omega}{l}\right)^2\right]^2 + \left(\frac{bm\omega}{l}\right)^2}}\,. \tag{9.68}$$

Taking into account in equation (9.68) that b is assumed to be small and using condition (9.53) it is possible to ignore the value $(bm\omega/l^2)$. When summing over index m for all $m > 1$, terms Ω^2 also can be neglected. Using this simplification, substituting (9.68) in (9.61) and summing the corresponding numerical series, we obtain

$$\frac{a_p\omega}{|\omega^2 - \Omega^2|} \geq -\frac{\Delta}{\chi_1(0)}\left\{\frac{bl}{\pi\omega}\left[\frac{1}{\left(1 - \frac{\Omega^2 l^2}{\omega^2}\right)^2} + \frac{\pi^2}{6} - 1\right] + \frac{1-R}{2(1+R)}\right\}, \tag{9.69}$$

where according to (6.75), the value $\chi_1(0)$ takes the form

$$\chi_1(0) = \frac{\exp\left(-\frac{\pi b}{\omega}\right)\sin\frac{2\pi\Omega_1}{\omega}}{\Omega_1\left[1 + \exp\left(-\frac{2\pi b}{\omega}\right) - 2\exp\left(-\frac{\pi b}{\omega}\right)\cos\frac{2\pi\Omega_1}{\omega}\right]} \qquad \left(\Omega_1 = \sqrt{\Omega^2 - \frac{b^2}{4}}\right). \tag{9.70}$$

For symmetrical double-sided impact pairs using a similar simplification we find from (9.62)

$$\frac{a_p\omega}{|\omega^2 - \Omega^2|} \geq -\frac{\Delta}{\chi_2(0)}\left\{\frac{2bl}{\pi\omega}\left[\frac{1}{\left(1 - \Omega^2 l^2/\omega^2\right)^2} + \frac{\pi^2}{8} - 1\right] + \frac{1-R}{2(1+R)}\right\}, \tag{9.71}$$

where according to (6.76), $\chi_2(0)$ is found from the equation

$$\chi_2(0) = \frac{\exp\left(-\frac{\pi b}{2\omega}\right)\sin\frac{\pi\Omega_1}{\omega}}{\Omega_1\left[1 + \exp\left(-\frac{2\pi b}{\omega}\right) + 2\exp\left(-\frac{\pi b}{2\omega}\right)\cos\frac{\pi\Omega_1}{\omega}\right]}\,. \tag{9.72}$$

In the case, when $\omega/l \gg \Omega$, equation (9.71) can be simplified by neglecting the value of Ω^2:

$$a_p \geq -\frac{\Delta}{\chi_2(0)}\left[\frac{\pi b l}{4}+\frac{\omega(1-R)}{2(1+R)}\right]. \tag{9.73}$$

From (9.73) it follows that as the excitation frequency increases, the relative effect of dissipation during impact grows by comparison with viscous damping. This result agrees with that presented during the analysis of exact solutions (see §3, section 5).

As an example we will find the limiting value of Δ in an arrangement of stops and the corresponding impact velocity $\dot{x}_-$ for a linear oscillator with double-sided collisions for the case of sub-periodic resonance of the third order ($l = 3$).Suppose the parameter system and excitation take the values $\Omega = 62.81\ /s$; $R = 0.5$; $b = 12.51\ /s$; $a_p = 4g \approx 40\ m/s^2$; $\omega = 236.1\ /s$.

From equation (9.71), we calculate

$$J = -\Delta/\chi_2(0) = 0.189\, kgm/s\ , \tag{9.74}$$

consequently, according to (9.65), $\dot{x}_- = 0.126 m/s$. In (9.72), if we neglect the value of viscous friction ($b \equiv 0$) we find $\chi_2(0) = -24.5.10^{-3} s/kg$. Finally, from (9.74), we find $\Delta = 4.65.10^{-3}\ m$. For comparison we will find the similar value of Δ from the exact expression (3.17). After a considerably more difficult calculation using (3.10), (3.16) we obtain $a = 0.82.10^{-3}\ m$; $\Delta = 4.43.10^{-3}\ m$. We can see that condition (9.67) is satisfied, which explains the high accuracy of the approximate solution.

In conclusion, we note that the approximate approach described can be used for other forms of dissipation and excitation force, guaranteeing resonant vibration of the system with arbitrary degrees of freedom (see §14, section 4), and, taking account of (6.90), with the use of any mathematical model of impact.

§10. Stability and transients of forced vibration

1. Let us now look at the stability of the steady-state solutions found in §9 for the equations of forced vibration in relation to small disturbances. For this it is convenient to make use of methods which correspond to the approximate character of the solution sought. The majority of such methods are based on the assumption that small disturbances do not change the form of the solution, but will only lead to the slow drift of its parameters. In the case of a harmonic solution this means that the disturbed motion takes the form of "harmonic" vibration with term $\widetilde{m}_x(t)$, with amplitude $\widetilde{a}_x(t)$ and phase $\widetilde{\varphi}(t)$ which change slowly with time:

$$x(t) = \widetilde{m}_x(t) + \widetilde{a}_x(t)\cos[\omega t - \widetilde{\varphi}(t)]. \tag{10.1}$$

For convenience, the phase from the expression for the disturbing force will in future be transferred to the solution, i.e. by using a shift of the starting time.

By presenting the disturbed motion in the form (10.1) we can conveniently develop calculation methods to study the stability of approximate solutions based on various modifications to the method of slowly changing amplitude [221].

These assumptions also make it possible to study stability directly from the harmonic linearisation equations for systems. The reduction of systems to the equivalent autonomous ones makes it possible to formulate the necessary conditions on the basis of the stability of systems at the vibration boundary.

The following convenient form for the stability conditions can be suggested:

$$[d\Theta(\widetilde{a}_x,\omega)/d\widetilde{a}_x]_{\widetilde{a}_x=a_x} > 0\,. \tag{10.2}$$

where $\Theta(\widetilde{a}_x,\omega)$ is the imaginary part of the *characteristic complex* $\mathcal{D}(j\omega)$ of the linearised system, reduced to the equivalent autonomous one, at the frequency of the periodic solution considered. The tilde above the corresponding value implies that when substituted into (10.2) it is considered as a slow variable, changing around the constant value, corresponding to the steady-state solution. We will show that condition (10.2) represents the disturbed balance of work of a non-conservative force for the period of stable motion. For example, for a linearised system described by equation (9.1) and the expressions (9.5), the work of the dissipative force $(b+\beta)\dot{x}$ for the period of motion (10.1) with "frozen" slow parameters is found from the expression

$$E_d = \int_0^{2\pi/\omega}(b+\beta)[\dot{x}(t)]^2\,dt = \pi(b+\beta)\omega\widetilde{a}_x^2\,. \tag{10.3}$$

The work done by the external force $P(t) = a_p\cos\omega t$ during this motion will be as follows:

$$E_e = \int_0^{2\pi/\omega}P(t)\dot{x}(t)dt = \pi a_p\widetilde{a}_x\sin\widetilde{\varphi}\,. \tag{10.4}$$

The *energy condition for stability* [221], reflects the fact that, during change of the disturbed amplitude $\widetilde{a}_x$ in the close neighbourhood to the steady-state value a_x, the balance of work due to the dissipative and disturbing force is disrupted in such a way so as to compensate this disturbance:

$$\frac{d}{d\widetilde{a}_x}(E_d - E_e)_{\widetilde{a}_x=a_x} > 0\,.$$

Substituting here (10.3), (10.4), we obtain

$$\frac{d}{d\widetilde{a}_x}\left(b+\beta-\frac{a_p\sin\widetilde{\varphi}}{\widetilde{a}_x\omega}\right)_{\widetilde{a}_x=a_x} > 0\,. \tag{10.5}$$

At the same time, by applying condition (10.2) to the characteristic equation (9.10), we obtain expression (10.5) directly. Thus, condition (10.2) represents a more convenient form for the energy conditions for stability.

2. In (10.1) three unknown functions appear instead of one, and we can connect them in turn to two arbitrary conditions. Let us assume that in the disturbed motion $\widetilde{m}_x(t)$ and $\widetilde{a}_x(t)$ are related in the same way as in the steady-state case, and that this applies also to $\widetilde{\varphi}(t)$ and $\widetilde{a}_x(t)$.

Differentiating (10.5) using the relationship for $\widetilde{\varphi}$ in terms of $\widetilde{a}_x$, we obtain

$$\left[\frac{d\beta}{d\widetilde{a}_x}+\frac{a_p}{\omega\widetilde{a}_x^2}\left(\sin\widetilde{\varphi}-\frac{d\widetilde{\varphi}}{d\widetilde{a}_x}\widetilde{a}_x\cos\widetilde{\varphi}\right)\right]_{\widetilde{a}_x=a_x} > 0\,. \tag{10.6}$$

The term $(d\widetilde{\varphi}/d\widetilde{a}_x)_{\widetilde{a}_x=a_x}$ is found by differentiating (10.12) as an implicit function $\widetilde{\varphi}$ of $\widetilde{a}_x$:

$$\left(\frac{d\varphi}{d\tilde{a}_x}\right)_{\tilde{a}_x=a_x}=-\frac{1}{a_p\sin\varphi}\left[\left(\Omega^2+k-\omega^2\right)+\tilde{a}_x\frac{dk}{d\tilde{a}_x}\right]_{\tilde{a}_x=a_x}. \qquad (10.7)$$

Substituting the expression for sinφ from (9.11) into (10.7), we obtain

$$\left(\frac{d\varphi}{d\tilde{a}_x}\right)_{\tilde{a}_x=a_x}=-\frac{1}{(b+\beta)\omega a_x}\left[\left(\Omega^2+k-\omega^2\right)+\tilde{a}_x\frac{dk}{d\tilde{a}_x}\right]_{\tilde{a}_x=a_x}. \qquad (10.8)$$

Using (10.8), (9.11) and (9.12), condition (10.6) takes the final form:

$$\left[\omega^2(b+\beta)\left(b+\beta+\tilde{a}_x\frac{d\beta}{d\tilde{a}_x}\right)+\right.$$
$$\left.+(\Omega^2+k-\omega^2)\left(\Omega^2+k-\omega^2+\tilde{a}_x\frac{dk}{d\tilde{a}_x}\right)\right]_{\tilde{a}_F=a_F}>0 \qquad (10.9)$$

Expression (10.9) coincides with the stability conditions obtained by Van der Pol's method using a much longer transformation [221]. In addition, the study of stability using Van der Pol's method gives another condition

$$[2(b+\beta)+\tilde{a}_x d\beta/d\tilde{a}_x]_{\tilde{a}_x=a_x}>0, \qquad (10.10)$$

which, however, is satisfied for all forms of dissipative force.

When using the method of harmonic linearisation by force it is possible to retain the above reasoning, bearing in mind that the linearisation coefficients k and β are implicitly related to $\tilde{a}_x$. Then, using the transformations

$$\frac{dk}{d\tilde{a}_x}=\frac{dk}{d\tilde{a}_F}\frac{d\tilde{a}_F}{d\tilde{a}_x}=\frac{dk}{d\tilde{a}_F}k,\ \frac{d\beta}{d\tilde{a}_x}=\frac{d\beta}{d\tilde{a}_F}\frac{d\tilde{a}_F}{d\tilde{a}_x}=\frac{d\beta}{d\tilde{a}_F}k$$

condition (10.9) leads to a convenient form for applying this method:

$$\left[(k-\omega^2)\left(k-\omega^2+\tilde{a}_F\frac{dk}{d\tilde{a}_F}\right)+\omega^2(b+\beta)(b+\beta+\tilde{a}_F\frac{d\beta}{d\tilde{a}_F})\right]_{\tilde{a}_F=a_F}>0. \qquad (10.11)$$

Here k and β are the force linearisation coefficients, dependent on $\tilde{a}_F$; a_F is the steady-state value of $\tilde{a}_F$, corresponding to the periodic solution studied.

It is shown in [221] that the boundaries of the stability domains satisfying condition (10.9), appear on the (ω, a_x) plane as the locus of points, in which the resonance curves constructed from equation (9.13), have vertical tangents. This implies that the branches of the amplitude characteristics for nonlinear regimes, which lie above the backbone curves, occur in stability domains. In double-valued regions below the backbone curve, the branches correspond to unstable solutions. We will demonstrate this using as an example vibro-impact systems in which the coefficient of harmonic linearisation k is described by expression (9.18). Assuming for simplicity $b+\beta\equiv 0$, we obtain from (10.9) two boundary branches for the stability domains

$$[\Omega^2+k(\tilde{a}_x)-\omega^2]_{\tilde{a}_x=a_x}=0, \qquad (10.12)$$

$$[\Omega^2+k(\tilde{a}_x)-\omega^2+\tilde{a}_x dk(\tilde{a}_x)/d\tilde{a}_x]_{\tilde{a}_x=a_x}=0. \qquad (10.13)$$

Expression (10.12) represents the equations of the backbone curves. Substituting (10.18) in (10.13), we find

$$\omega = \sqrt{3}\Omega . \tag{10.14}$$

This expression defines the asymptotes of the backbone curve (see (9.21)). Analysing inequality (10.9) we conclude that the stable branches of the amplitude-frequency characteristics are arranged above the backbone line, and the unstable branches between the backbone curves and their asymptotes, which proves the point. Fig. 10.1 shows the unstable domains hatched, over the curves of the amplitude-frequency characteristics for systems with single-sided stops having either gap (Fig. 10.1, *a*) or interference (Fig. 10.1, *b*). Of course, the results obtained are valid only for nonlinear solutions, since Fig. 10.1, *a* shows another boundary $a_x = \Delta$, which separates existence domains for linear and nonlinear solutions. Similar results are obtained using condition (10.11). Note that the stability conclusions agree well with the known conditions for nonlinear systems and with the results obtained when applying exact methods to the analysis of the vibro-impact systems considered (see §2).

3. We shall now look at transient processes for forced vibration, from some initial conditions under the action of a harmonic force $P(t)$. The study of such effects is important in the use of resonant processes, which we often try to fulfil in vibro-impact systems. At the same time the dynamics of such processes is interesting, in terms of the conditions studied, for the passing of resonance in machines during run-up or run-down, i.e. during growth or decay of the excitation frequency. Note also, that the approach used allows the limits of application of the methods of studying stability of periodic motion described above to be further refined, and also to examine stability of such systems "at large" with finite values of disturbance.

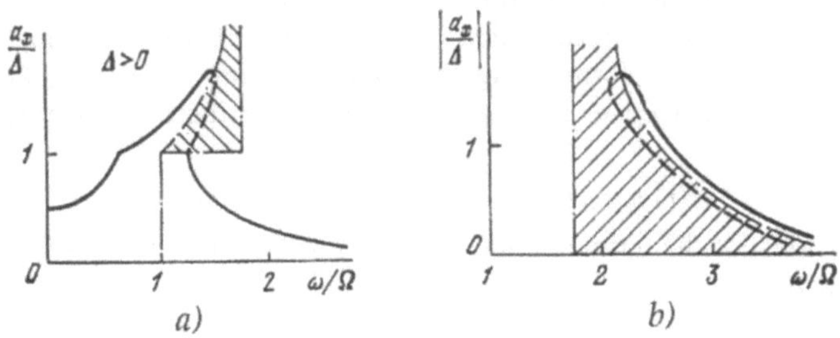

Fig. 10.1

We shall now turn to an example of a second order symmetrical system, described by the equation

$$(s^2 + bs + \Omega^2)x + \Phi(x, sx) = a_p \cos(\omega t + \tilde{\varphi}), \tag{10.15}$$

where $\tilde{\varphi}$ is a variable, found during the solution process and inserted on the right hand side for convenience of analysis. The solution required for $x(t)$ will be approximated by a single frequency function with variable amplitude $\tilde{a}_x$ and phase

$\tilde{\varphi}$:

$$x(t) = \tilde{a}_x(t)\cos\omega t\,. \tag{10.16}$$

By including phase on the right hand side, the study of process $x(t)$ leads strictly to systems of time measurement displaced relative to the external excitation by a phase shift $\tilde{\varphi}(t)$.

In contrast to the transient of process (10.1), considered during the study of stability, we will now discard the assumption that amplitude $\tilde{a}_x(t)$ changes slowly, since resonance can be realised for a small number of periods of the disturbing force even from zero initial conditions. With this in mind, it is convenient to approximate the change in $\tilde{a}_x(t)$ by the following relationship

$$d\tilde{a}_x dt = \tilde{a}_x\tilde{\xi}\,, \tag{10.17}$$

where $\tilde{\xi}$ represents a slowly changing quantity, found from the solution process. As a result the rate of change of amplitude will depend principally on the absolute value of the term $\tilde{\xi}$, and not on its rate of change, since even for a small change in $\tilde{\xi}$ the expression (10.17) can describe a rapid change of amplitude through the expression

$$\tilde{a}_x(t) = C\exp\int_0^t \tilde{\xi}dt\,, \tag{10.18}$$

where C is defined by the initial conditions.

Using the above equation we will search for an approximate solution to equation (10.15) of the form (10.16), assuming $\tilde{\xi}$ and $\tilde{\varphi}$ are change slowly.

To find a single-frequency periodic process of the type described, occurring in a nonlinear autonomous system, E P Popov proposed an effective approach [181] based on the asymptotic method of Krylov-Bogolyubov

Applying harmonic linearisation to equation (10.15) to bring it to the equivalent autonomous form using the form of solution (10.16), (10.17), this method can be extended to the class of system considered. Since according to (10.16), (10.17)

$$\dot{x}(t) = sx(t) = -\tilde{a}_x\omega\sin\omega t + \tilde{a}_x\tilde{\xi}\cos\omega t\,, \tag{10.19}$$

the expressions for the harmonic linearisation coefficients according to (7.16) will take the form

$$k = \frac{1}{\pi\tilde{a}_x}\int_0^{2\pi}\Phi(\tilde{a}_x\cos\psi, -\tilde{a}_x\omega\sin\psi + \tilde{a}_x\tilde{\xi}\cos\psi)\cos\psi d\psi\,, \tag{10.20}$$

$$\beta = \frac{1}{\pi\tilde{a}_x}\int_0^{2\pi}\Phi(\tilde{a}_x\cos\psi, -\tilde{a}_x\omega\sin\psi + \tilde{a}_x\tilde{\xi}\cos\psi)\sin\psi d\psi\,, \tag{10.21}$$

i.e. they are functions of three unknowns $\tilde{a}_x$, ω and $\tilde{\xi}$. In cases when the value of $\tilde{\xi}$ is small or nonlinear or $\Phi(x)$ is independent of the derivatives of the process, the usual equations for the coefficients of harmonic linearisation can be used: in the first case in the form (7.24), (7.25), in the second case in the form

$$k = \frac{1}{\pi a_x} \int_0^{2\pi} \Phi(a_x \cos\psi)\cos\psi d\psi \tag{10.22}$$

substituting a_x for $\tilde{a}_x$ in the final expression.

To convert equation (10.15) to the equivalent homogeneous form we will rewrite the right hand side in the form

$$a_P \cos(\omega t + \tilde{\varphi}) = a_P \cos\omega t \cos\tilde{\varphi} - a_P \sin\omega t \sin\tilde{\varphi}\,, \tag{10.23}$$

expressing $\cos\omega t$ and $\sin\omega t$ in terms of the solution sought for $x(t)$. From (10.16) we find

$$\cos\omega t = x / \tilde{a}_x\,; \tag{10.24}$$

from (10.19) using (10.16) we find

$$\sin\omega t = -\frac{sx}{\tilde{a}_x\omega} + \frac{\tilde{\xi}x}{\tilde{a}_x\omega}\,. \tag{10.25}$$

Substituting (10.24), (10.25) into (10.23), we find

$$a_P \cos(\omega t + \tilde{\varphi}) = \frac{a_P \sin\tilde{\varphi}}{\tilde{a}_x\omega} sx + \frac{a_P}{\tilde{a}_x}\left(\cos\tilde{\varphi} - \frac{\tilde{\xi}}{\omega}\sin\tilde{\varphi}\right)x\,. \tag{10.26}$$

Carrying out harmonic linearisation of the nonlinear function $\Phi(x, sx) = kx + \beta sx$ in (10.15) and substituting in (10.26) we obtain the following autonomous system:

$$\left[s^2 + \left(b + \beta - \frac{a_P \sin\tilde{\varphi}}{\tilde{a}_x\omega}\right)s + \Omega^2 + k - \frac{a_P}{\tilde{a}_x}\left(\cos\tilde{\varphi} - \frac{\tilde{\xi}}{\omega}\sin\tilde{\varphi}\right)\right]x = 0\,. \tag{10.27}$$

Transient processes of type (10.16), (10.17) correspond to the existence of pairs of complex roots $s = \tilde{\xi} \pm j\omega$ to the characteristic equation of the linearised system (10.27). Substituting $s = \tilde{\xi} + j\omega$ in the latter and setting the real and imaginary parts to zero, we obtain

$$\tilde{\xi}^2 + (b + \beta)\tilde{\xi} + \Omega^2 + k - \omega^2 - \frac{a_P \cos\tilde{\varphi}}{\tilde{a}_x} = 0\,, \tag{10.28}$$

$$\tilde{\xi} = \frac{1}{2}\left(\frac{a_P \sin\tilde{\varphi}}{\tilde{a}_x\omega} - b - \beta\right). \tag{10.29}$$

Equations (10.28), (10.29) define how $\tilde{\varphi}$ depends on $\tilde{a}_x$ and $\tilde{\xi}$ on $\tilde{a}_x$. When $\tilde{\xi} \equiv 0$ we obtain the existence conditions for steady-state regimes ($\tilde{a}_x = a_x = \text{const}$), which coincide, naturally, with (9.11), (9.12).

In the case when expression (10.16), (10.17) represent the disturbed motion of the system, assuming $C = a_x$ in (10.18), we see that as amplitude $\tilde{a}_x$ varies from the steady-state value a_x, stability will be maintained as $\tilde{\xi}$ changes sign according to the inequality

$$(d\tilde{\xi} / d\tilde{a}_x)_{\tilde{a}_x = a_x} < 0\,. \tag{10.30}$$

Substituting in (10.30) the expression for $\tilde{\xi}$ from (10.29), we obtain the condition for stability in the form (10.5). Hence, conditions (10.2) and (10.30) are equivalent.

Expression (10.28) also shows that the calculated expression $(d\tilde{\varphi}/d\tilde{a}_x)_{\tilde{a}_x=a_x}$ used in section 2 based on (9.12), satisfying the steady-state case, appears correct only for sufficiently small values of the damping factor $b+\beta$. This also defines the limits of application for methods developed in sections 1 and 2 to investigate stability of forced vibration in nonlinear systems with small disturbances. With larger dissipative losses, the expression $(d\tilde{\xi}/d\tilde{a}_x)_{\tilde{a}_x=a_x}$ must be calculated on the basis of the complete expressions (10.28), (10.29). We will now find this value, assuming for simplification that $\beta \equiv 0$. In this case, neglecting $\tilde{\xi}^2$ in (10.28) as of second order accuracy and differentiating (10.28), (10.29) as implicit functions of $\tilde{a}_x$, in the vicinity of the steady-state value a_x we find

$$\left[b\frac{d\tilde{\xi}}{d\tilde{a}_x}+\frac{dk}{d\tilde{a}_x}+\frac{a_P}{\tilde{a}_x^2}\left(\tilde{a}_x\frac{d\tilde{\varphi}}{d\tilde{a}_x}\sin\tilde{\varphi}+\cos\tilde{\varphi}\right)\right]_{\tilde{a}_x=a_x}=0\,,$$

$$\left(\frac{d\tilde{\xi}}{d\tilde{a}_x}\right)_{\tilde{a}_x=a_x}=\left[\frac{a_P}{2\tilde{a}_x^2\omega}\left(\tilde{a}_x\frac{d\tilde{\varphi}}{d\tilde{a}_x}\cos\tilde{\varphi}-\sin\tilde{\varphi}\right)\right]_{\tilde{a}_x=a_x}.$$

Eliminating the value $(d\tilde{\xi}/d\tilde{a}_x)_{\tilde{a}_x=a_x}$ from these expressions, we obtain using (9.11), (9.12):

$$\left(\frac{d\tilde{\varphi}}{d\tilde{a}_x}\right)_{\tilde{a}_x=a_x}=-\left[\frac{\Omega^2+k-\omega^2+\tilde{a}_x dk/d\tilde{a}_x-b^2/2}{\tilde{a}_x b\omega[1+(\Omega^2+k-\omega^2)/(2\omega^2)]}\right]_{\tilde{a}_x=a_x}. \tag{10.31}$$

Comparing equations (10.31) and (10.8) and noting that in the domains considered, the range of the system parameters is given by

$$(\Omega^2+k-\omega^2)/(2\omega^2)\ll 1$$

since they are close to resonance, we find that, for small b they differ only to second order accuracy. Consequently, it is unsuitable to change from the simplified expression (10.8) to the full expression (10.31), during the study of qualitative principles, as indicated.

4. The transient process is clearly illustrated by the dependence of $\tilde{\xi}$ on $\tilde{a}_x$. To find it we can eliminate phase $\tilde{\varphi}$ from the system of equations (10.28), (10.29). We find

$$\tilde{\xi}^2+(b+\beta)\tilde{\xi}+\Omega^2+k-\omega^2-\sqrt{a_P/\tilde{a}_x^2-\omega^2(2\tilde{\xi}+b+\beta)^2}=0\,. \tag{10.32}$$

Expression (10.32) represents an algebraic equation of the fourth degree in $\tilde{\xi}$. To simplify further analysis we will put $b+\beta=0$. In this case equation (10.32) is transformed to the equation

$$\tilde{\xi}^2+\Omega^2+k-\omega^2-\sqrt{a_P/\tilde{a}_x^2-4\omega^2\tilde{\xi}^2}=0$$

which can be converted to the biquadratic equation

$$.\,\tilde{\xi}^4+2\tilde{\xi}^2(\Omega^2+k-\omega^2)+(\Omega^2+k-\omega^2)-a_P/\tilde{a}_x^2=0\,. \tag{10.33}$$

From equation (10.33) we find eventually

$$\tilde{\xi}^2 = -(\Omega^2 + k - \omega^2) + (\Omega^2 + k - \omega^2)\sqrt{1 + \frac{a_P / \tilde{a}_x^2 - (\Omega^2 + k - \omega^2)^2}{(\Omega^2 + k - \omega^2)^2}} . \quad (10.34)$$

We shall look at the case when the transient process satisfies the resonance condition

$$\Omega^2 + k - \omega^2 = 0 . \quad (10.35)$$

Substituting (10.35) into (10.34) and assuming that a_p is a small value, we can expand the radical as a power series, limited to the linear terms. As a result we obtain

$$\tilde{\xi}^2 = \frac{a_P^2}{2\tilde{a}_x^2(\Omega^2 + k - \omega^2)} = \frac{a_P^2}{4\tilde{a}_x^2\omega^2} .$$

whence

$$\tilde{\xi} = a_P / (2\tilde{a}_x \omega) . \quad (10.36)$$

Substituting (10.36) into (10.17), we find

$$d\tilde{a}_x / dt = a_p / (2\omega) \quad (10.37)$$

and consequently using (10.35)

$$\tilde{a}_x = \frac{a_P}{2\omega} t = \frac{a_P}{2\sqrt{\Omega^2 + k}} t . \quad (10.38)$$

In the absence of nonlinearity ($k \equiv 0$) we obtain the known formula for the development of resonance of a linear system. Substituting in (10.38) the value k appropriate to various nonlinearities, we find how the resonance amplitude changes during run-up and run-down of a nonlinear system. Since expression (10.38) is obtained from condition (10.35), it corresponds to the most stressed transient process.

As an example we shall look at the character of passing of resonance band of frequencies of an oscillator, colliding against a stop. We shall make use of the expressions for the linearisation coefficients k in the form (9.18). When $m_P \equiv 0$, we find

$$k = 2\Omega^2(1 - \Delta / \tilde{a}_x) . \quad (10.39)$$

Substituting (10.39) in (10.38), we obtain

$$\tilde{a}_x = \frac{a_P}{2\Omega\sqrt{3 - 2\Delta / \tilde{a}_x}} t . \quad (10.40)$$

From (10.40) it follows that, when the stops are arranged with initial gap ($\Delta > 0$), with increase in $\tilde{a}_x$ the rate of growth of the vibration amplitude during run-up of the system within the limits of the resonance zone decreases from the rate of linear resonance (when $a_x \approx \Delta$) to the value

$$\frac{d\tilde{a}_x}{dt} \sim \frac{a_P}{2\sqrt{3}\Omega} \quad (10.41)$$

when $\tilde{a}_x \to \infty$. When the stops are arranged with interference ($\Delta < 0$), then during run-down an increase in the velocity of growth of the system occurs from the infinitely small value for low values of $\tilde{a}_x$ to a value (10.41) (when $\tilde{a}_x \to \infty$). Hence, the velocity of growth of resonant vibration during run-up is in principle higher than during run-down. When $\Delta = 0$ the system swings at resonance with velocity (10.41), almost half that of the linear case.

§11. Parametric vibration.

1. Parametric vibration of mechanisms, machines and structures is an essential factor of their vibration activity, often causing complete break down of their normal operation due to *parametric resonance*. A range of such examples is presented in [91]. As a rule, this study leads to a linear formulation based on the analysis of Hill or Mathieu equations [203] and the analysis is restricted to finding the conditions which cause parametric resonance to emerge.

However, an estimate of the level of vibration for such systems can be obtained only by evaluating a series of additional *nonlinear* factors, accompanying the actual vibration. These include fuller characteristics for elastic and dissipative features of systems, gaps, stops, etc.

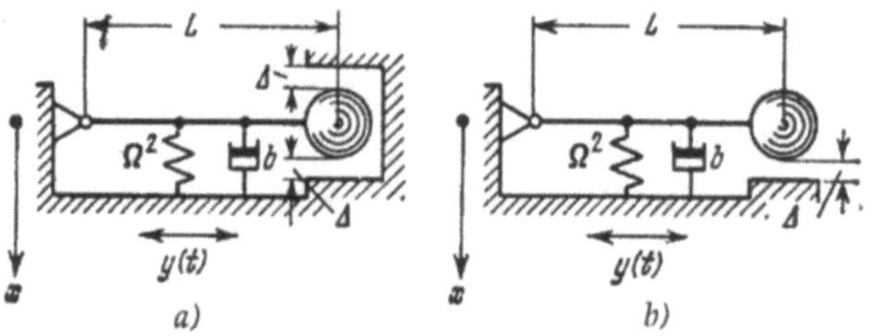

Fig. 11.1

A study of nonlinear parametric effects makes it necessary to use approximate methods of nonlinear mechanics. In particular, N N Bogolyubov and N M Krylov applied asymptotic methods [138], and in the monograph by V V Bolotin [65], the study of nonlinear parametric effects is combined with the problem of dynamic stability of structures using Van der Pol's method and by complete expansion of Fourier series. These methods give identical first approximations, and it is normal to study nonlinear parametric effects within such a framework. In addition, another investigation, described by V O Kononenko [132] and based on asymptotic methods, consisted of crossing through the parametric resonance domains of nonlinear systems when studying interaction with the exciting vibrator.

The existence of gaps and rigid stops leads to the onset of collisions, considerably influences the character of parametric vibration. It is quite natural to extend the methods explained to the process indicated. This section is devoted to the description of such an approach [42]. By using frequency methods and the ideas of harmonic linearisation, we can obtain a clear physical interpretation of transformation, which will lead to solutions, and will give a much shorter route to the basic expressions for the first approximation.

We shall look at a second degree system with one periodically changing parameter, writing the nonlinear equation in the form

$$\ddot{x} + b\dot{x} + \Omega^2[1 - h\cos(\omega t + \varphi)]x + \Phi(x, \dot{x}) = 0 . \tag{11.1}$$

Here h is the amplitude of modulation, where, in the general case $h = h(\omega)$; φ is the unknown phase of the parametric excitation; $\Phi(x, \dot{x})$ is the nonlinear elastic and dissipative characteristic of the system. Such a type of system is representative of a wide class of problems in machine dynamics: vibration of mechanisms arranged on oscillating bases [117, 186], vibration of anisotropic shafts in bearings [67], vibration of mechanisms with nonlinear transformation functions [220], etc.

As an example, Fig. 11.1 shows two dynamic models of mechanisms with vibrating bases, having double-sided (Fig. 11.1, *a*) and single-sided (Fig. 11.1, *b*) rigid stops for the motion of an impacting mass $M = 1$,with a ball-end attached to the base by a weightless rigid bar. The single-sided stop can be arranged (Fig. 11.1, b) either with initial gap or interference Δ. The zero position of the system with a single-sided stop coincides with the equilibrium of its linear part and we find that situations of gap or interference differ only in the sign of Δ.

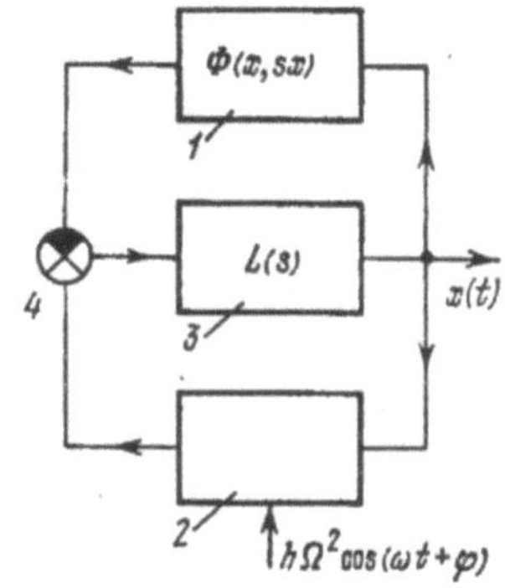

Fig. 11.2

Let us assume that the base vibrates according to a harmonic law of the form

$$y(t) = y_0 \cos(\omega t + \varphi)\,. \qquad (11.2)$$

We will show that the equation for small relative vibration of the impacting mass of the mechanism takes the form (11.1). Let us call the angle of rotation of the mechanism relative to the base, ψ, and let us assume that the projections of the viscous and elastic forces on the x-axis are proportional to the velocity $\dot{x}$ and displacement x of the impacting mass in this direction through the coefficients b and Ω^2. Applying the inertia force of translational motion to the mass as $\ddot{y}(t)$, we can write Newton's second law for the force projected onto the x-axis:

$$\ddot{x} = -b\dot{x} - \Omega^2 - \Phi(x,\dot{x}) - \ddot{y}\sin\varphi \qquad (11.3)$$

Here $\Phi(x,\dot{x})$ is the force characteristic for the impacting pair. Substituting (11.2) into (11.3) and using the expression

$$h = y_0\omega^2 /(\Omega^2 L)\,, \qquad 11.4)$$

where L is the distance from the centre of inertia of the mass to the joint (see Fig. 11.1), we obtain, assuming $\sin\varphi \approx \varphi$, equation (11.1).

Let us rewrite equation (11.1) in the following form:

$$(s^2 + bs + \Omega^2)x = xh\Omega^2 \cos(\omega t + \varphi) - \Phi(x, sx)\,. \qquad (11.5)$$

Equation (11.5) corresponds with the schematic diagram for the signal transformation shown in Fig. 11.2. Here *1* is the nonlinear block with characteristic $\Phi(x, sx)$; *2* is the multiplying block, in which the product of process $x(t)$ with a parametric effect, $h\Omega^2\cos(\omega t + \varphi)$ is produced; *3* is the linear block with operator $L(s) = (s^2 + bs + \Omega^2)^{-1}$; *4* is the summer. From Fig. 11.2, it can be seen that the solution sought for $x(t)$ exists as the output from the linear block, which represents an oscillator with viscous damping. Using the obvious filtering effects of this block, it is natural to search for an approximate solution of equation (11.1) in harmonic form. Such an approach allows us to make use of the output from blocks *1* and *2* only for the basic harmonic components and to develop a solution using harmonic linearisation. A description for such treatment of parametric vibration was given by A Leongard in [143].

The appearance of the arbitrary harmonic signal $a_x \cos \nu t$ at the output of the linear filter *3* (Fig. 11.2), after its passage through the multiplying block *2*, leads to a pattern combining tones with frequency ω − ν and ω + ν as the result of the transformation

$$a_x \cos \nu t \cdot h\Omega^2 \cos(\omega t + \varphi) = \tfrac{1}{2} h\Omega^2 a_x \{\cos[(\omega - \nu)t + \varphi] + \cos[(\omega + \nu)t + \varphi]\}$$

which moves further through the linear blocks without altering the frequency. As a result, the frequency condition for steady-state vibration at the fundamental to exist takes the form

$$\omega - \nu = \nu \tag{11.6}$$

and consequently ν = ω/2. Such behaviour is called *primary parametric resonance*. By studying mutual passage of the signal throughout the loop of the schematic, we can obtain the frequency conditions for secondary parametric resonances to exist

$$\nu = m\omega / 2 \;\; (m = 2,3,\ldots)\,. \tag{11.7}$$

To form the corresponding *m-th* resonance, higher harmonics, occurring in the vibration spectrum, must pass through the loop *m* times until the conditions of frequency balance arise, similar to (11.6). Noting that for each passage through the linear block, a weakening of the amplitude of the higher harmonics occurs, so that as the number of resonances increases so each requires higher levels of modulation *h* to maintain them. It is obvious that any form of parametric resonance occurs only when ν is close to some natural frequency of the system.

Let us find the existence conditions in the system for the principal parametric resonance, occurring at minimum level of modulation. Vibration on output from the linear part of the system using its filtering effect will be expressed in the form

$$x(t) \approx m_x + a_x \cos(\omega t / 2) = m_x + x^0(t) \tag{11.8}$$

Process (11.8) exists at the entry of blocks *1* and *2* (Fig. 11.2). Let us carry out harmonic linearisation for the characteristics of these blocks. For the multiplier we calculate directly

$$\begin{aligned} F_1(t) &= [m_x + a_x \cos(\omega t / 2)] h\Omega^2 \cos(\omega t + \varphi) = \\ &= \tfrac{1}{2} h\Omega^2 a_x \cos(\omega t / 2 + \varphi) + h\Omega^2 m_x \cos(\omega t + \varphi) + \\ &\qquad + \tfrac{1}{2} h\Omega^2 a_x \cos(3\omega t / 2 + \varphi) \end{aligned} \tag{11.9}$$

Noting that the second and third harmonics in (11.9) are suppressed due to the filtering effect of the linear system, the transformation of process *x*(*t*) in the multiplier can be approximately written using the complex coefficient of amplification k_1, carrying out the change of amplitude and phase of the fundamental harmonic of vibration of system *x*(*t*) according to (11.9). Formally such a coefficient can be obtained in the following way. Using Euler's formula, we can represent the basic harmonic input and output signals of the multiplier in the form of the sum of an exponential function

$$x^0(t) = a_x \cos \frac{\omega t}{2} = \frac{a_x}{2}\left[\exp\left(j\frac{\omega t}{2}\right) + \exp\left(-j\frac{\omega t}{2}\right)\right], \tag{11.10}$$

$$F_1(t) \approx \frac{1}{2} h\Omega^2 a_x \cos\left(\frac{\omega t}{2} + \varphi\right) = \\ = \frac{h\Omega^2 a_x}{4}\left\{\exp\left[j\left(\frac{\omega t}{2} + \varphi\right)\right] + \exp\left[-j\left(\frac{\omega t}{2} + \varphi\right)\right]\right\}. \tag{11.11}$$

Identifying the input and output signals with the first harmonics in equations (11.10), (11.11) we finally obtain

$$k_1 = \frac{F_1(t)}{x^0(t)} = \frac{h\Omega^2}{2}\exp j\varphi = \frac{h\Omega^2}{2}(\cos\varphi + j\sin\varphi)\,. \tag{11.12}$$

Hence, the linearisation coefficients of the multiplier depend only on the unknown phase for the process.

Using harmonic linearisation for the function $\Phi(x, xs)$ we find

$$\Phi(x, sx) \approx m_F + (k + \beta s)x^0\,. \tag{11.13}$$

Substituting from the linearised expression for the function on the right hand side in (11.5), we find, using (11.12), (11.13)

$$[s^2 + (b+\beta)s + \Omega^2 + k - k_1]x^0 + \Omega^2 m_x + m_F = 0\,. \tag{11.14}$$

Separating the constant and centred components in (11.14), we find

$$m_F + \Omega^2 m_x = 0 \tag{11.15}$$

$$[s^2 + (b+\beta)s + \Omega^2 + k - k_1]x^0 = 0 \tag{11.16}$$

Expression (11.15) is similar to (9.7). It gives the connection between m_x and a_x. Substituting this into (11.16), we obtain the equation to find a_x.

From (11.15), it can be seen that the constant components of parametric vibration only arise because of the non-symmetry of the function $\Phi(x, sx)$. In the symmetrical case $m_F \equiv 0$, and we can find a_x from equation (11.16) directly. Let us look at how to solve it.

As in the case of forced vibration, equation (.11.16) represents some equivalent autonomous systems, undergoing harmonic vibration similar to the parametric. The existence conditions for periodic regimes of type (11.8) for such systems are represented by the presence of pairs of imaginary roots $s = \pm j\omega/2$ for their characteristic equations

$$\mathcal{D}(s) = s^2 + (b+\beta)s + \Omega^2 + k - k_1 = 0\,. \tag{11.17}$$

Substituting $s = j\omega/2$ in (11 17) and taking account of (11.12), we obtain the characteristic equation of the systems in the form

$$\mathcal{D}\left(j\frac{\omega}{2}\right) = \Omega^2 + k - \frac{\omega^2}{4} - \frac{h\Omega^2}{2}\cos\varphi + j\left[(b+\beta)\frac{\omega}{2} - \frac{h\Omega^2}{2}\sin\varphi\right]. \tag{11.18}$$

From (11.18) it follows that parametric excitation causes "negative damping," which explains the effects it causes. Setting the real and imaginary parts of (11.18) to zero, we find

$$\sin\varphi = (b+\beta)\omega/(h\Omega^2)\,, \tag{11.19}$$

$$\cos\varphi = 2(\Omega^2 + k - \omega^2/4)/(h\Omega^2)\,. \tag{11.20}$$

Eliminating φ from (11.19), (11.20), we obtain the final expression for a_x. It is convenient to write it in the following form:

$$\Omega^2 + k = \left(\frac{\omega}{2}\right)^2 \pm \frac{1}{2}\sqrt{h^2\Omega^2 - (b+\beta)^2\omega^2}\,. \tag{11.21}$$

To calculate a_x from this equation, it is necessary to substitute into it the values of linearisation coefficients k and β.

As can be seen from (11.21) it is possible that two different periodic solutions may exist for certain values of the system parameters. The existence conditions for parametric vibration according to (11.21) take the form

$$h \geq (b+\beta)\omega / \Omega^2 \tag{11.22}$$

This defines the modulation for which the periodic regime of motion is sustained. Of course, this value is proportional to damping in the system.

2. It is possible to use the method described in §10 to study the stability of parametric vibration, based on analysis of the stability of solutions for the equivalent linearised autonomous system. In the case considered, the imaginary part of the characteristic complex, according to (11.18), takes the form

$$\Theta(a_x, \omega) = (b+\beta)\frac{\omega}{2} - \frac{h\Omega^2}{2}\sin\varphi\,. \tag{11.23}$$

Applying condition (10.2) to (11.23) and using the connection between $\tilde{\varphi}$ and $\tilde{a}_x$, we obtain

$$\left(\frac{d\beta}{d\tilde{a}_x} - \frac{h\Omega^2}{\omega}\frac{d\varphi}{d\tilde{a}_x}\cos\varphi\right)_{\tilde{a}_x = a_x} > 0\,. \tag{11.24}$$

The expression $d\tilde{\varphi}/d\tilde{a}_x$ is found by differentiating (11.20) as an implicit function $\tilde{\varphi}$ of $\tilde{a}_x$. Bearing in mind (11.19), we find

$$\left(\frac{d\tilde{\varphi}}{d\tilde{a}_x}\right)_{\tilde{a}_x = a_x} = -\frac{2}{\omega(b+\beta)}\left(\frac{dk}{d\tilde{a}_x}\right)_{\tilde{a}_x = a_x}. \tag{11.25}$$

Substituting the values (11.20) and (11.25) into (11.24), we obtain finally

$$\left[\frac{d\beta}{d\tilde{a}_x} + \frac{\omega^2}{4(b+\beta)}\frac{dk}{d\tilde{a}_x}\left(\Omega^2 + k - \frac{\omega^2}{4}\right)\right]_{\tilde{a}_x = a_x} > 0\,. \tag{11.26}$$

To use force harmonic linearisation, the corresponding conditions for stability are obtained by differentiating the equivalent force linearisation coefficients in (11.26), k (instead of $\Omega^2 + k$) and β, as implicit functions of $\tilde{a}_x$. Since, for force characteristics of impact pairs

$$d\tilde{a}_F / d\tilde{a}_x = k > 0$$

we find,

$$\left[\frac{d\beta}{d\tilde{a}_F} + \frac{\omega^2}{4(b+\beta)}\frac{dk}{d\tilde{a}_F}\left(k - \frac{\omega_2}{4}\right)\right]_{\tilde{a}_F = a_F} > 0 \tag{11.27}$$

where a_F is the steady-state value of $\tilde{a}_F$, corresponding to the periodic regime considered.

In cases when the coefficient β is independent of the process parameters or is zero, the conditions of stability can be considerably simplified. For example, from (11.26) we find when $\beta \equiv 0$

$$\left[\frac{dk}{d\tilde{a}_x}\left(\Omega^2 + k - \frac{\omega^2}{4}\right)\right]_{\tilde{a}_x = a_x} > 0\,, \tag{11.28}$$

i.e. for a hard system ($dk/da_x > 0$) the condition for stability becomes

$$\Omega^2 + k > (\omega/2)^2\,, \tag{11.29}$$

which corresponds to the positive sign in (11.21); for soft systems ($dk/da_x < 0$) the inequality reverses to:

$$\Omega^2 + k < (\omega/2)^2 \tag{11.30}$$

and the regime is stable, corresponding to the negative sign in (11.21). A similar relationship is obtain from (11.27).

Note that, so far, we have not in practice used the specific of vibro-impact systems, since the results obtained for the dynamics and stability of parametric vibration are suitable for all nonlinear systems, described by equations of the type (11.1), the nonlinear terms of which allow harmonic linearisation. The conditions for applying the methods considered coincide with the usual requirements set out for harmonic linearisation (see §7). In particular it is satisfactory to use this for systems with filtering effects in their linear part.

Secondary parametric resonances can be studied by a similar scheme to determine the corresponding harmonics at the output of the linear part of the system. The effects of higher harmonics can be found in the same way.

As an example let us look at the case of cubic nonlinearity $\Phi(x, sx) = \Phi(x) = \gamma x^3$, for which the dynamics solution is described, using asymptotic methods for the first approximation, in reference [64]. For these functions the harmonic linearisation coefficients take the form

$$k = \frac{3}{4}\gamma a_x^2,\ \beta = 0\,. \tag{11.31}$$

Substituting (11.31) into (11.21), we find

$$a_x^2 = \frac{4}{3\gamma}\left[\left(\frac{\omega}{2}\right)^2 - \Omega^2 \pm \frac{1}{2}\sqrt{h^2\Omega^4 - b^2\omega^2}\right], \tag{11.32}$$

which coincides with the solution specified. We will evaluate its stability using condition (11.28). Noting that $dk/da_x = 1.5\gamma a_x$, we obtain a stable solution when $\gamma > 0$, corresponding to the positive sign of (11.32), and to the negative sign when $\gamma < 0$. The conclusions regarding stability have obvious physical meanings [64].

3. Let us now look at how parametric vibration becomes established. For this, we can use the method similar to the study of forced vibration stability in §10. With this aim, the value of φ in equation (11.1) will be replaced by a slowly changing value $\tilde{\varphi}$ and we shall search for a solution for the principal resonance in the form

$$x(t) \approx \tilde{a}_x \cos(\omega t/2) \tag{11.33}$$

where

$$d\tilde{a}_x / dt = \tilde{a}_x \tilde{\xi}\,, \tag{11.34}$$

in which $\tilde{\xi}$ is a slowly changing variable. The existence of a constant component in $x(t)$ has no specific influence on the calculation scheme to be described later, hence we shall assume for brevity that the system is symmetrical $\Phi(-x, sx) = -\Phi(x, sx)$ and consequently $m_x \equiv 0$.

Carrying out the transformation of the signal (11.33) in the multiplier similar to (11.9), it can be seen that the same gain coefficient (11.12) will occur for harmonic linearisation, but, in contrast to the steady-state process, coefficient k_1 is now a slowly changing function of time

$$k_1 = \tfrac{1}{2} h\Omega^2 (\cos\tilde{\varphi} + j\sin\tilde{\varphi})\,, \tag{11.35}$$

The linearisation coefficients of function $\Phi(x, sx)$ can be found from formulae (10.20), (10.21), while for the nonlinear form $\Phi(x)$ they coincide with the usual harmonic linearisation coefficients, using $\tilde{a}_x$ instead of a_x.

As a result, we can use the characteristic equation in the form (11.17) to study the establishment of the process, treating the linearisation coefficients which appear in it in a similar manner.

Transient processes of the linearised system (11.16) of the type (11.33), (11.34) correspond to the existence of pairs of complex roots $s = \tilde{\xi} \pm j\omega/2$ for its characteristic equation (11.17). Substituting $s = \tilde{\xi} + j\omega/2$ in the latter and setting the real and imaginary parts to zero, we obtain using (11.35)

$$\tilde{\xi}^2 + (b+\beta)\tilde{\xi} + \Omega^2 + k - \frac{1}{2} h\Omega^2 \cos\tilde{\varphi} - \left(\frac{\omega}{2}\right)^2 = 0\,, \tag{11.36}$$

$$\tilde{\xi} = \frac{1}{2}\left(\frac{h\Omega^2}{\omega}\sin\tilde{\varphi} - b - \beta\right). \tag{11.37}$$

The system of equations (11.36), (11.37) determines the relationships between $\tilde{\varphi}$ and $\tilde{a}_x$, $\tilde{\xi}$ and $\tilde{a}_x$. When $\tilde{\xi} \equiv 0$ we obtain the existence condition for steady-state regimes ($\tilde{a}_x = a_x = \text{const}$), which coincide with (11.19), (11.20). Introducing the stability condition (10.30) into (11.37), we obtain condition (11.24). Let us find the error arising in the value of $(d\tilde{\varphi} / d\tilde{a}_x)_{\tilde{a}_x = a_x}$ when using the abbreviated equation (11.19), using the method described in section 2 to study stability. We shall find the value of the specified term on the basis of the full relationships (11.36), (11.37) differentiating them as implicit functions of $\tilde{a}_x$ in the vicinity of the steady-state value a_x. For simplification, assuming $\beta = 0$ and neglecting the square of small values in (11.36) near the specified value of $\tilde{\xi}$, we find

$$\left(b\frac{d\tilde{\xi}}{d\tilde{a}_x} + \frac{dk}{d\tilde{a}_x} + \frac{1}{2}h\Omega^2 \sin\tilde{\varphi}\frac{d\tilde{\varphi}}{d\tilde{a}_x}\right)_{\tilde{a}_x = a_x} = 0, \quad \left(\frac{d\tilde{\xi}}{d\tilde{a}_x} = \frac{1}{2}\frac{h\Omega^2}{\omega}\cos\tilde{\varphi}\frac{d\tilde{\varphi}}{d\tilde{a}_x}\right)_{\tilde{a}_x = a_x}.$$

Eliminating the value $(d\tilde{\xi}/d\tilde{a}_x)_{\tilde{a}_x=a_x}$ from these expressions, we obtain using (11.19), (11.20)

$$\left(\frac{d\tilde{\varphi}}{d\tilde{a}_x}\right)_{\tilde{a}_x=a_x} = \left\{\frac{-dk/d\tilde{a}_x}{\frac{1}{2}b\omega[1+2(\Omega^2+k-\omega^2/4)/\omega^2}\right\}_{\tilde{a}_x=a_x}. \tag{11.38}$$

Comparing expressions (11.38) and (11.25) and noting that for parametric resonance domains the value

$$2(\Omega^2+k-\omega^2/4)/\omega^2 << 1,$$

we find that they practically coincide.

To find the relationship of $\tilde{\xi}$ to $\tilde{a}_x$, which characterises the transient process, we shall eliminate phase $\tilde{\varphi}$ from equations (11.36), (11.37). We find

$$\tilde{\xi}^2+(b+\beta)\tilde{\xi}+\Omega^2+k-\left(\frac{\omega}{2}\right)^2+\sqrt{\frac{h^2\Omega^2}{4}-\frac{\omega^2}{4}(2\tilde{\xi}+b+\beta)^2}.$$

To simplify this equation, let us assume $b+\beta\equiv 0$. As a result we obtain a biquadratic equation, whose real solution takes the following form:

$$\tilde{\xi}_2 = -\left[\Omega^2+k+\left(\frac{\omega}{2}\right)^2\right]+\sqrt{\left[\Omega^2+k+\left(\frac{\omega}{2}\right)^2\right]^2+\frac{h^2\Omega^2}{4}-\left[\Omega^2+k-\left(\frac{\omega}{2}\right)^2\right]^2}. \tag{11.39}$$

According to (11.39) the conditions for soft excitation of parametric vibration ($\tilde{\xi}\geq 0$) lead to the following relationships:

$$\frac{h^2\Omega^2}{4}-\left[\Omega^2+k-\left(\frac{\omega}{2}\right)^2\right]^2\geq 0,$$

from which we find

$$\left|1+\frac{k}{\Omega^2}-\left(\frac{\omega}{2\Omega}\right)^2\right|\leq\frac{h}{2}. \tag{11.40}$$

Assuming that when $\tilde{a}_x=0$ the system is not nonlinear ($k\equiv 0$), we obtain the well-known relationships for self-excitation of parametric vibration for a linear system (see, for example, [65])

$$\sqrt{1-\frac{h}{2}}\leq\frac{\omega}{2\Omega}\leq\sqrt{1+\frac{h}{2}}. \tag{11.41}$$

Of course, this relationship can also be obtained from (11.21) when $a_x=0,\ b+\beta=0$. Expression (11.41) is obtained in reference [65] as the result of the first approximation from the equation for critical frequency, which represents Hill's infinite determinant. Assuming that h has a small value and expanding the radical as a series of linear terms in h, we obtain the formula of N M Beliayev [54]

$$1-\frac{h}{4}\leq\frac{\omega}{2\sqrt{c}}\leq 1+\frac{h}{4}. \tag{11.42}$$

Let us examine the case of a transient process when resonance conditions are supported throughout the entire process

$$\Omega^2 + k - (\omega/2)^2 = 0. \tag{11.43}$$

Substituting (11.43) into (11.39) and expanding the radical as a series in terms of the small parameter h, we obtain, limiting the expansion to linear terms, $\tilde{\xi}^2 = \dfrac{h^2\Omega^2}{16(\omega/2)^2}$, whence, using (11.43) we find

$$\tilde{\xi} = \frac{h\Omega^2}{4\sqrt{\Omega^2 + k}} \tag{11.44}$$

When $k \equiv 0$ this corresponds to the well-known *characteristic index* for a linear system (see, for example, [65]). Substituting the value of k for various nonlinearities, we find the character of the change of resonant amplitude for start-up or run-down conditions.

Taking $k = \frac{3}{4}\gamma a_x^2$ (cubic nonlinearity), then

$$\tilde{\xi} = \frac{h\Omega^2}{4\sqrt{\Omega^2 + \frac{3}{4}\gamma a_x^2}}, \tag{11.45}$$

i.e. pulling during run-down of the system with soft nonlinearity $(\gamma < 0)$ occurs as $\tilde{\xi}$ increases, characterised by the rate of oscillation (see (10.18)), but run-up of a system with hard non linearity $(\gamma > 0)$ is accompanied by a decrease in this term.

4. Let us now turn to examine parametric *vibro-impact* regimes. From the formulas obtained above, calculation of energy dissipation due to impact is equivalent to increasing the equivalent coefficient of viscous damping b to a value β, thus without impairing the qualitative side of the problem the coefficient of restitution can be taken as unity $(\beta \equiv 0)$.

To study the case of symmetrical double-sided collisions we will make use of the harmonic force linearisation. Substituting the expression for the linearisation coefficients in the form (9.38) in (11.21) in place of Ω^2+k, we obtain

$$a_F = \frac{4\Delta}{\pi}\left[\left(\frac{\omega}{2}\right)^2 - \Omega^2\left(1 - \frac{\pi}{4}\right) \pm \frac{1}{2}\sqrt{h^2\Omega^4 - b^2\omega^2}\right], \tag{11.46}$$

where, because the nonlinearity considered is of hardening type, the negative sign before the radical corresponds to an unstable solution.

Since impacts correspond to the value $a_F > \Omega^2\Delta$, to find the existence conditions for periodic vibro-impact regimes we shall use in (11.46)

$$a_F \geq \Omega\Delta. \tag{11.47}$$

As a result we obtain the relationship, defining the frequency range of the vibration examined as

$$\left(\frac{\omega}{2}\right)^2 \geq \Omega^2 - \frac{1}{2}\sqrt{h^2\Omega^4 - b^2\omega^2}. \tag{11.48}$$

Because of the structure of the system (Fig. 11.1, *a*) the amplitude is constant in the range shown and equals Δ.

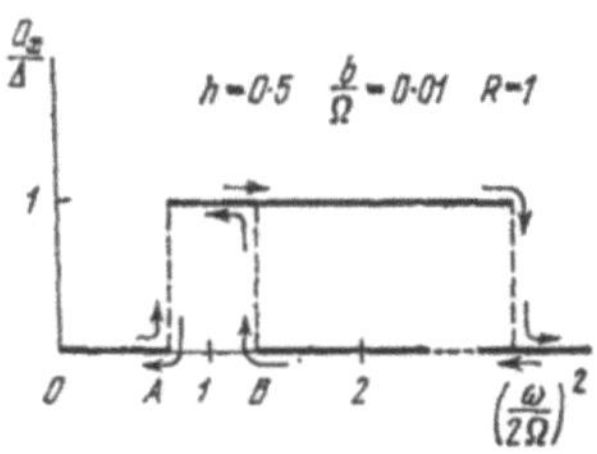

Fig. 11.3

Fig. 11.3 shows the amplitude-frequency characteristics for symmetrical systems. It can be seen that as the excitation frequency increases, pulling can occur up to the break-down frequency defined by (11.22). By moving the stops apart when excitation frequency is constant, it is possible to obtain amplitude pulling. It is interesting to note that in contrast to forced vibration, amplitude pulling for parametric vibration can not be limited by the dissipative factors considered. In actual systems limitation occurs due to the fact that at large amplitudes additional nonlinear factors occur and, moreover, the loading of the vibratory source of energy leads to a decrease in the level of modulation and, consequently existence conditions (11.22) are infringed.

In the frequency range *AB*. (Fig. 11.3), when the relationship

$$\Omega^2 - \frac{1}{2}\sqrt{h^2\Omega^4 - b^2\omega^2} < \left(\frac{\omega}{2}\right)^2 < \Omega^2 + \frac{1}{2}\sqrt{h^2\Omega^4 - b^2\omega^2}, \tag{11.49}$$

obtained using (11.21) when $k \equiv 0$, is satisfied, the system's state of equilibrium is unstable, and the system develops into a regime with impacts. This instability domain for a linear parametric system is disposed between the stops, consequently, even taking into account the loss of energy during collisions, vibro-impact regimes will always exist in this frequency range. With the help of hard excitation at the pulling frequency, situated between point *B* and the break-down frequency, the behaviour of systems can change from stable equilibrium to vibro-impact.

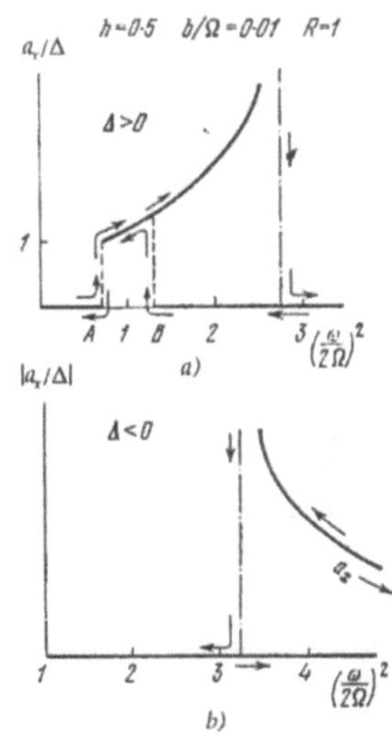

Fig. 11.4

Let us now look at the case of single-sided limited motion (Fig. 11.1, b), using the expressions for the linearisation coefficients in the form (9.18). When $\Delta > 0$ the nonlinearity is hard, so regimes corresponding to the positive sign in (11.21) are stable; when $\Delta < 0$ the nonlinearity becomes soft and the sign for stable regimes changes to negative. Substituting the known value of k in (11.21), we obtain when $\beta \equiv 0$

$$a_x = -2\Delta\left[\left(\frac{\omega}{2\Omega}\right)^2 - 3 \pm \sqrt{h^2 - \frac{4b^2}{\Omega^2}\left(\frac{\omega}{2\Omega}\right)^2}\,\right]^{-1}. \tag{11.50}$$

Fig. 11.4 shows amplitude-frequency characteristics for non-symmetrical systems according to formula (11.50), corresponding to stops arranged with a gap (Fig. 11.4, *a*) and with interference (Fig. 11.4, *b*). Only the branches for stable regimes are shown.

For systems with a gap, the same zone of unstable equilibrium occurs (11.49) close to the excitation frequency, at almost double the natural frequency. With interference the equilibrium state is always stable and the shift to a vibration regime can happen only with additional excitation in the frequency range found from the relationship

$$\left(\frac{\omega}{2\Omega}\right)^2 \geq 3+\frac{1}{2}\sqrt{h^2-\frac{4b^2}{\Omega^2}\left(\frac{\omega}{2\Omega}\right)^2} \tag{11.51}$$

In both cases the vibration of the system can be pulled to infinite amplitude. When the system is arranged with a gap this corresponds to the smooth increase of the excitation frequency, while with interference, with decreasing frequency. For example, the latter can lead to dangerous vibration of systems during run-down when force-locked components are linked. According to (11.50) infinite oscillation of systems is observed as the frequency nears the value given by

$$\left(\frac{\omega}{2\Omega}\right)^2 = 3\pm\frac{1}{2}\sqrt{h^2-\frac{4b^2}{\Omega^2}\left(\frac{\omega}{2\Omega}\right)^2}\,. \tag{11.52}$$

Here the positive sign corresponds to resonant frequency of a system with interference and the negative sign to that with a gap. As the system passes through the resonant frequency (11.52) the vibration breaks down and dies away completely. Note that in the examples studied damping does not limit the maximum value of amplitude. However, according to (11.22) when damping is increased, parametric effects are eliminated.

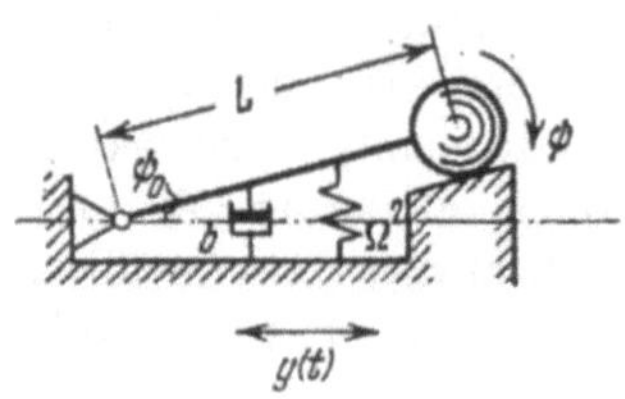

Fig. 11.5

The resonance amplitude is not limited as is the increase of interference Δ (see (11.50)). The latter does not even affect the resonant frequency. Hence increasing the force locking mechanism (Fig. 11.1, *b*) leads to a growth in the value needed for the starting impulse, but does not eliminate the possibility of parametric resonance. The situation shown is particularly dangerous, since it can not be discovered even during machine testing.

To estimate how the vibrations grow for regimes with pulling while maintaining resonance conditions (11.43), when we substitute (9.18) in (11.44), we find

$$\tilde{\xi} = \frac{h\Omega}{4\sqrt{3-2\Delta/a_x}}\,. \tag{11.53}$$

This means that for a stable system with gap $(\Delta > 0)$, as the excitation frequency increases, index $\tilde{\xi}$ decreases from the corresponding value for a linear system to the value

$$\tilde{\xi} \approx h\Omega/(4\sqrt{3}), \tag{11.54}$$

almost twice as small. During run-down of a system with interference $(\Delta < 0)$ an increase in ξ occurs from a small value to a value given by (11.54). This same value characterises the rate of parametric resonance when $(\Delta = 0)$. In the latter case the system can not develop stable steady-state regimes.

5. The schematic diagram of Fig. 11.2 represents a wide class of systems which can be studied in a similar way to the above presentation. To conclude the description of methods to study periodic motion of non-autonomous systems, we shall use as an example the problems associated with the simultaneous effects of external and parametric excitation [44]. Such excitation is exhibited by the simplest form of mechanism with force locking, fixed on the base (Fig. 11.5), which moves according to the law $y(t) = y_0 \cos(\omega t + \varphi)$ with a non-zero angle ψ_0 between the axis of the bar and the direction of the vibration [117].

Keeping the same assumptions of mechanism properties and parameter designation as in section 1, we arrive, taking $x = \psi L$, at equation (11.3). Substituting (11.2) in (11.3), we find using (11.4)

$$(s^2 + bs + \Omega^2)x = hL\Omega^2 \cos(\omega t + \varphi)\sin\psi - \Phi(x, sx). \tag{11.55}$$

Expanding $\sin\psi$ as a Taylor series near the reference angle ψ_0, and limiting the expansion to linear terms we find

$$\sin\psi \approx \sin\psi_0 + (\psi - \psi_0)\cos\psi_0. \tag{11.56}$$

Substituting (11.56) into (11.55), we obtain

$$(s^2 + bs + \Omega^2)x = hfx\cos(\omega t + \varphi) + hd\cos(\omega t + \varphi) - \Phi(x, sx), \tag{11.57}$$

where $f = \Omega^2 \cos\psi_0$; $d = \Omega^2(L\sin\psi_0 - \Delta\cos\psi_0)$; $\Delta = \psi_0 L$. The first term on the right hand side of equation (11. 57) represents a parametric excitation, and the second term external excitation. It can be seen that as ψ_0 increases the external excitation grows but the parametric excitation decreases. Equation (11.57) corresponds to the schematic diagram of Fig. 11.6. It differs from the arrangement of Fig. 11.2 by the addition of external excitation $P(t) = hd\cos(\omega t + \varphi)$ and the introduction of coefficient f in the parametric excitation.

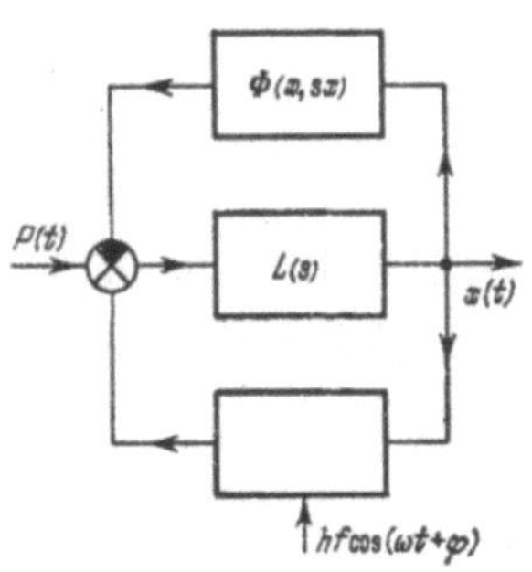

Fig. 11.6

An interesting feature of this system is that when it is installed with pre-arranged interference $(\psi_0 < 0)$ due to accidental break away of the mass from the stop, stable equilibrium can be violated and an intensive vibro-impact regime is established. In this case the amplitude of the exciting force can be below the level of force due to the initial strain of the spring.

Let us find the existence conditions for the specified regimes. The joint effects of parametric and external excitation produce the main dynamic effects on the frequency of the principle and secondary parametric resonances. This occurs because of the development of a strong "damping decreasing" parametric effect at these frequencies causing favourable conditions for the external force to produce excitation corresponding to sub-periodic resonance of ratio 2 and fundamental resonance. A method to find an approximate solution with frequency $\omega/2$ was developed above and recurs in this case in its principle feature, hence we will continue by studying vibration at frequency ω. In addition, with external and parametric vibration interaction, resonance at frequency ω is more vigorous.

An approximate periodic solution of (11.57) will be sought in the form

$$x(t) \approx m_x + a_x \cos \omega t = m_x + x^0(t). \tag{11.58}$$

The action (11.58) appears at entry to the multiplier and nonlinear blocks. Let us establish the harmonic linearisation characteristics of these blocks. In the first case we find directly

$$\begin{aligned} F_1(t) &= (m_x + a_x \cos \omega t) hf \cos(\omega t + \varphi) = \\ &= \tfrac{1}{2} hfa_x \cos\varphi + hfm_x \cos(\omega t + \varphi) + \tfrac{1}{2} hfa_x \cos(2\omega t + \varphi). \end{aligned} \tag{11.59}$$

Discarding higher harmonics in (11.59) by taking account of the filtering effects of the system, we can write the transformation of the harmonic process in the multiplier using a complex gain factor

$$F_1(t) \approx \tfrac{1}{2} hfa_x \cos\varphi + k_1 x^0(t). \tag{11.60}$$

where

$$k_1 = \frac{hfm_x}{a_x} \exp j\varphi. \tag{11.61}$$

The nonlinear block will be linearised, as usual, using formula (11.13).

Using (11.58), the external excitation $P(t)$ is transformed to the form

$$P(t) = hd(\cos\omega t \cos\varphi - \sin\omega t \sin\varphi) = \frac{hd}{a_x}\left(\cos\varphi + \frac{\sin\varphi}{\omega} s\right) x^0(t). \tag{11.62}$$

Substituting expressions (11.60), (11.13), (11.62) in the right hand side of (11.57) and separating the constant and centred components we find

$$\Omega^2 m_x + m_F - \tfrac{1}{2} hfa_x \cos\varphi = 0, \tag{11.63}$$

$$\left[s^2 + \left(b + \beta - \frac{hd \sin\varphi}{a_x \omega}\right) s + \Omega^2 + k - k_1 - \frac{hd}{a_x}\right] x^0(t) = \mathcal{D}(s) x^0(t) = 0. \tag{11.64}$$

The existence conditions for periodic regimes of type (11.58) of the linearised system (11.64) are the presence of roots $s = \pm j\omega$ in the characteristic equation. Substituting $s = j\omega$ and setting the real and imaginary parts to zero we obtain, using (11.61)

$$(b + \beta)\omega - \frac{h(d + m_x f)}{a_x} \sin\varphi = 0, \tag{11.65}$$

$$\Omega^2 + k - \omega^2 - \frac{h(d + m_x f}{a_x} \cos\varphi = 0. \tag{11.66}$$

Eliminating the trigonometric functions from (11.65), (11.66), we find

$$a_x = h|d + m_x f| [(\omega_0^2 - \omega^2)^2 + \omega^2 (b + \beta)^2]^{-1/2}, \tag{11.67}$$

where $\omega_0 = \sqrt{\Omega^2 + k}$. Comparing (11.67) with the similar expression (9.13) for forced vibration, it is easy to establish the additional effects on the amplitude due to parametric excitation. From (11.67) it follows that vibro-impact regimes exist for any small values of h, and, as a consequence, when the conditions provide a "force locking" mechanism

$$|\Delta| > hL\sin\psi_0 \,. \tag{11.68}$$

Obviously, such a regime can only result with hard excitation.

Let us re-write (11.63) with the help of (11.66), (11.67) in the following form:

$$\begin{aligned} m_F &= -\Omega^2 m_x + \frac{f(\omega_0^2-\omega^2)a_x^2}{2(d+m_x f)} = \\ &= -\Omega^2 m_x + \frac{h^2 f(d+m_x f)(\omega_0^2-\omega^2)}{2[(\omega_0^2-\omega^2)^2+\omega^2(b+\beta)^2]} \end{aligned} \,. \tag{11.69}$$

Equations (11.67), (11.69) make it possible to define the values m_x and a_x using the equations for the linearisation coefficients m_F and k. Subsequently, from the relationship

$$\varphi = \arctan[\omega(b+\beta)/(\omega_0^2-\omega^2)]\,, \tag{11.70}$$

obtained from (11.65), (11.66), the phase of the process can be found.

We will use the method described in §10 to study the stability of the periodic solutions. Applying condition (10.2) to the characteristic complex $\mathcal{D}(j\omega)$ for the linearised system (11.64), we find, using (11.61)

$$\frac{d}{d\tilde{a}_x}\left[b+\beta-\frac{h(d+m_x f)\sin\varphi}{\tilde{a}_x\omega}\right]_{\tilde{a}_x=a_x} > 0\,. \tag{11.71}$$

We will assume that slowly changing functions $\tilde{m}_x(t), \tilde{a}_x(t)$ and $\tilde{\varphi}(t)$ are connected during the disturbed motion by equations (11.63), (11.66).

Differentiating (11.71), we find

$$\left[\frac{d\beta}{d\tilde{a}_x} - \frac{hf\tilde{a}_x\sin\tilde{\varphi}\dfrac{d\tilde{m}_x}{d\tilde{a}_x} + h(d+\tilde{m}_x f)\left(\tilde{a}_x\cos\tilde{\varphi}\dfrac{d\tilde{\varphi}}{d\tilde{a}_x} - \sin\tilde{\varphi}\right)}{\tilde{a}_x^2\omega}\right]_{\tilde{a}_x=a_x} > 0\,. \tag{11.72}$$

The derivative $d\tilde{\varphi}/d\tilde{a}_x$ is found by differentiating (11.66) as an implicit function $\tilde{\varphi}$ of $\tilde{a}_x$. Using (11.67) we find

$$\frac{d\tilde{\varphi}}{d\tilde{a}_x} = \frac{\omega\dfrac{d\beta}{d\tilde{a}_x}(\omega_0^2-\omega^2) - \dfrac{dk}{d\tilde{a}_x}\omega(b+\beta)}{(\omega_0^2-\omega^2)+\omega(b+\beta)^2}\,. \tag{11.73}$$

To simplify the relationship we will put $\beta \equiv 0$. Substituting expression (11.73) in (11.72) we obtain after transformation and using the steady-state values $\sin\varphi$, $\cos\varphi$ and $h(d+m_x f)$ found from equations (11.65) to (11.67),

$$\left[\begin{array}{l}\tilde{a}_x(\omega_0^2-\omega^2)\dfrac{dk}{d\tilde{a}_x}-hf\sqrt{(\omega_0^2-\omega^2)^2+\omega^2b^2}\,\dfrac{d\tilde{m}_x}{d\tilde{a}_x}\\ \qquad\qquad\qquad\qquad\qquad +(\omega_0^2-\omega^2)+\omega^2 b\end{array}\right]_{\tilde{a}_x=a_x}>0. \quad (11.74)$$

When $b\equiv 0$ condition (11.74) can be separated into two factors

$$\left[(\omega_0^2-\omega^2)\left(\omega_0^2-\omega^2+\tilde{a}_x\frac{dk}{d\tilde{a}_x}-hf\frac{d\tilde{m}_x}{d\tilde{a}_x}\right)\right]_{\tilde{a}_x=a_x}>0. \quad (11.75)$$

Comparing (11.75) with the similar equation (10.11) when $b+\beta\equiv 0$, obtained for forced vibration, we see that the additional effects of parametric excitation led to the appearance in the condition of a term proportional to the value of modulation.

In the resulting expressions, describing the conditions for existence and stability of periodic solutions, linearisation coefficient k exists due to the dependence of the natural frequency ω_0 on amplitude a_x of the fundamental. Noting that the motion considered has a resonant character and consequently appears close to the natural frequency of the conservative system, for the relationship $\omega_0(a_x)$, it is possible to apply the exact equations for the backbone curve of the fundamental. Expanding the law of motion for a conservative system (2.6) found by the "stitching" method, as a Fourier series, we find for the coefficients the two first terms of the series

$$m_x=\frac{\Delta\omega_0}{\pi\Omega}\tan\frac{\pi\Omega}{\omega_0} \quad (11.76)$$

$$a_x=-\frac{2\Delta\omega_0\Omega}{\pi(\omega_0^2-\omega^2)}\tan\frac{\pi\Omega}{\omega} \quad (11.77)$$

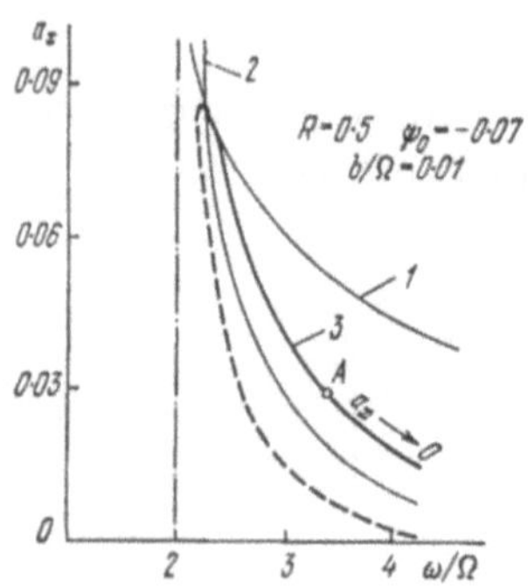

Fig. 11.7

Obviously, we could use the general equation for the harmonic linearisation coefficients obtained in §7 to obtain a solution for the problem considered. We simply wanted to demonstrate here yet another method to calculate the linearisation coefficients, based on a fuller description of the particular problem. According to (11.67), (11.70) the consideration of energy losses during impact does not provide any qualitative peculiarities, hence we estimate these losses with the simplified relationship (7.61). Greater accuracy can be obtained by using in (7.57) the value of the pre-impact velocity $\dot{x}_-$, found for free vibration of a conservative system.

Using expressions (11.76), (11.77) allows us to solve equation (11.67) independently and to construct amplitude-frequency characteristics for the system. Fig. 11.7 shows examples of such construction for systems with interference $(\Delta<0)$. Since $a_x\geq 0$, then in this case according to (11.77) $\omega_0\geq 2\Omega$. Substituting in (11.77) $\omega=\omega_0$, we find the expression for the energy boundary of vibration using (11.76)

$$a_x = \frac{h}{\omega(b+\beta)} \left| d + \frac{\Delta\omega_0 f}{\pi\Omega} \tan\frac{\pi\Omega}{\omega_0} \right|. \tag{11.78}$$

The intersection of the energy boundary *1* with the backbone curve *2* found from expression (11.77) gives the highest point of the resonance curve. The remaining points (curve 3) are conveniently obtained as described in §9, section 1, constructed graphically as the intersections of the family of curves $a_x(\omega_0)$, produced from (11.67) using (11.76), (7.61) for various values ω = const and from the backbone curves.

To investigate stability according to condition (11.75) we find from (11.76)

$$\begin{aligned} \frac{d\tilde{m}_x}{d\tilde{a}_x} &= \frac{\Delta dk / d\tilde{a}_x}{2\pi\omega_0\Omega} \left(\tan\frac{\pi\Omega}{\omega_0} - \frac{\pi\Omega}{\omega_0} \Big/ \cos^2\frac{\pi\Omega}{\omega_0} \right) = \\ &= \frac{\Delta dk / d\tilde{a}_x}{4\pi\omega_0\Omega\cos^2(\pi\Omega/\omega_0)} \left(\sin\frac{2\pi\Omega}{\omega_0} - \frac{2\pi\Omega}{\omega_0} \right) < 0. \end{aligned} \tag{11.79}$$

since the sign of Δ coincides with the sign of $dk / d\tilde{a}_x$, but the factor in curved brackets is of negative sign. Thus, in the example considered the two final in the second factor of condition (11.75) take the same sign. Using an argument similar to that explained in §10, it can be shown that in Fig. 11.7 these correspond to stable regimes, lying on branch *3* of the curve, above the backbone line for the system. The dotted line shows unstable solutions.

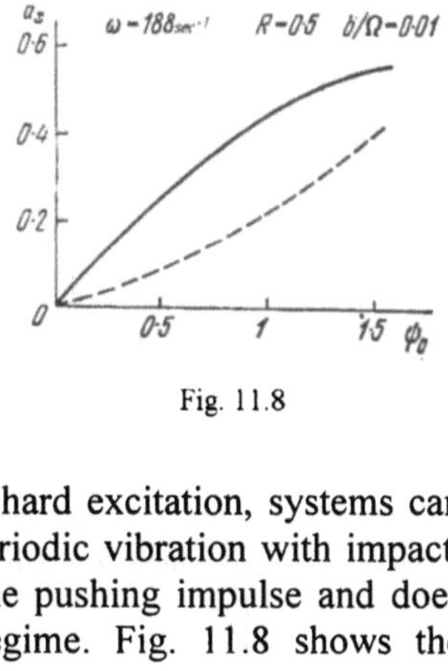

Fig. 11.8

From Fig. 11.7 it can be seen that as a result of hard excitation, systems can shift from an equilibrium condition to regimes of periodic vibration with impact. Increasing interference also causes an increase in the pushing impulse and does not eliminate the possibility of a vibro-impact regime. Fig. 11.8 shows the relationship between the amplitude of steady vibration a_x and the value of the initial angular displacement ψ_0. The maximum amplitude occurs for forced vibration ($\psi_0 = \pi/2$).

6. To demonstrate some of the effects described an experimental arrangement was constructed as shown in Fig. 11.9. The beam *1*, fixed in the support *2*, is installed on base *3* of a horizontal vibration test-bed, and is subjected to vibration in the direction of the arrows shown. At the free end of the beam a ball *4* is firmly attached, and its motion is limited by one or two stops, 5.

The motion of the ball was measured by a strain-gauge *6* fixed on the beam and whose signal was recorded on an oscillograph; the vibration of the table was recorded by a similar method using an inductance transducer. The instant when the ball and the stop touched was recorded by a contact.

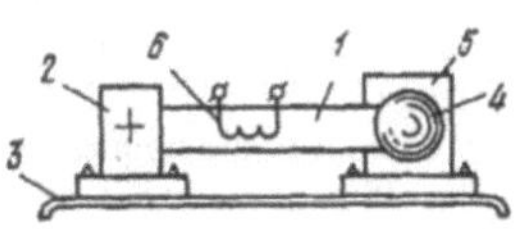

Fig. 11.9

With $\psi_0 = 0$ an oscillogram was recorded of the fundamental parametric resonance of a system with a natural linear frequency $\Omega/2\pi = 10Hz$. Fig. 11.10, *a* shows the oscillation of the system (curve *1*) without stops at a fixed excitation frequency $\omega/2\pi = 20Hz$. Curve *2* in the oscillogram corresponds to the vibration velocity of the table when

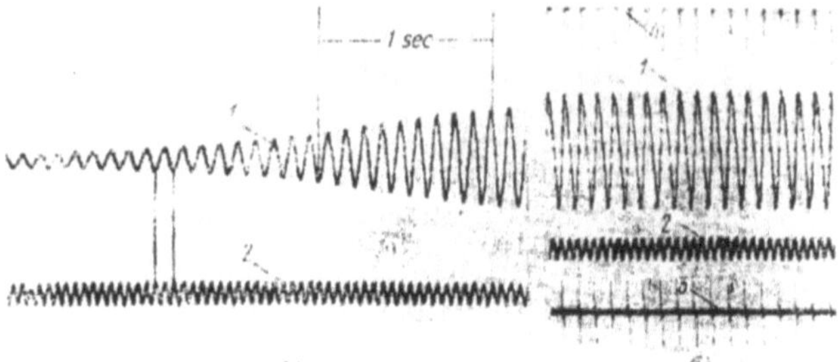

(a) (b)

Fig. 11.10

$y_0 \approx 10^{-3}m$, at a frequency of $50Hz$. Noting that $L \approx 7.1\times10^{-3}m$, we find from (11.4) that $h = 0.56$. Substituting the value of h in (11.44) we find when $k \equiv 0, \xi = h\Omega/4 \approx 0.88/s$. Direct measurement of

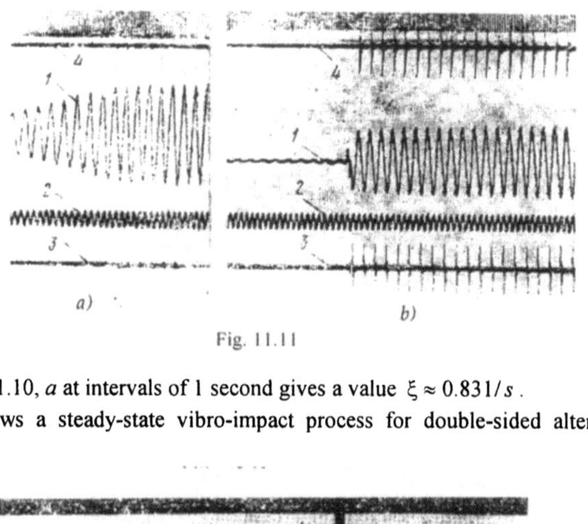

a) b)

Fig. 11.11

the amplitude in Fig. 11.10, *a* at intervals of 1 second gives a value $\xi \approx 0.831/s$.

Fig. 11.10, *b* shows a steady-state vibro-impact process for double-sided alternate collisions

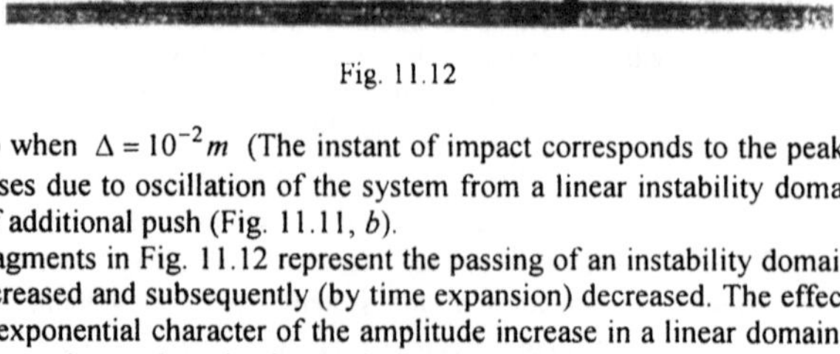

Fig. 11.12

against the stop when $\Delta = 10^{-2}m$ (The instant of impact corresponds to the peaks in curves *3* and *4*.) The process arises due to oscillation of the system from a linear instability domain (Fig. 11.11, *a*), or with the help of additional push (Fig. 11.11, *b*).

The two fragments in Fig. 11.12 represent the passing of an instability domain as the frequency of excitation is increased and subsequently (by time expansion) decreased. The effect of pulling is clearly shown and the exponential character of the amplitude increase in a linear domain. After breakdown of the vibro-impact regime, a beat is observed caused by the closeness of the natural frequency of the system.

Fig. 11.13, *a* shows the process of establishing a vibro-impact regime for a single-sided stop with

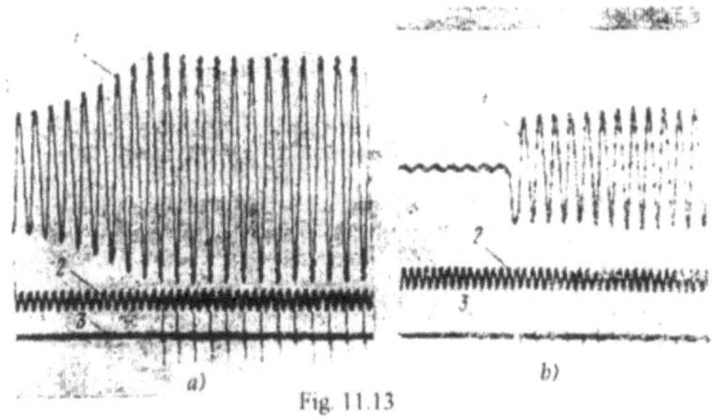

Fig. 11.13

gap $\Delta = 10^{-2} m$. Fig. 11.13, *b* shows the shift to a similar regime assisted by an additional push. Finally, Fig.11.14 shows the effects considered in section 5 of the loss of stability of the system with pre-arranged interference as the result of an initial push. Here curve *3* represents the position of the ball without

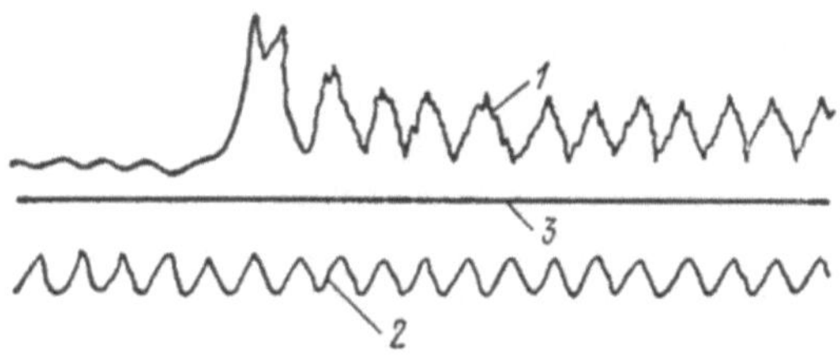

Fig. 11.14

interference. The case described corresponds to point *A* of the amplitude characteristics in Fig. 11.7.

§12. Vibration of self-sustaining systems

1. Let us look at solution methods and the basic principles of self-sustaining vibration of systems with collisions. A necessary condition for steady-state vibration in self-sustaining systems is the existence of a source of energy with which the vibrating system interacts, as the result of which compensating dissipative losses arises in it and vibration at the established level is maintained. Considering vibrating systems with stationary stops, we shall deal with various dynamic models of vibration excitation. The simplest approach is to introduce a special force, nonlinearly related to the velocity $B(\dot{x})$, into the differential equation of the system. In this case, the equation for a vibro-impact system will contain not less than two nonlinear terms, as a result of which it is not only effective to use approximate methods, but frequently the only possible way.

A more general model of a self-sustaining system can be made by using a description for the source of the excitation force which allows us to find the strain due to the loading of the source by the vibrating system. This extension of the dynamic model for the system leads to the introduction of additional equations, which describe the condition of the energy source, and consequently increasing the basic degree of the system of differential equations. A feature of the system of

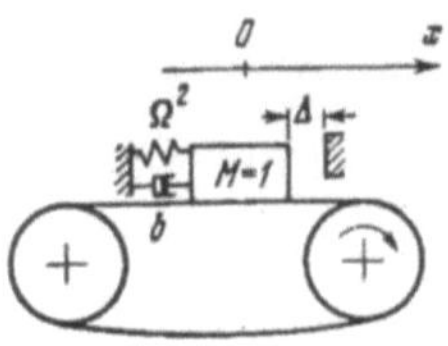
Fig. 12.1

equations described is the existence of nonlinear terms which represent the interaction of vibrating systems with the excitation source, usually, in the form of a product of functions of generalised coordinates which determine the state of the sub-system described.

To study these systems we will examine the effects of resonance types (*autoresonance*) in self-sustaining systems which exist near the natural frequencies of the impacting oscillator, assuming that the latter are significantly different from the natural frequencies of the energy source. As a result the equations for the source serve principally to define the level of excitation of the vibrator, which then brings us conveniently to the study of the process described in the present chapter.

The classical example of a mechanism of vibration excitation in self-sustaining systems is represented by frictional interaction, caused by various types bodies sliding relative to one another (Fig. 12.1). Depending on the character of the contact, the interaction force $B(\dot{x})$ takes various functional forms, however the principal feature is the existence of a range of velocities over which the force linking the rubbing surfaces decreases as the velocity of sliding increases. This occurs, for example, due to the appearance of additional adhesion with relatively "dry" surfaces (static friction) or with low sliding velocity between greased surfaces [134].

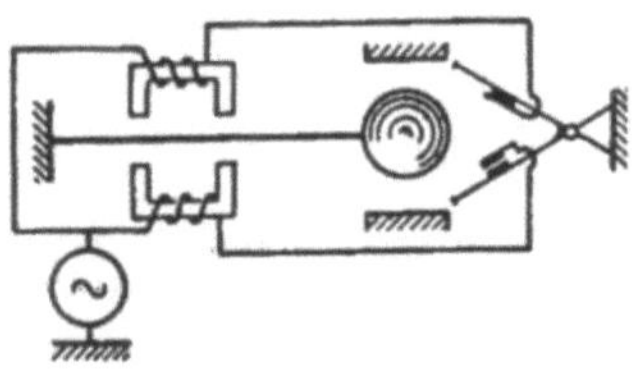
Fig. 12.2

For vibro-impact systems with one degree of freedom the operator equation for self-sustained vibration in similar cases can be written as:

$$(s^2 + bs + \Omega^2)x + \Phi_E(x, sx) + \Phi_D(x, sx) = B(sx), \tag{12.1}$$

where $(\partial B / \partial \dot{x})_{\dot{x}=0} > 0$. For sliding of greased surfaces this is usually taken as

$$B(\dot{x}) = h\dot{x} - b_0 sign\dot{x}, \tag{12.2}$$

where h and b_0 are given positive values.

Equation (12.1) can be used to describe other mechanisms for self-sustaining vibration with impacts. For example, in the construction of machines with electro-, pneumatic, and hydro-impact effects to obtain greater effectiveness of the excitation one tries to produce

$$B(\dot{x}) = U\text{sign}\dot{x}, \tag{12.3}$$

where U is the maximum gain, developed by the energy source. Excitation (12.3) guarantees the maximum contribution of the energy source to the vibrating system (see Chapter 5). Similar excitation can be used when modelling electromagnetic contact breakers (Fig. 12.2), for the construction of vibrating relays, bells, etc. Finally, for several types of gas-dynamic processes, self-sustaining vibration can be produced by a force like

$$B(\dot{x}) = H(\dot{x} - \sigma\dot{x}^3), \tag{12.4}$$

where H, σ are positive quantities.

Fig. 12.3 shows a schematic diagram for the transformation of signals corresponding to equation (12.1). Existing in the linear part of the system $L(s) = (s^2 + bs = \Omega^2)^{-1}$ a filtering effect gives the basis for carrying out harmonic linearisation of all nonlinear elements of the system. It is preferable to use the identical linearisation method for all nonlinearities, however it can be expedient to use a different method, when the impact pair characteristics are linearised by force, but the vibration exciter characteristics $B(\dot{x})$ by displacement. In this case it is necessary to add to the resulting expression conditions connecting the amplitude of the contact force and displacement $a_F = ka_x$.

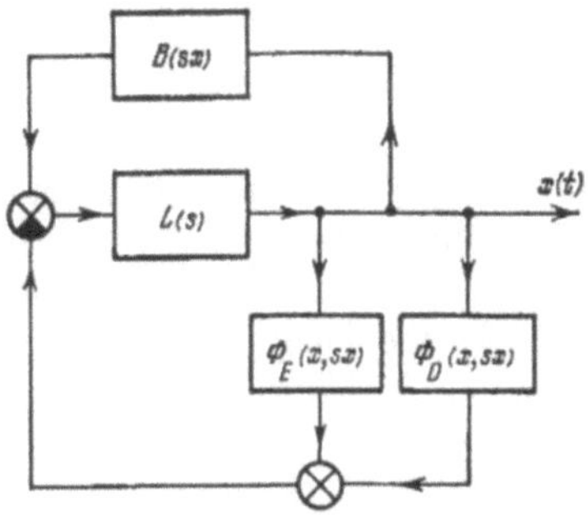

Fig. 12.3

As in previous studies, the solution $x(t)$ of equation (12.1) is searched for approximately in the form (9.2), and replacing the force characteristics for the impact pair $\Phi_E(x,sx)$ and $\Phi_D(x,sx)$ with the linearisation expression (9.5). Also applying the harmonic linearisation function $B(sx)$ using the form

$$B(sx) \approx \lambda sx, \tag{12.5}$$

where λ is the harmonic linearisation coefficient which depends on a_x and ω in the general case, we obtain from (12.1)

$$[s^2 + (b + \beta - \lambda)s + \Omega^2 + k]x^0 + m_F + \Omega^2 m_x = 0 \qquad (x^0 = x - m_x). \tag{12.6}$$

Separating the constant and centred components in (12.6), we obtain the equations for self-sustaining vibration in the standardised form studied earlier (see, for example (9.7), (9.8) or (11.15), (11.16))

$$m_F + \Omega^2 m_x = 0, \tag{12.7}$$

$$[s^2 + (b + \beta - \lambda)s + \Omega^2 + k]x^0 = 0. \tag{12.8}$$

The relationship (12.7) defines the connection $m_x(a_x)$, which after substitution into equation (12.8) makes the latter independent.

Existence conditions for steady-state vibration at frequency ω appear as a pair of imaginary roots to the characteristic equation of the system

$$\mathscr{D}(s) = s^2 + (b + \beta - \lambda)s + \Omega^2 + k = 0. \tag{12.9}$$

When $s = j\omega$ we obtain the characteristic complex system in the form

$$\mathscr{D}(j\omega) = \Omega^2 + k - \omega^2 + j(b + \beta - \lambda). \tag{12.10}$$

Setting the real and imaginary parts of (12.10) to zero we find

$$\omega = \sqrt{\Omega^2 + k}, \tag{12.11}$$

$$\lambda = b + \beta. \tag{12.12}$$

Thus, according to (12.11) the connection between frequency and the amplitude of steady-state vibration is found from the equations for the backbone curves of the system, i.e. steady-state vibration in the system considered can be treated as free vibration supported by excitation forces. Condition (12.12) represents the equation for the energy boundary of vibration and defines the level of free vibration at which an energy balance is achieved between the excitation force and damping. Examining (12.11), (12.12) as implicit expressions $a_x(\omega)$, it is convenient to obtain a graphical solution for this system, by locating the intersections between these curves on the (ω, a_x) plane.

To study stability of self-sustaining vibration found by harmonic linearisation, we can apply different criteria, based on the behaviour of the system considered near the steady-state. In particular, it is convenient to make use of condition (10.2). In the case considered, according to (12.10), the imaginary part of the characteristic complex takes the form

$$\Theta(a_x, \omega) = b + \beta - \lambda . \tag{12.13}$$

Applying condition (10.2) to (12.13), we obtain

$$\left(\frac{d\beta}{d\tilde{a}_x} - \frac{d\lambda}{d\tilde{a}_x} \right)_{\tilde{a}_x = a_x} > 0 . \tag{12.14}$$

Condition (12.14) has a simple physical meaning, implying that as the amplitude of stable vibration increases then the dissipative forces should perform more work than the excitation forces but the reverse is true with decreasing amplitude. This is shown in many works devoted to the study of self-sustaining vibration.

Together with a study of the stability of self-sustaining vibration, it is also necessary to examine stability of equilibrium when studying autonomous systems, since solutions of the type $x = m_x = \text{const}$ also satisfy equation (12.1). This problem can be dealt with by considering the linear equations with constant coefficients, obtained from (12.1) by replacing all nonlinear functions by their Taylor's series expansions to linear accuracy. For these, stability conditions are generated using the Routh-Hurwitz criterium or its equivalent.

2. As an example, let us look at vibro-impact regimes of systems as described by expressions (12.1), (12.4). The harmonic linearisation coefficient λ for function (12.4) takes the following form

$$\lambda = H(1 - 3/4\, \sigma a_x^2 \omega^2) . \tag{12.15}$$

Substituting (12.15) in (12.12) and assuming $\beta = \rho\omega$ (see (7.61), (7.62)), we obtain the following equation for the energy boundaries of vibration

$$a_x = \frac{2}{\omega} \sqrt{\frac{1}{3\sigma} \left(1 - \frac{b + \rho\omega}{H} \right)} . \tag{12.16}$$

From this we find the existence condition for steady-state vibration in the form

$$H > b + \rho\omega . \tag{12.17}$$

With the stops arranged to have a gap, satisfying condition (12.17) leads to unstable equilibrium for the system and soft excitation of vibration. With interference, self-sustaining vibration can only be established through an initial push, which allows it to overcome the initial force of the spring $\Omega^2\Delta$..

Superposing expressions (12.16) (curve *1* in Fig. 12.4) on the graph of the backbone curve *2* for the system found from equation (12.11), we find the values of amplitude and frequency for steady-state vibration. Fig. 12.4, *a* to *c* correspond to different variations of single-sided stops, and Fig. 12.4, *d* to symmetrical double-sided stops. From Fig. 12.4, *a* it can be seen that with a sufficiently large gap, vibration can occur without touching the stops.The vibration frequency in this case is $\omega = \Omega$. Substituting this frequency into (12.16), we find, for example, that the following condition for excitation of vibration with impacts occurs for the case of symmetrically arranged stops ($a_x = \Delta$) (Fig. 12.4, *d*)

$$H > \frac{b + \rho\Omega}{1 - \frac{3}{4}\sigma\Omega^2\Delta^2}. \tag{12.18}$$

According to Fig. 12.4, *b*, when $\Delta = 0$ the vibro-impact regime is isochronous.

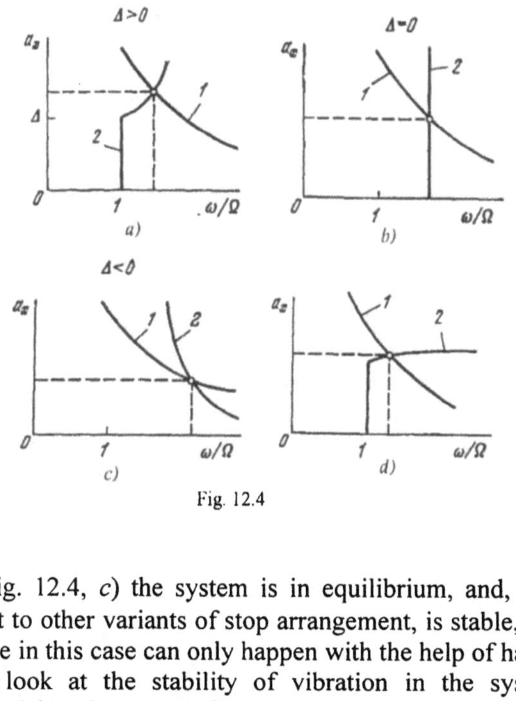

Fig. 12.4

When $\Delta < 0$ (Fig. 12.4, *c*) the system is in equilibrium, and, as can be easily found, in contrast to other variants of stop arrangement, is stable, so that shift into a vibration regime in this case can only happen with the help of hard excitation.

Let us now look at the stability of vibration in the system considered. Substituting (12.15) into (12.14) we find

$$\frac{3}{2}\sigma H\omega^2 a_x > 0. \tag{12.19}$$

Comparing expressions (12.19) and (12.17) shows that when $\sigma > 0$ the regime is stable for all existence domains.

Let us now look at the case when self-sustaining vibration is excited in the system described by equation (12.1) with the function $B(\dot{x})$ taking the form (12.3). The harmonic linearisation coefficient for this function is

$$\lambda = \frac{4U}{\pi a_x \omega}. \tag{12.20}$$

Substituting (12.20) into (12.12) we find the energy boundary for vibration

$$a_x = \frac{4U}{\pi(\rho\omega^2 + b\omega)}. \tag{12.21}$$

Superposing expression (12.21) on the backbone curves we find the amplitude and frequency for self-sustaining vibration. For symmetrical double-sided stops, assuming $a_x = \Delta$ in (12.21), we obtain the following expression for the vibration frequency:

$$\omega = \frac{b}{2\rho}\left(\sqrt{1+\frac{16\rho U}{\pi b^2 \Delta}} - 1\right), \tag{12.22}$$

where the formula is only valid in the natural frequency range of the oscillator $(\omega \geq \Omega)$. Using this we find from (12.22) the existence condition for a vibro-impact regime

$$U \geq \tfrac{1}{4}\pi\Delta\Omega(b + \rho\omega). \tag{12.23}$$

Substituting (12.20) in (12.14) we find the conditions for stable self-sustaining vibration $U > 0$. Thus vibration is stable for all existence domains in the given system.

3. Let us turn to the study of self-sustaining vibration of vibro-impact systems with a non-ideal energy source. The work of many authors has been devoted to the study of interaction of elastic systems with the energy source causing vibration [61, 62, 101, 133, 216, 227 and others]. The traditional method for analytic study of similar problems is the asymptotic methods of nonlinear mechanics. We shall extend the harmonic linearisation method to the class of problem shown, turning our attention specifically to the behaviour of vibro-impacts systems with "limited" excitation [25].

We shall start by looking at a system (Fig. 12.5), which consists of an electric or hydraulic motor producing a variable torque L, driving a shaft with moment of inertia I, which excites vibration of an impacting oscillator having mass M, through a crank of radius r attached to the shaft, and a connecting-rod mechanism which is an elastically sprung element of stiffness c. The damping in the guide and other losses, will be represented by a viscous force proportional to the absolute velocity of the striker by a factor n. The zero position of the striker corresponds when the spring is undeformed and the crank arm horizontal; from this position, the displacement of the striker is limited by the gap (interference) Δ. An example of a similar system are shown in Fig. 4.2.

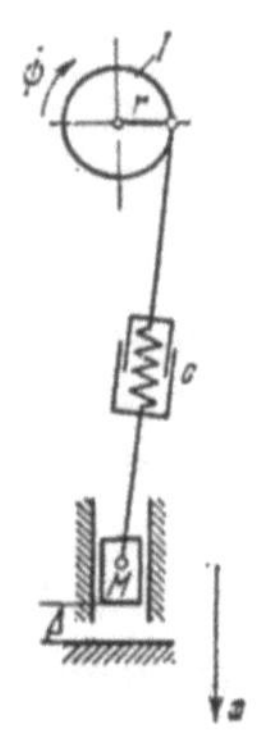

Fig. 12.5

Measuring displacement of the vibrating mass as $x(t)$ and rotation angle of the shaft as $\psi(t)$ and assuming that the length of the crank is essentially less than that of the connecting rod, we can write the differential equation in terms of these basic coordinates for a unit oscillator mass and shaft moment of inertia, in the form

$$\ddot{x} + b\dot{x} + \Omega^2 x + \Phi_E(x,\dot{x}) + \Phi_D(x,\dot{x}) = r\Omega^2 \sin\psi, \tag{12.24}$$

$$\ddot{\psi} = L + r\Omega^2(x - r\sin\psi)\cos\psi, \tag{12.25}$$

where $b = n/M$; $\Omega^2 = c/M$. The second term on the right hand side of (12.25) represents the load which the vibrating system transmits to the motor shaft.

To find the torque L we shall use the equation for the *dynamic* characteristics of the motor [215]

$$\dot{L} + \tau L = \mu\tau(\omega^0 - \psi). \tag{12.26}$$

Here $\tau = T_M^{-1}$: T_M is the time constant of the motor (electromagnetic for an electric motor and hydraulic for a hydraulic motor); μ is a coefficient of proportionality $(\mu > 0)$; ω^0 is the idling velocity.

Assuming that r and L are proportional to the small parameter ε: we establish from (12.25) that

$$\psi = \omega t + \varepsilon v(t) \tag{12.27}$$

where ω = const is the unknown average angular velocity of the crank, while $v(t)$ is an oscillatory function of time. When $\dot{L}$ is small we may assume a small variation in the angular velocity of the motor by comparison to its average value.

Substituting (12.27) in equations (12.24), (12.25) and expanding their right hand sides as series in terms of the small parameter ε, and limiting the expansion to linear terms, we obtain:

$$\ddot{x} + b\dot{x} + \Omega^2 x + \Phi_E(x, \dot{x}) + \Phi_D(x, \dot{x}) = \varepsilon r\Omega^2 \sin\omega t, \tag{12.28}$$

$$\ddot{\psi} = \varepsilon(L + r\Omega^2 x \cos\omega t). \tag{12.29}$$

Hence, (12.28) can be considered independently. It represents the vibration of an impacting oscillator under the influence of harmonic excitation as represented by the right hand side. Thus, we can use the periodic solution for the problem indicated, found by using either the approximation described above or by using an exact method (see §2, §9). An important feature of such a solution in this case is that it depends on the unknown parameter ω, which must be found subsequently from (12.26), (12.29) using this solution.

To obtain the final expression, we will take advantage of the simplified approximate solution of (9.1), similar to (12.24), found in §9 section 1 using harmonic linearisation, bearing in mind that here $a_p = \varepsilon r\Omega^2$, $m_p \equiv 0$, while phase φ now represents the time lag of the approximate solution

$$x(t) \approx m_x + a_x \sin(\omega t - \varphi) \tag{12.30}$$

relative to the excitation. As a result, similar to (9.11) to (9.13) we find

$$a_x = \varepsilon r\Omega^2[(\Omega^2 + k - \omega^2) + (b + \beta)^2\omega^2]^{-1/2}, \tag{12.31}$$

$$\sin\varphi = \frac{(b + \beta)\omega a_x}{\varepsilon r\Omega^2}, \tag{12.32}$$

$$\cos\varphi = \frac{(\Omega^2 + k - \omega^2)\omega a_x}{\varepsilon r\Omega^2}, \tag{12.33}$$

while for the linearisation coefficients m_F, k and β we will use their forms (9.17), (9.18), (7.61). In this case the connection between m_x and a_x is defined by (7.63), but expressions of type (9.7) for the constant components

$$m_F + \Omega^2 m_x = 0 \tag{12.34}$$

coincide with (9.17) in view of (7.63).

Resonant vibration of the oscillator, which will be of interest to us later, occurs for small values of the detuning $\Omega^2 + k - \omega^2$ and small coefficients of dissipation $b + \beta$, since no matter how small the excitation, amplitude a_x at resonance (see (12.31)) is not a small value. The value of m_x follows the same order.

Let us find the corresponding periodic solution for the system of equations (12.26), (12.29). Using (12.30) the equations described can be considered as the equation of a linear non-autononmous system subjected to the effects of a constant moment and biharmonic excitation at frequencies ω and 2ω. Consequently, substituting (12.30) into (12.29) and transforming the resulting trigonometric functions on the right hand side, we find

$$\ddot{\psi} - \varepsilon L = \varepsilon r \Omega^2 \left[-\frac{a_x}{2} \sin\varphi + m_x \cos\omega t + \frac{a_x}{2} \sin(2\omega t - \varphi) \right]. \tag{12.35}$$

Searching for a solution to the system of equations (12.26), (12.35) in the form

$$\begin{aligned} \psi &= \omega + a_\omega^{\mathrm{I}} \sin(\omega t - \vartheta_1) + a_\omega^{\mathrm{II}} \sin(2\omega t - \vartheta_2), \\ L &= m_L - a_L^{\mathrm{I}} \sin(\omega t - \Theta_1) - a_L^{\mathrm{II}} \sin(2\omega t - \Theta_2), \end{aligned} \tag{12.36}$$

we find using harmonic balance, neglecting small values of $a_\omega^{\mathrm{I}} / \omega$ *and* $a_\omega^{\mathrm{II}} / \omega$ as explained above in the description of (12.27) since after multiplying by ε they become second order small quantities:

$$m_L - \frac{1}{2} r \Omega^2 a_x \sin\varphi = 0 \qquad (m_L = \mu(\omega^0 - \omega)), \tag{12.37}$$

$$a_\omega^{\mathrm{I}} = -\frac{\varepsilon r \Omega^2}{\omega} m_x, \; a_\omega^{\mathrm{II}} = \frac{\varepsilon r \Omega^2}{4\omega} a_x, \tag{12.38}$$

$$\vartheta_1 = \pi, \tan\vartheta_2 = -\frac{\Omega^2 + k - \omega^2}{(b+\beta)\omega}, \tag{12.39}$$

$$a_L^{\mathrm{I}} = -\frac{\varepsilon\tau\mu r \Omega^2}{\omega\sqrt{\tau^2 + \omega^2}} m_x, \; a_L^{\mathrm{II}} \frac{\varepsilon\tau\mu r \Omega^2}{4\omega\sqrt{\tau^2 + 4\omega^2}} a_x, \tag{12.40}$$

$$\tan\Theta_1 = \frac{\omega}{r}, \; \tan\Theta_2 = \frac{\tau\sin\vartheta_2 + 2\omega\cos\vartheta_2}{\tau\cos\vartheta_2 - 2\omega\sin\vartheta_2}. \tag{12.41}$$

Note that harmonic vibration of the shaft with frequency ω occurs only as a result of the non-symmetry of the law of motion of the oscillator ($m_x \neq 0$), arising due to single-sided collisions, and proportional to this non-symmetry (drift), while at the same time harmonic double frequency is formed in accordance with the law for change of amplitude of vibration of the oscillator a_x.

Expression (12.37) establishes the relationship between the average torques and load on the motor shaft, caused by the vibrating system. This defines the energy boundary of vibration. Substituting the value of $\sin\varphi$ from (12.32) into (12.37), we obtain eventually

$$\mu(\omega^0 - \omega) - \tfrac{1}{2}(b+\beta) w a_x^2 = 0. \tag{12.42}$$

The components of the torque proportional to b and β, determine the load produced by energy dissipation in the vibrating system and by the work process causing collisions. The power control of the motor leads to the shift of idling value of the angular velocity ω^0.

The system of finite equations (12.31), (12.42) is similar to the corresponding ones (12.11), (12.12) considered in section 1, and enables us to find the amplitude a_x and frequency ω for self-sustaining vibration of the impact oscillator, from which all the remaining parameters for motion of the system can the found. Similar equations were obtained in [132] using the asymptotic method.

The character of possible values for ω in relation to the arrangement of the stops and motor power, proportional to the value of ω^0, is conveniently found graphically from the solution of equation (12.42) (Fig. 12.6) by substituting (12.31) into it and using the equations for the harmonic linearisation coefficients (9.18), (7.61). In Fig. 12.6

$$m_s = \tfrac{1}{2}(b+\beta)\omega a_x^2 \qquad (12.43)$$

is the average moment of the force in the motor, caused by vibration of the oscillator. The intersection of the lines for moments $m_L(\omega)$ and $m_s(\omega)$ set the required value of ω. The remaining unknown values can then be found from formulas (12.31) to (12.33), (12.38) to (12.41). Note that for the given system

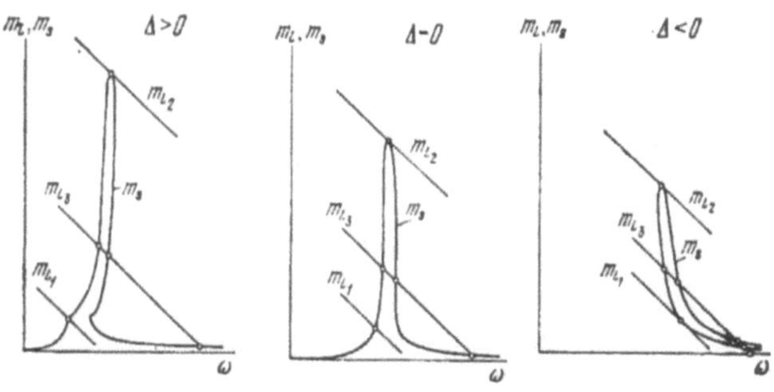

Fig. 12.6

when $\Delta < 0$, a vibro-impact regime can occur only when an additional hard excitation is supplied.

4. Let us look at the stability of the periodic solutions obtained in terms of small disturbances. Representing the disturbance of motion of the oscillator in the form

$$x(t) = \tilde{m}_x(t) + \tilde{a}_x(t)\sin[\tilde{\omega}(t)t - \tilde{\varphi}(t)], \qquad (12.44)$$

we shall as before use the stability condition (10.2), which, when we apply this to the imaginary parts of the characteristic complex of equation (12.28), gives, similar to (10.5)

$$\frac{d}{d\tilde{a}_x}\left(b+\beta-\frac{\varepsilon r\Omega^2\sin\tilde{\varphi}}{\tilde{\omega}\tilde{a}_x}\right)_{\tilde{a}_x=a_x} > 0. \qquad (12.45)$$

However, an important feature of the case considered is the fact that now frequency must also be considered as a slowly changing function $\tilde{\omega}(t)$. Hence, assuming $\tilde{a}_x(t)$ is an independent function it is necessary to apply *three* arbitrary conditions to its connections with the remaining slowly changing functions in (12.44). Retaining the conditions (12.34), (12.33) for the connections between $\tilde{m}_x(t)$ and $\tilde{a}_x(t)$, $\tilde{\varphi}(t)$ and $\tilde{a}_x(t)$ we add expression (12.42) as the connection between $\tilde{\omega}(t)$ and $\tilde{a}_x(t)$.

Differentiating (12.45) using the expressions above, we find

$$\left[\frac{d\beta}{d\tilde{a}_x}+\frac{\varepsilon r\Omega^2}{\tilde{\omega}^2\tilde{a}_x^2}\left(\tilde{\omega}\sin\tilde{\varphi}+\frac{d\tilde{\omega}}{d\tilde{a}_x}\tilde{a}_x\sin\tilde{\varphi}-\frac{d\tilde{\varphi}}{d\tilde{a}_x}\tilde{\omega}\tilde{a}_x\cos\tilde{\varphi}\right)\right]_{\tilde{a}_x=a_x}>0. \qquad (12.46)$$

Differentiating (12.33) as an implicit function $\tilde{\varphi}$ of $\tilde{a}_x$, we find

$$\left(\frac{d\tilde{\varphi}}{d\tilde{a}_x}\right)_{\tilde{a}_x=a_x}=-\frac{1}{\varepsilon r\Omega^2\sin\varphi}\left(\Omega^2+k-\tilde{\omega}^2+\frac{dk}{d\tilde{a}_x}\tilde{a}_x-2\frac{d\tilde{\omega}}{d\tilde{a}_x}\tilde{\omega}\tilde{a}_x\right)_{\tilde{a}_x=a_x}. \qquad (12.47)$$

As the result of differentiating (12.42) we find

$$\left(\frac{d\tilde{\omega}}{d\tilde{a}_x}\right)_{\tilde{a}_x=a_x}=-\left\{\frac{\mu\omega^0\tilde{a}_x\left[\frac{d\beta}{d\tilde{a}_x}\tilde{a}_x+2(b+\beta)\right]}{2[\mu+\frac{1}{2}(b+\beta)\tilde{a}_x^2]^2}\right\}_{\tilde{a}_x=a_x} \qquad (12.48)$$

To simplify the final relationship, we assume $\beta\equiv 0$, and we find from (12.46) using (12.31), (12.32), (12.47), (12.48)

$$\begin{aligned}&b^2\omega^2-\frac{\mu\omega^0 b\omega a_x^2}{(\mu+\frac{1}{2}ba_x^2)}[b^2-2(\Omega^2+k-\omega^2)]+\\&+(\Omega^2+k-\omega^2)\left[\Omega^2+k-\omega^2+a_x(dk/d\tilde{a}_x)_{\tilde{a}_x=a_x}\right]>0.\end{aligned} \qquad (12.49)$$

Comparing (12.49) with the stability condition for forced vibration (10.9) we can see that when energy dissipation in the system is neglected, they coincide; consequently in the system considered, instability will occur only in the lower branch of the double-valued domain of the amplitude-frequency characteristics drawn using (12.31). In the general case the stability picture changes. Substituting the value of k from (9.18) in (12.49), when $m_p\equiv 0$, we obtain eventually

$$\begin{aligned}&b^2\omega^2-\frac{\mu\omega^0 b\omega a_x^2}{(\mu+\frac{1}{2}ba_x^2)}[b^2-2(3\Omega^2+\frac{2\Omega^2\Delta}{a_x}-\omega^2)]+\\&+(3\Omega^2+\frac{2\Omega^2\Delta}{a_x}-\omega^2)(3\Omega^2-\omega^2)>0.\end{aligned} \qquad (12.50)$$

The unstable solution found using condition (12.50) is drawn in Fig. 12.7 as a dotted line. In addition, notice that when the stops are arranged with initial interference, satisfying the stability conditions depends strongly on parameter μ, which is defined by the "rigidity" of the static characteristic of the motor. As the value of μ changes from an infinitely large value to zero four distinctive domains for the value of μ are developed similar to [132]. These domains are represented in the form of sectors in Fig. 12.7, *c*. In sector *1*, corresponding to steep characteristics, the system behaves similarly to the analogous non-autonomous system, excited by a periodic force with fixed frequency ω. The stability picture closely resembles this condition, since the lower branch of $m_s(\omega)$ remains unstable in this case.

Another picture is observed when $m_L(\omega)$ lies in sector *3*. This case is also

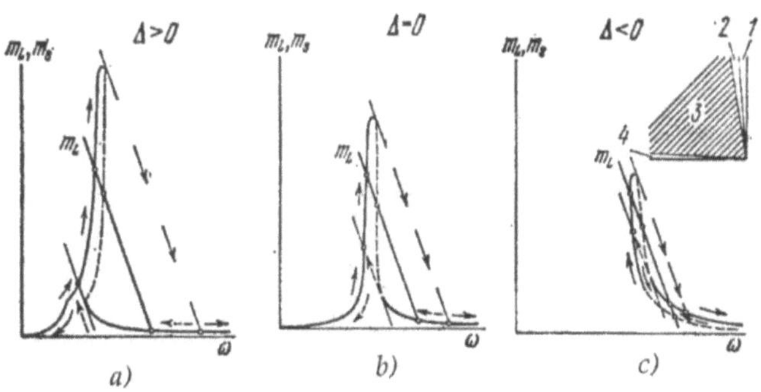

Fig. 12.7

shown in Fig. 12.7, *c*. Here an unstable solution appears on the fragment of the upper branches of expression $m_s(\omega)$.

In sector *2* condition (12.50) is satisfied by all steady-state solutions which produces some ambiguity. The experiment described in [132] for a similar situation, showed the absence of stable periodic motion. Finally, in cases when $m_L(\omega)$ lies in sector *4*, calculation using condition (12.50) shows that the system is near the boundaries of stable domains. According to [132], steady-state regimes were not achieved experimentally in such situations.

Assuming that power control of the motor corresponds to a shift of $m_L(\omega)$ parallel to itself (because of a proportional change in ω^0), we find that, as we increase the power, the applied moment $m_s(\omega)$ and correspondingly, the angular velocity ω will change in the direction shown in Fig. 12.7 by continuous arrows, but when the power is decreased, the angular velocity changes in the direction shown by dotted arrows. At the instant of jumps of the working point of characteristic $m_L(\omega)$ there will be transient processes leading to the corresponding final change in the value of ω. Note that when $\Delta < 0$, increasing the motor power can sometimes lead to a decrease in the average angular velocity ω, because of rapid growth in the vibration amplitude of the oscillator. As a result, for any stop arrangement the velocity of the motor changes negligibly as the amplitude increases during resonance, no matter how the power is controlled, but changes significantly after the oscillator passes maximum amplitude (the *Sommerfeld effect*).

Fig. 12.8, in agreement with the previous construction (compare to Fig. 12.7) shows with full lines the branches of the amplitude-frequency characteristics for the oscillator corresponding to run-up and slowing down of the motor. Fig. 12.9 shows similar branches for the amplitude-frequency characteristics according to (12.38), (12.40) for the fundamental harmonic component of vibration of the angular velocity and moment for an electric motor.

5. Using the methods outlined in §10, section 3, we shall now look at the

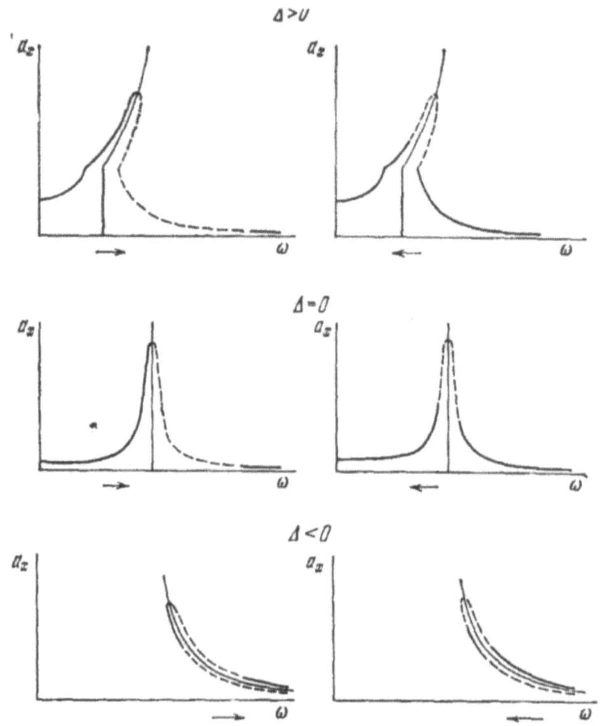

Fig. 12.8

process by which steady-state vibration of the system considered (Fig. 12.5) is established. Rejecting the assumption of slowly changing function $\widetilde{a}_x(t)$ in the transfer process, we shall approximate it with expression (10.17) and will look for an approximate solution of equation (12.28) in the form (12.30), (10.17), assuming $\widetilde{\xi}(t), \widetilde{\varphi}(t)$ and $\widetilde{\omega}(t)$ are slowly changing functions. In this situation the linearisation coefficients (9.17), (9.18) (7.61) retain the same structure as in the steady-state case.

Taking this into account, and substituting (12.30), (10.17) in the harmonically linearised equation (12.28), we obtain similar to (10.28), (10.29)

$$\tilde{\xi}^2 + (b+\beta)\tilde{\xi} + \Omega^2 + k - \tilde{\omega}^2 - \frac{\varepsilon r \Omega^2 \cos\tilde{\varphi}}{\tilde{a}_x} = 0. \quad (12.51)$$

$$\tilde{\xi} = \frac{1}{2}\left(\frac{\varepsilon r \Omega^2 \sin\tilde{\varphi}}{\tilde{\omega}\tilde{a}_x} - b - \beta\right). \quad (12.52)$$

For constant components expression (12.34) is retained. The angular velocity $\tilde{\omega}(t)$ is found from the static characteristic of the motor

$$\tilde{m}_L(t) = \mu[\omega^0 - \tilde{\omega}(t)]. \quad (12.53)$$

This assumes a sufficiently slow rate of transient processes in the shaft of the motor so that the Sommerfeld effect is observed. When $\tilde{\xi} \equiv 0$, we obtain the existence conditions for a steady-state regime ($\tilde{a}_x = a_x = \text{const}$) from (12.51), (12.52), which coincide, of course, with (12.32), (12.33).

The establishment of vibration is defined by the dependence of $\tilde{\xi}$ on $\tilde{a}_x$. To find this, let us eliminate phase $\tilde{\varphi}$ from equation (12.51), (12.52). We find

$$\tilde{\xi}^2 + (b+\beta)\tilde{\xi} + \Omega^2 + k - \tilde{\omega}^2 - \sqrt{(\varepsilon r \Omega^2 / \tilde{a}_{x^2})^2 - (2\tilde{\xi} + b + \beta)^2 \tilde{\omega}^2} = 0. \quad (12.54)$$

Equation (12.54) represents an algebraic equation of the fourth degree in $\tilde{\xi}$. For simplicity we shall assume for further analysis that $b + \beta \equiv 0$, and from equation (12.54) we find eventually

$$\tilde{\xi} = -(\Omega^2 + k - \tilde{\omega}^2) + (\Omega^2 + k - \tilde{\omega}^2)\sqrt{1 + \frac{(\varepsilon r \Omega^2 / \tilde{a}_x)^2 - (\Omega^2 + k - \tilde{\omega}^2)^2}{(\Omega^2 + k + \tilde{\omega}^2)^2}}. \quad (12.55)$$

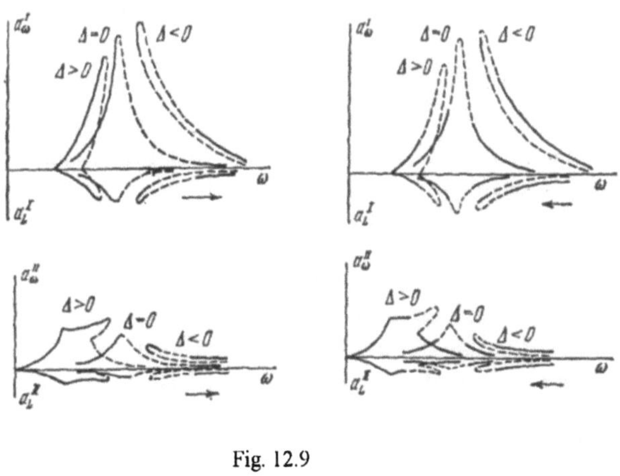

Fig. 12.9

Let us find now the acceleration of the motor shaft during the transfer process. Substituting (12.44) in (12.29) and bearing in mind (12.27) we obtain, averaged over rapidly oscillating terms,

$$\frac{d\widetilde{\omega}}{dt} = \varepsilon\left[\mu(\omega^0 - \widetilde{\omega}) - \frac{1}{2}r\Omega^2\widetilde{a}_x \sin\widetilde{\varphi}\right]. \tag{12.56}$$

From equation (12.56) using (12.52) we find when $b + \beta \equiv 0$

$$\frac{d\widetilde{\omega}}{dt} = \varepsilon\left[\mu(\omega^0 - \widetilde{\omega}) - \frac{\widetilde{\xi}\widetilde{\omega}\widetilde{a}_x^2}{\varepsilon}\right] \tag{12.57}$$

or, substituting (10.17),

$$\frac{d\widetilde{\omega}}{dt} = \varepsilon\left[\mu(\omega^0 - \widetilde{\omega}) - \frac{\widetilde{\omega}\widetilde{a}_x}{\varepsilon}\frac{d\widetilde{a}_x}{dt}\right], \tag{12.58}$$

i.e. the average torque of the motor load caused by vibration of the system is proportional to the product of speed of amplitude variation and maximum vibration velocity in the period $T = 2\pi / \widetilde{\omega}$.

Let us look at the case when the resonance correlation during the transfer process

$$\Omega^2 + k - \widetilde{\omega}^2 = 0. \tag{12.59}$$

is satisfied. Substituting (12.59) into (12.55) and expanding the radical as a series in terms of parameter ε, we obtain, limiting the expansion to linear terms,

$$\widetilde{\xi} = \varepsilon r\Omega^2 / (2\widetilde{\omega}\widetilde{a}_x). \tag{12.60}$$

Substituting (12.60) into (10.17), we find

$$d\widetilde{a}_x / dt = \varepsilon r\Omega^2 / (2\widetilde{\omega})$$

and consequently

$$\widetilde{a}_x = \frac{\varepsilon r\Omega^2}{2\widetilde{\omega}}t = \frac{\varepsilon r\Omega^2}{2\sqrt{\Omega^2 + k}}t. \tag{12.61}$$

Comparing (12.61) with (10.38), we can see that the transfer process in this case evolves similarly to forced vibration. Using (12.60), equation (12.57) takes the form

$$\frac{d\widetilde{\omega}}{dt} = \varepsilon\left[\mu(\omega^0 - \widetilde{\omega}) - \frac{1}{2}r\Omega^2\widetilde{a}_x\right]. \tag{12.62}$$

Contrasting (12.62) with (12.56) we can see that the torque on the motor is a maximum in the transfer process considered. Hence, expressions (12.61), (12.62) correspond to the most stressed transient process.

6. Vibration is significantly influenced by the characteristics of the excitation source in pneumatic impact systems. Let us look at a vibro-impact system with single-sided pneumatic action (Fig. 12.10), assuming that the connection between the working chamber of the cylinder at pressure p_1 and the atmosphere at pressure p_2 occurs through ports *1* or *2* depending on the motion of the piston.

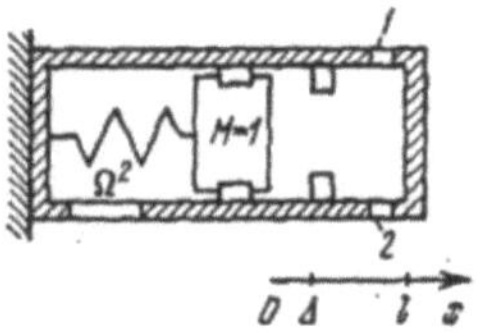

Fig. 12.10

Taking the origin for the motion of the piston, x, from its position when the spring is undeformed, we can write the equation of motion for a unit mass piston in the form

$$\ddot{x} + b_0 \mathrm{sign}\dot{x} + \Omega^2 x + \Phi_E(x,\dot{x}) + \Phi_D(x,\dot{x}) = -(p - p_2)S. \quad (12.63)$$

Here S is the cross-sectional area of the piston; Ω^2 is the elasticity of the spring; b_0 is the amplitude of the dry friction force; p is the pressure inside the cylinder.

An equation describing the change of pressure in the cylinder in relation to the vibration of the piston can be found based on the equation for an ideal gas

$$pV = R_0 \mathrm{WT}, \quad (12.64)$$

where V is the volume of the working chamber of the cylinder; T is the absolute temperature of the gas; W is the weight of gas; R_0 is the specific gas constant.

For simplification later we shall assume that expansion and compression of the gas in the cylinder is isothermal. In [13] it is shown that with this assumption the calculated time for change of pressure in the cylinder is increased by 10 to 20% compared to reality. Differentiating equation (12.64) with respect to time, using the assumption stated, we find

$$\dot{p}V + p\dot{V} = R_0 \mathrm{T}\dot{\mathrm{W}}$$

Noting that $V = (l - x)S$, where l is the coordinate of the right end of the cylinder (Fig. 12.10), we find $\dot{V} = -S\dot{x}$. Denoting the gas flow rate $\dot{W} = G_1 - G_2$ where G_1, G_2 are the flow rates for entry to and exit from the cylinder we find

$$\dot{p}(l - x) = \frac{R_0 \mathrm{T}}{\mathrm{S}}(G_1 - G_2) + p\dot{x}. \quad (12.65)$$

Let us find the existence conditions for autoresonant tuning of pneumatic vibro-impact systems. Such tuning assumes significant vibration of the piston for small variation of the pressure in the cylinder relative to the average value. As result, compression and exhaust occur with large drops in pressure. This allows us to assume in both cases that the gas flow is close to being *supercritical* and to describe the appropriate flow rate by the expressions [106]

$$G_1 = nS_1 p_1, \quad G_2 = nS_2 p, \quad (12.66)$$

where S_1, S_2 are the effective areas of the inlet and outlet ports, $n \approx 0.350/\sqrt{\mathrm{T}}$. Apart from the simplicity of calculating this expression, the assumption suggested decreases the time for pressure change in the cylinder by comparison with the real value and also compensates to some measure for errors arising by accepting the isothermal character of process.

The control of overlap for the corresponding ports will be characterised using logic functions $\eta_1(x,\dot{x})$, $\eta_2(x,\dot{x})$ which are assumed zero for a closed port and one for an open port. To accommodate this and bearing in mind (12.66), we can write equation (12.65) in the following final form:

$$\dot{p}(l - x) = n_1[S_1 p_1 \eta_1(x,\dot{x}) - S_2 p \eta_2(x,\dot{x})] + p\dot{x} \quad (12.67)$$

where $n_1 = R_0 \mathrm{T}n/S$.

Let us look for approximate solutions to the system of equations (12.63), (12.67) in the form

$$x(t) \approx m_x + a_x \cos\omega t, \quad (12.68)$$

$$p(t) \approx m_p - a_p \cos(\omega t + \varphi), \tag{12.69}$$

where ω is the unknown frequency of self-sustaining vibration; φ is the phase lag between vibration of the piston and pressure in the cylinder.

Carrying out harmonic linearisation of the dry friction force in (12.63), we find

$$b_0 \operatorname{sign}\dot{x} \approx \beta_1 \dot{x} \tag{12.70}$$

where the harmonic linearisation coefficient β_1 by analogy with (12.20) will be as follows:

$$\beta_1 = 4b_0 /(\pi a_x \omega) \tag{12.71}$$

Substituting (12.69), (12.70) into equation (12.63), we obtain

$$\begin{aligned} &\ddot{x} + \beta_1 \dot{x} + \Omega^2 x + \Phi_E(x,\dot{x}) + \Phi_D(x,\dot{x}) = \\ &\qquad = -(m_p - p_2)S + a_p S \cos(\omega t + \varphi), \end{aligned} \tag{12.72}$$

i.e. we obtain once again the equivalent equation for forced vibration, similar to (9.1).

Carrying out harmonic linearisation of the impact force characteristic in the form (9.5) we obtain, similar to (9. 11) to (9.13)

$$a_x = a_p S[(\Omega^2 + k - \omega^2)^2 + (\beta_1 + \beta)^2 \omega^2]^{-1/2} \tag{12.73}$$

$$\sin\varphi = \frac{(\beta_1 + \beta)\omega a_x}{a_p S}, \qquad \cos\varphi = \frac{(\Omega^2 + k - \omega^2) a_x}{a_p S} \tag{12.74}$$

where, for the present, we shall use the forms (9.17), (9.18), (7.61) for the harmonic linearisation coefficients m_F, k and β.

By analogy with (9. 7), we find from (12.72) for the constant terms

$$m_F + \Omega^2 m_x = -(m_p - p_2)S\ . \tag{12.75}$$

This expression also follows from (9.17), since now

$$m_F = -(m_p - p_2)S \tag{12.76}$$

and, moreover, condition (7.63) is satisfied. Comparing (12.73) to (12.75) with similar solutions (12.31) to (12. 34) for the problem above, we notice that in this case, not only is the frequency of vibration ω unknown, but also the value of m_p and a_p, which represent the average value and amplitude of the excitation for an impact oscillator. These values can be found using the gas-dynamic equation (12.67). When we substitute the solution (12.68) into it, we obtain a linear non-homogeneous equation of the first order relative in p with periodic coefficients and periodic "external" excitation.

Under conditions of autoresonance, when the pressure is in phase with the piston velocity, the term $p\dot{x}$ of the equation considered prevents abrupt change of pressure (the piston enlarges the working volume during inlet and decreases it during expansion). As the piston vibration intensifies then its smoothing effects become stronger. Similar influences on pressure are shown during operation by the term $l - x$, representing time for the pressure drop to change. As vibration of the piston increases, the average value of these coefficients increases (similar to the time constant of the response), because $m_x < 0$. As a result, according to (12.67) it is possible to ensure low vibration of the pressure in the chamber relative to the average value when the inlet and exhaust ports have small area, no matter how intensive the vibration of the piston. This shows that the assumptions taken when introducing this equation are valid.

The solution is conveniently carried out in the following manner. Performing harmonic linearisation of the logic functions, we find

$$\begin{aligned} n_1 S_1 \eta_1(x,\dot{x}) &\approx m_{\eta_1} + \alpha_1 (x - m_x) + \lambda_1 \dot{x}, \\ n_1 S_2 \eta_2(x,\dot{x}) &\approx m_{\eta_2} + \alpha_2 (x - m_x) + \lambda_2 \dot{x}, \end{aligned} \tag{12.77}$$

where the linearisation coefficients depend on the parameters (12.68). Replacing the nonlinear terms in equations (12.62), (12.67) by their linearised approximations (9.5), (12.76) we obtain, substituting (12.68), (12.69):

$$\begin{aligned} &- a_x\omega^2 \cos\omega t - \beta_1 a_x \omega \sin\omega t + \Omega^2 (m_x + a_x \cos\omega t) + m_F + \\ &\qquad + k a_x \cos\omega t - \beta a_x \omega \sin\omega t = -[m_p - a_p \cos(\omega t + \varphi) - p_2] S, \end{aligned} \tag{12.78}$$

$$\begin{aligned} &-a_p \omega \sin(\omega t + \varphi)(l - m_x - a_x \cos\omega t) = p_1 (m_{\eta_1} + \alpha_1 a_x \cos\omega t - \\ &\quad - \lambda_1 a_x \omega \sin\omega t) - [m_p - a_p \cos(\omega t + \varphi)](m_{\eta_2} + \alpha_2 a_x \cos\omega t - \\ &\qquad\qquad - \lambda_2 a_x \omega \sin\omega t - a_x \omega \sin\omega t. \end{aligned} \tag{12.79}$$

Differentiating (12.79) and eliminating $\dot{p}$ and p from (12.79) using the expressions obtained and (12.78), we receive expressions which depend only on the motion parameters of the piston. Using harmonic balance we can find, correct on the basis of the main harmonic component

$$m_F = \frac{\left[m_{\eta_2}(p_2 S - \Omega^2 m_x) - p_1 S m_{\eta_1} + \tfrac{1}{2}(\beta_1 + \beta)\lambda_2 a_x^2 \omega^2 + \tfrac{1}{2}\alpha_2 a_x^2 (\Omega^2 + k - \omega^2) \right]}{m_{\eta_2}}, \tag{12.80}$$

$$\omega^2 = \frac{m_{\eta_2}(\Omega^2 + k) + \alpha_2 (p_2 S - m_F - \Omega^2 m_x) - \alpha_1 p_1 S}{m_{\eta_1} + (l - m_x)(\beta_1 + \beta)}, \tag{12.81}$$

$$\omega^2 = \frac{\left[(l - m_x)(\Omega^2 + k) + m_{\eta_2}(\beta_1 + \beta) + (\lambda_2 + 1)(p_2 S - m_F - \Omega^2 m_x) - \lambda_1 p_1 S_1 \right]}{l - m_x}. \tag{12.82}$$

The system of finite equations (12.81), (12.82) is similar to the expressions (12.11), (12.12): from (12.81) the connection between amplitude and frequency of "natural" vibration of the piston in the system examined (using gas-dynamic processes) is established, (12.82) defines the energy boundaries of vibration. As a result the joint solution of equations (12.80) to (12.82), using the equations for the harmonic linearisation coefficients, gives the frequency of vibration ω, the amplitude of the piston a_x and the value of m_F. Using these values it is possible to find the process parameters a_p, φ and m_p (12.69), subsequently from (12.73) to (12.75).

To complete the solution of this problem it is necessary to settle the structure of the control arrangement which defines the cyclic reversal of the inlet and exhaust ports during motion of the piston. Here we can reverse the problem so that, given the requirements of the motion parameters, we can look for the linearisation coefficients of the logic functions and from them synthesise the control system. Chapter 5 will be devoted to the problems of structural synthesis of self-sustaining vibro-impact systems. In particular, §20 will consider the synthesis of auto-resonant systems based on harmonic linearisation methods.

The strategy of port switching has some arbitrariness but does not always lead to the establishment of autoresonance. Based on the simplified scheme described in section 2, in which control was described by equation (12.3), it is possible, for example, by analogy to choose the switching sequence for the ports, defining them by the following logic functions:

$$\eta_1(x,\dot{x}) = \eta(-\dot{x}), \quad \eta_2(x,\dot{x}) = \eta(\dot{x}), \tag{12.83}$$

where $\eta(\dot{x})$ is a unit step function, i.e. during motion of the piston away from the stops, only the inlet port is open, and only the outlet port during motion towards the stop. For greater effectiveness of excitation, it follows of course, that we must adjust the advance of switching to compensate for the inertia of the gas-dynamic process, however for simplification we will not dwell on this point.

The harmonic linearisation coefficients for expression (12.77) take the following form, using (12.83):

$$m_{\eta_1} = \frac{n_1 S_1}{2}, \; m_{\eta_2} = \frac{n_1 S_2}{2}, \; \lambda_1 = \frac{2n_1 S_1}{\pi a_x \omega}, \; \lambda_2 = -\frac{2n_1 S_2}{\pi a_x \omega}, \tag{12.84}$$

$$\alpha_1 \equiv 0, \; \alpha_2 \equiv 0.$$

7. Substituting $\alpha_1 \equiv 0, \alpha_2 \equiv 0$ in (12.81), we obtain

$$\omega^2 = \frac{\Omega^2 + k}{1 + (\beta_1 + \beta)(l - m_x)/m_{\eta_2}}. \tag{12.85}$$

Noting that autoresonance regimes arise with relatively small values of the dissipation coefficients, and, in the denominator of (12.85), neglecting values which are small compared to unity, we obtain again expression (12.11). Hence, even in the system studied, autoresonance occurs near to the natural frequency of the impacting oscillator. In practice, when conditions (12.11) are satisfied, then according to (12.74) the excitation phase remains equal to $\pi/2$.

From (12.80), when $\alpha_2 \equiv 0$,we obtain

$$m_F = \left(p_2 - p_1 \frac{m_{\eta_1}}{m_{\eta_2}}\right) S - \Omega^2 m_x + \frac{\beta_1 + \beta}{2m_{\eta_2}} \lambda^2 a_x^2 \omega^2. \tag{12.86}$$

Substituting (12.86) into (12.75) using the value of the linearisation coefficients (12.84), we find

$$m_p = p_1 \frac{S_1}{S_2} + \frac{2(\beta_1 + \beta)}{\pi S} a_x \omega. \tag{12.87}$$

Expression (12.87) is similar in appearance to that considered above when solving the previous equation (12.42) for the balance of average torques of the shaft of an electric motor.

Let us rewrite (12.87) in the form

$$(m_p - p_1 S_1 / S_2) S = 2(\beta_1 + \beta) a_x \omega / \pi. \tag{12.88}$$

The right hand side of this expression defines the value of the average force in the pneumatic system during vibration of the impact oscillator. As before, such a force is proportional to the loss of energy in the vibrating system due to damping and the performance of the working process, but also to the intensity of the vibration. The left-hand side of the expression represents the average force generated by the pneumatic system to compensate the force indicated.

In cases when the vibration load is negligibly small, we find from (12.87)

$$m_p = p_1 S_1 / S_2 . \tag{12.89}$$

The excitation at which condition (12.89) is satisfied will be called *idling*. Since $m_p < p_1$ the port areas must satisfy the condition

$$S_2 > S_1 . \tag{12.90}$$

As the load increases, m_p also increases, and as a result the average force developed by a pneumatic motor increases, and according to (12.88) tends in the limit to the value

$$p_1 S(1 - S_1 / S_2) , \tag{12.91}$$

which by analogy with the electric motor, will be called *stalling*. It must be understood that this value is only hypothetical, since by letting m_p approach to p_1, the initial assumption of flow character is infringed, since the approach of m_p to p_1 indicates insufficient power in the pneumatic actuator to develop autoresonance. According to (12.91), a radical method to increase the intensity of actuation is to increase the inlet pressure p_1. Limited opportunity can also be achieved by decreasing the ratio of area of the ports S_1 / S_2. According to (12.88) for large power reserves in the pneumatic system:

$$p_1 S_1 / S_2 >> 2(\beta_1 + \beta) a_x \omega / \pi ,$$

the vibration load can be neglected, using (12.89) to calculate the average pressure. In this case one talks of *ideal* sources of excitation.

For an actual excitation source with limited power, substituting the resonance value found from (12.73) when satisfying condition (12.11), for a_x in (12.87), the connection between m_p and a_p for autoresonance conditions is established

$$m_p = p_1 S_1 / S_2 + 2a_p / \pi . \tag{12.92}$$

In addition to the assumption already made, the flow must be supercritical

$$m_p - a_p > p_2, \quad m_p + a_p < p_1 , \tag{12.93}$$

and we obtain from (12.92) the following limits:

$$a_p < 2.74(p_1 S_1 / S_2 - p_2), \quad a_p < 0.61 p_1 (1 - S_1 / S_2) \tag{12.94}$$

Given the required value of a_p, it is possible to select suitable values of the pressures p_1, p_2 and the port areas S_1, S_2 from these inequalities.

Let us examine relationship (12.82), which defines the energy boundaries of vibration. Assuming, using (12.85), $\omega^2 \approx \Omega^2 + k$, we find

$$m_{\eta_2}(\beta_1 + \beta) + (\lambda_2 + 1)(p_2 S - m_F - \Omega^2 m_x) - \lambda_1 p_1 S = 0 .$$

Substituting here the value of m_F from (12.86) for the case of an ideal exciter

(neglecting small terms proportional to the dissipation coefficient)

$$m_{\eta_2}(\beta_1+\beta)+(\lambda_2+1)Sp_1S_1/S_2-\lambda_1 p_1 S=0 \qquad (12.95)$$

Expressing the harmonic linearisation coefficients appearing in (12.95) using (12.71), (7.61), (12.84), we obtain after transformation

$$a_x=\frac{2(2p_1S-b_0S_2/S_1)}{\left(\pi\frac{\rho}{2}\frac{S_2}{S_1}\omega^2+\frac{p_1}{n_1}\frac{S}{S_2}\omega\right)} \quad (\rho=0.16(1-R^2)). \qquad (12.96)$$

Comparing (12.96) with the similar relationship (12.21) obtained for a suddenly increasing excitation (12.3) in the direction of the velocity of the impacting oscillator, we can establish that gasdynamic processes are basically similar to additional viscous damping, the parameters of which are reflected in the second term of the denominator.

According to (12.96), increasing the dry friction force b_0 leads to a decrease in the amplitude of vibration of the piston, from which there must be a critical value of friction, when vibration ceases $(a_x=0)$. It is given by

$$b_0=2p_1SS_1/S_2. \qquad (12.97)$$

Decreasing the coefficient of restitution R also leads to a decrease in a_x, however, for all values of R, regimes do exist.

Superposing the curves for the vibration energy boundaries onto the backbone curves for the system, we obtain at the intersection points amplitude a_x and frequency ω similar to that found in Fig. 12.4. Let us find the equation for the backbone curves. Substituting the values for the linearisation coefficients (9.18) using (12.76), (12.89) into the approximate relationship (12.11), we find

$$a_x=\frac{2[\Omega^2\Delta+(p_1S_1/S_2-p_2)S]}{3\Omega^2-\omega^2}, \qquad (12.98)$$

where as before, gap (interference) Δ is measured from the piston position when the spring is undeformed. Setting the numerator in (12.98) to zero, we find

$$\Delta=(p_2-p_1S_1/S_2)S\Omega^{-2} \qquad (12.99)$$

To satisfy this condition, the vibration frequency ω is independent of amplitude a_x and must equal $\sqrt{3}\Omega$. This corresponds to the set of parameters for which the piston, under the effects of average force due to air pressure and elasticity of the spring, just touches the stops with neither gap or interference. As Δ is increased from this value the nonlinearity of the system remains of a soft type; as Δ is decreased, the system becomes harder.

Finding the values of a_x and ω from the system of finite equations (12.96), (12.98), we can find from these the value a_p, assuming $\Omega^2+k-\omega^2=0$ in (12.73). Using (12.71), (7.61) we obtain

$$a_p=\frac{a_x\omega}{S}\left[0.16(1-R^2)\omega+\frac{4b_0}{\pi a_x\omega}\right]. \qquad (12.100)$$

To verify the accuracy of the assumption regarding the ideal excitation source, used to construct the energy boundaries for vibration, it follows that the value of

m_P must be close to (12.89) and this must be satisfied with the help of (12.92). When using the simplified expressions (12.96), (12.98), it is necessary, in the same way, to check the results of calculation of small order of the second term in the denominator of (12.85). Above all, inequality (12.94) must always be satisfied.

Finally, note that the frequency of autoresonance, similar to (12.85), can be obtained from other logic structures for the switching arrangement. Really, assuming in (12.81) that

$$\alpha_2(p_2S - m_F - \Omega^2 m_x) - \alpha_1 p_1 S \equiv 0$$

and substituting (12.86), we find, for example, neglecting values proportional to the dissipation coefficient (the ideal excitation source):

$$\alpha_1 / m_{\eta_1} = \alpha_2 / m_{\eta_2} . \tag{12.101}$$

Linearisation coefficients in form (12.84) represent only one of the possible ways of achieving condition (12.101).

Reference [3] presents a numerical method which can be used to find solutions for the general set of equations which describe the dynamics of pneumatically driven strikers.

§13. Random vibration

1. Let us turn to consider statistical dynamics for impact oscillators, comparing statistical motion and force linearisation methods as examples and comparing the results with exact solutions [37, 38]. The study will be carried out for a system with double-sided stops (Fig. 3.1), assuming now that the applied external excitation $P(t)$ is a stationary normal random process. The dynamic model described has numerous technical applications, so it is worthwhile to study its statistical dynamics from this point of view also.

Assuming that $P(t) = m_P + P^0(t)$, where $m_P = \text{const}$; $P^0(t)$ is a centred random process, we shall write the equations for the system for an oscillator of unit mass for the statistical motion linearisation method in the form

$$(s^2 + bs + \Omega^2)x + \Phi_E(x, sx) + \Phi_D(x, sx) = m_P + P^0(t) . \tag{13.1}$$

Here the impact interaction force characteristics $\Phi_E(x, sx)$ and $\Phi_D(x, sx)$ are found from expressions similar to (6.15), (6.16).

Equation (13.1) corresponds to the structural scheme shown in Fig. 13.1. From this it can be seen that the additive component $\tilde{\chi}(t)$, acting on the input to the nonlinear block, appears as the output signal from the linear filter, of high factor when $\Omega >> b/2$. In agreement with the presentation of §8, section 2, this allows us to consider the situation described as approximately normal and consequently, to assume normality for the complete process at entry to the nonlinear block. As a result, we can describe its statistical linearisation according to the formulae obtained in §8.

Assuming the solution of equation (13.1) in the form

$$x(t) = m_x + x^0(t) , \tag{13.2}$$

where $m_x = \text{const}$; $x^0(t) = x(t) - m_x$ is the centred random process, we find the linearised force characteristics for impact interaction

$$\Phi_E(x,s) \approx m_F + k(x - m_x) = m_F + kx^0, \tag{13.3}$$

$$\Phi_D(x,sx) \approx \beta sx. \tag{13.4}$$

Here m_F, k and β are expressed in terms of unknown statistical characteristics of the solution with the help of (8.75) to (8.77), where $y(t) = \dot{x}(t)$.

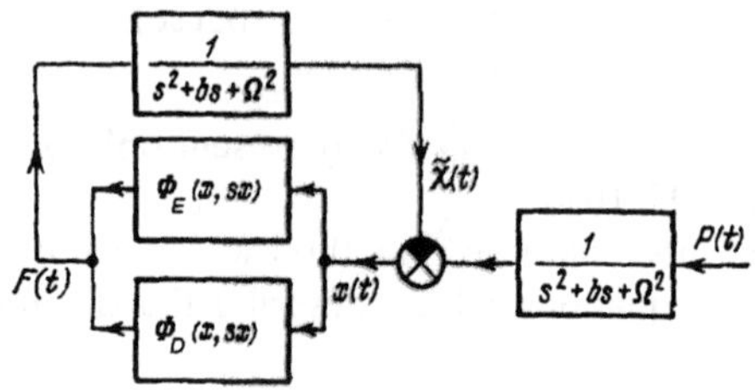

Fig. 13.1

Substituting (13.3), (13.4) into (13.1) and separating the constant and centred components, we find

$$m_F + \Omega^2 m_x = m_P, \tag{13.5}$$

$$[s^2 + (b+\beta)s + \Omega^2 + k]x^0(t) = P^0(t). \tag{13.6}$$

From (13.6) using formula (8.27) for the transformation by the linear system of a stationary random process, we find the spectral density of the linearised system

$$S_x(\omega) = \frac{S_P(\omega)}{\left|\Omega^2 + k - \omega^2 + (b+\beta)j\omega\right|^2}, \tag{13.7}$$

where $S(\omega)$ represents the spectral density of the process indicated by the index. Using (13.7) we find from (8.15)

$$\sigma_x^2 = \frac{1}{2\pi}\int_{-\infty}^{\infty} \frac{S_P(\omega)d\omega}{\left|\Omega^2 + k - \omega^2 + (b+\beta)j\omega\right|^2}, \tag{13.8}$$

$$\sigma_y^2 = \frac{1}{2\pi}\int_{-\infty}^{\infty} \frac{\omega^2 S_P(\omega)d\omega}{\left|\Omega^2 + k - \omega^2 + (b+\beta)j\omega\right|^2}. \tag{13.9}$$

For simplification of the later relationships we will assume that $P_0(t)$ has constant spectral density $S_P(\omega) = S_0 = \text{const}$, i.e. white noise appears. Applying this and using the values of integrals of type (13.8), (13.9) [184], we find

$$\sigma_x^2 = \frac{S_0}{2(b+\beta)(\Omega^2 + k)}, \tag{13.10}$$

$$\sigma_y^2 = \frac{S_0}{2(b+\beta)}. \tag{13.11}$$

The relationships (13.5), (13.10), (13.11) together with the values of the linearisation coefficients, complete the system of equations to find the unknowns m_x, σ_x, σ_y.

Let us look at some particular cases of solution for the equations shown. Assuming $m_P \equiv 0$, then $m_x \equiv 0$ and consequently $m_F \equiv 0$. Using this with (8.76), (8.77) we find when $M = 1$

$$k = \frac{\Delta(1+R^2)}{4\sqrt{2\pi}\sigma_x} \frac{\exp[-\Delta^2/(2\sigma_x^2)]}{\mathcal{F}(\Delta/\sigma_x)} \frac{\sigma_y^2}{\sigma_x^2}, \tag{13.12}$$

$$\beta = \frac{1-R^2}{2\pi} \frac{\exp[-\Delta^2/(2\sigma_x^2)]}{\mathcal{F}(\Delta/\sigma_x)} \frac{\sigma_y}{\sigma_x}, \tag{13.13}$$

where $\mathcal{F}(u)$ is the probability integral.

Dividing (13.11) by (13.10) we find

$$\sigma_y^2/\sigma_x^2 = \Omega^2 + k. \tag{13.14}$$

Substituting (13.14) into (13.12), (13.13), we obtain

$$k = \Omega^2 \left[\frac{4\sqrt{2\pi}\sigma_x \mathcal{F}(\Delta/\sigma_x)}{\Delta(1+R^2)\exp[-\Delta^2/(2\sigma_x^2)]} \right]^{-1} \tag{13.15}$$

$$\beta = \frac{1-R^2}{2\pi} \frac{\exp[-\Delta^2/(2\sigma_x^2)]}{\mathcal{F}(\Delta/\sigma_x)} \sqrt{\Omega^2 + k} =$$

$$= \frac{(1-R^2)\Omega}{2\pi} \frac{\exp[-\Delta^2/(2\sigma_x^2)]}{\mathcal{F}(\Delta/\sigma_x)} \left[1 - \frac{\Delta}{4\sqrt{2\pi}\sigma_x} \frac{\exp[-\Delta^2/(2\sigma_x^2)]}{\mathcal{F}(\Delta/\sigma_x)} \right]^{-1/2}. \tag{13.16}$$

Substituting (13.15), (13.16) into (13.10), we obtain an equation to find σ_x. When $R = 1$ ($\beta = 0$) it can be simplified significantly and takes the form

$$\sigma_x^2 = \frac{S_0}{2b\Omega^2} \left[1 - \frac{\Delta}{4\sqrt{2\pi}\sigma_x} \frac{\exp[-\Delta^2/(2\sigma_x^2)]}{\mathcal{F}(\Delta/\sigma_x)} \right]. \tag{13.17}$$

When $\Delta \to \infty$ from (13.17) we find, noting that $\mathcal{F}(\infty) = ½$:

$$\sigma_x^2 = \sigma_{x0}^2 = S_0/(2b\Omega^2), \tag{13.18}$$

which coincides with the the expression for the mean square value of the linear oscillator, excited by white noise [77]. As Δ increases, as anticipated, the influence of the stops remains of less significance.

Let us find the average frequency ω_0 of $x(t)$,i.e. the mathematical expectation of the frequency when the process will intersect the level $x = 0$ with $\dot{x} > 0$. In agreement with formula (8.25), we find

$$\omega_0 = 2\pi \int_0^\infty v w(0,v)\, dv, \tag{13.19}$$

where $w(u,v)$ is the joint probability density of $x(t)$ and $y(t) = \dot{x}(t)$. Using (8.61), (8.65) we find from (13.19)

$$\omega_0 = 2\pi C_2 \int_0^\infty v \exp\left(-\frac{v^2}{2\sigma_y^2} \right) dv = \frac{\sigma_y}{2\sigma_x \mathcal{F}(\Delta/\sigma_x)} = \frac{\sqrt{\Omega^2 + k}}{\mathcal{F}(\Delta/\sigma_x)} \tag{13.20}$$

According to (13.15) when $\Delta \to \infty$ the value of $k \to 0$ and consequently the average frequency tends to the natural frequency of the linear system $\omega \to \Omega$.

Let us find the mathematical expectation for the impact frequency as double the number of crossings when $x(t)$ intersects level Δ with $\dot{x}(t) > 0$. Using (8.25), we find using (8.61), (8.65)

$$f = 2\int_0^\infty v w(\Delta, v) dv = 2C_2 \exp\left(-\frac{\Delta^2}{2\sigma_x^2}\right)\int_0^\infty v \exp\left(-\frac{v^2}{2\sigma_y^2}\right) dv =$$

$$= \frac{\exp\left(-\frac{\Delta^2}{\sigma_x^2}\right)}{2\pi \mathscr{F}\left(\frac{\Delta}{\sigma_x}\right)}\frac{\sigma_y}{\sigma_x} = \frac{\exp\left(-\frac{\Delta^2}{\sigma_x^2}\right)}{2\pi \mathscr{F}\left(\frac{\Delta}{\sigma_x}\right)}\sqrt{\Omega^2 + k}. \tag{13.21}$$

2. Let us now look at the solution of the problem of forced random vibration of an impacting oscillator using the statistical *force* linearisation method. In this case, by analogy with (8.28) the equation for the system can be written in the following form:

$$[s^2 + (b+\beta)s]x + \Phi(x) = m_P + P^0(t), \tag{13.22}$$

where β is found from (8.54); $P_0(t)$ is a stationary centred normal random process as before with constant spectral density.

Let us linearise equation (13.22), assuming

$$\Phi(x) \approx m_F + k(x - m_x) = m_F + kx^0. \tag{13.23}$$

Here $k = 1/q$ is found from (8.85) and depends on the basic moment of the random process when $F(t) = \Phi[x(t)]$. Substituting (13.23) into (13.22) we obtain the condition

$$m_F \equiv m_P.$$

It can be used directly in the formula for k, which as a result depends on motion only through the value σ_F.

From (8.15), (8.27), for the system studied, we find similar to (13.8)

$$\sigma_x^2 = \frac{1}{2\pi}\int_{-\infty}^{\infty}\frac{S_0 d\omega}{\left|k - \omega^2 + (b+\beta)j\omega\right|^2}.$$

Multiplying both sides of this expression by k^2, and noting that

$$k^2\sigma_x^2 \approx \sigma_F^2, \tag{13.24}$$

we find

$$\sigma_F^2 = \frac{1}{2\pi}\int_{-\infty}^{\infty}\frac{k^2 S_0 d\omega}{\left|k - \omega^2 + (b+\beta)j\omega\right|^2} = \frac{kS_0}{2(b+\beta)}. \tag{13.25}$$

For numerical analysis of the quantity β, as we already found in §8, we can assign approximate prior values according to (8.54), (8.57). Then using the equation for k, equation (13.25) contains a single unknown quantity σ_F, which defines its solution. Knowing σ_F, it is possible to find all the remaining parameters. In general, coefficient β can be found using the more complete expression (8.56), however the calculation process in this case is significantly more difficult.

Let us now study in detail the case when $m_P \equiv 0$, and consequently, $m_F \equiv 0$, while the coefficient of restitution R = 1 (β = 0). In this case, after substituting (8.85), equation (13.25) takes the form

$$\sigma_F^2 = \frac{\Omega^2 S_0}{4b\mathcal{F}(\Omega^2\Delta/\sigma_F)}. \tag{13.26}$$

When $\Delta \to \infty$ we find for the linear case

$$\sigma_F^2 = \Omega S_0/(2b)$$

Calculating σ_F from equation (13.26), we can find σ_x^2 from this using (13.24)

$$\sigma_x^2 = \frac{S_0\mathcal{F}(\Omega^2\Delta/\sigma_F)}{b\Omega^2}. \tag{13.27}$$

We will find the average frequency of $F(t)$ as the mathematical expectation of the frequency intersecting $F(t)$ at $F = 0$ when $\dot{F}(t) > 0$. Obviously this coincides with the average frequency ω_0 of $x(t)$, since $\Phi(0) = 0$. By analogy with (13.19) we find

$$\omega_0 = 2\pi\int_0^\infty \dot{F}w(0,\dot{F})d\dot{F}, \tag{13.28}$$

where $w(F,\dot{F})$ is the joint probability density of $F(t)$ and $\dot{F}(t)$. Assuming the simplification

$$w(F,\dot{F}) = \frac{1}{2\pi\sigma_F\sigma_{\dot{F}}}\exp\left[-\frac{(F-m_F)^2}{2\sigma_F^2} - \frac{\dot{F}^2}{2\sigma_{\dot{F}}^2}\right], \tag{13.29}$$

we find from (13.28)

$$\omega_0 = \frac{\sigma_{\dot{F}}}{\sigma_F}. \tag{13.30}$$

For $\sigma_{\dot{F}}$ we find, similarly to (13.25)

$$\sigma_{\dot{F}}^2 = \frac{1}{2\pi}\int_{-\infty}^{\infty}\frac{k^2\omega^2 S_0 d\omega}{\left|k - \omega^2 + (b+\beta)j\omega\right|^2} = \frac{S_0 k^2}{2(b+\beta)}. \tag{13.31}$$

Substituting the values (13.31), (13.25) in (13.30), we find using (8.85)

$$\omega_0 = \sqrt{k} = \Omega[2\mathcal{F}(\Omega^2\Delta/\sigma_F)]^{-1/2}. \tag{13.32}$$

The average number of collisions in unit time is found as twice the mathematical expectation of the frequency at exit from $F(t)$ at level $\Omega^2\Delta$ when $\dot{F}(t) > 0$. Similarly to (13.21) using (13.29), (8.85) we find

$$f = 2\int_0^\infty \dot{F} w(\Omega^2\Delta, \dot{F}) d\dot{F} = \frac{1}{\pi}\exp\left[-\frac{(\Omega^2\Delta)^2}{2\sigma_F^2}\right]\frac{\sigma_{\dot{F}}}{\sigma_F} =$$
$$= \frac{\sqrt{k}}{\pi}\exp\left[-\frac{(\Omega^2\Delta)^2}{2\sigma_F^2}\right] = \frac{\Omega}{\sqrt{2\pi}}\frac{\exp[-(\Omega^2\Delta)^2/(2\sigma_F^2)]}{\sqrt{\mathcal{F}(\Omega^2\Delta/\sigma_F)}}. \tag{13.33}$$

3. To estimate the accuracy of the approximate solutions found we use the fact that when $R = 1$ an exact solution for the problem considered can be constructed for the case when excitation $P(t)$ exists as centred white noise $(P(t) = P^0(t))$. As before, assuming the system is symmetrical (Fig. 3.1) and noting that in the interval between collisions the vibration of the oscillator is defined by the equation

$$\ddot{x} + b\dot{x} + \Omega^2 x = P^0(t), \tag{13.34}$$

it can be considered as a *two-dimensional Markov process* [204], in terms of vector phase coordinates $\{x(t), y(t) = \dot{x}(t)\}$, since under conditions of δ-correlation of the process, the excitation of the vector components in any time instant $t_0\{x(t_0), y(t_0)\}$ completely defines the statistical state when $t \geq t_0$.

The common probability density $w(u,v)$ of $x(t)$ and its derivative with respect to time $y(t)$ for the stationary solution of equation (13.34) satisfies the Fokker-Planck equation [204]

$$\frac{\partial}{\partial u}(vw) - \frac{\partial}{\partial v}[(bv + \Omega^2 u)w] = \frac{1}{2}S_0\frac{\partial^2 w}{\partial v^2}. \tag{13.35}$$

This partial differential equation is of the *parabolic* type. To find an unambiguous solution it is necessary to assign boundary conditions for $w(u,v)$, which is non-zero in regions where the arguments take the values $-\Delta \leq u \leq \Delta,\ -\infty < v < \infty$. Taking into account the law which represents rebound of the mass from the stops ($R. = 1$), we arrive at the following boundary conditions:

$$w(\Delta, v) \equiv w(\Delta, -v),\quad w(-\Delta, v) \equiv w(-\Delta, -v). \tag{13.36}$$

When $v \to \pm\infty$ the density function must reduce to zero:

$$w(u, -\infty) \equiv 0,\quad w(u, \infty) \equiv 0. \tag{13.37}$$

The known solution of (13.35) is satisfied with conditions (13.36), (13.37), [204]

$$w(u,v) = C\exp\left[-\frac{2b}{S_0}\left(\frac{\Omega^2 u^2}{2} + \frac{v^2}{2}\right)\right], \tag{13.38}$$

where C is a constant found from the condition of normalisation

$$\int_{-\Delta}^{\Delta} du \int_{-\infty}^{\infty} w(u,v)dv = 1. \tag{13.39}$$

Substituting (13.38) in the left hand side of (13.39) we find after integration

$$C\int_{-\Delta}^{\Delta}\exp\left(-\frac{b\Omega^2 u^2}{S_0}\right)du\int_{-\infty}^{\infty}\exp\left(-\frac{bv^2}{S_0}\right)dv = \frac{2\pi S_0}{b\Omega}\mathcal{F}\left(\frac{\sqrt{2b}\Delta\Omega}{\sqrt{S_0}}\right).$$

Consequently,

$$C = \frac{b\Omega}{2\pi S_0 \mathscr{F}(\sqrt{2b}\Delta\Omega/\sqrt{S_0})} = \frac{1}{4\pi\Omega\sigma_{x0}^2 \mathscr{F}(\Delta/\sigma_{x0})}, \tag{13.40}$$

where according to (13.18) σ_{x0}^2 represents the standard deviation of $x(t)$ in the absence of stops. Hence corresponding to (13.38), (13.40) the distribution of velocity for any coordinate is normal, the distribution for any coordinateis limited to the Gaussian curve bounded by the values $u = \pm\Delta$ on the graph of density for the coordinate.

Let us find the mean square value of $x(t)$:

$$\sigma_x^2 = C\int_{-\Delta}^{\Delta} u^2 \exp\left(-\frac{b\Omega^2 u^2}{S_0}\right)du \int_{-\infty}^{\infty} \exp\left(-\frac{bv^2}{S_0}\right)dv = \\ = \sigma_{x0}^2\left[1 - \frac{\Delta\exp[-\Delta^2/(2\sigma_{x0}^2)]}{\sqrt{2\pi}\sigma_{x0}\mathscr{F}(\Delta/\sigma_{x0})}\right]. \tag{13.41}$$

For the mean square value of velocity we find

$$\sigma_y^2 = C\int_{-\Delta}^{\Delta} \exp\left(-\frac{b\Omega^2 u^2}{S_0}\right)du \int_{-\infty}^{\infty} v^2 \exp\left(-\frac{bv^2}{S_0}\right)dv = \Omega^2\sigma_{x0}^2. \tag{13.42}$$

Let us find the average frequency ω_0 of $x(t)$. Using (13.38), (13.40), we find from (13.19)

$$\omega_0 = 2\pi C\int_0^{\infty} v\exp\left(-\frac{bv^2}{S_0}\right)dv = \frac{\Omega}{2\mathscr{F}(\Delta/\sigma_{x0})}. \tag{13.43}$$

Finally, by analogy with (13.21) let us find the mathematical expectancy for the collision frequency

$$f = 2C\exp\left(-\frac{\Delta^2}{2\sigma_{x0}^2}\right)\int_{-\infty}^{\infty} v\exp\left(-\frac{bv^2}{S_0}\right)dv = \\ = \frac{\Omega}{4\pi\mathscr{F}(\Delta/\sigma_{x0})}\exp\left(-\frac{\Delta^2}{2\sigma_{x0}^2}\right). \tag{13.44}$$

Let us compare the results of approximate and exact solutions. Fig. 13.2 shows expressions for the standard deviation of $x(t)$ for the systems considered dependent on the similar value with the absence of stops. The curves correspond to the formulae found by the exact method (13.41) and the methods of statistical motion (13.17) and force (13.27) linearisation. The gradual "saturation" of standard deviation is clearly seen for the exiting process $x(t)$ as the intensity of the entry excitation $P(t)$ increases. Comparison shows that the statistical force linearisation method produces higher accuracy here than motion linearisation.

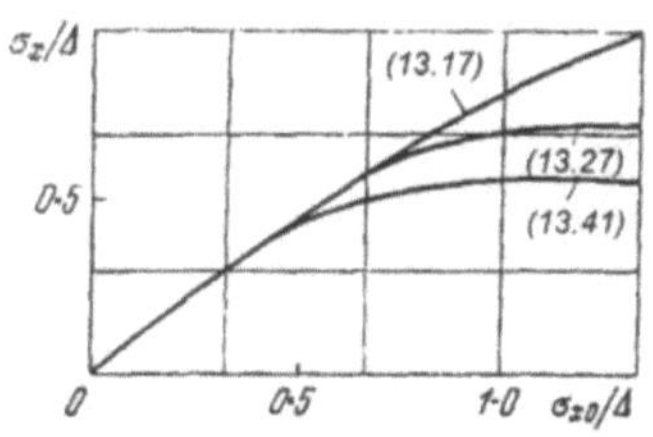

Fig. 13.2

The curves for the expressions $\omega_0(\Delta/\sigma_{x0})$ and $f(\Delta/\sigma_{x0})$ (Fig. 13.3) obtained by different methods almost coincide. It can be seen that as Δ increases, the

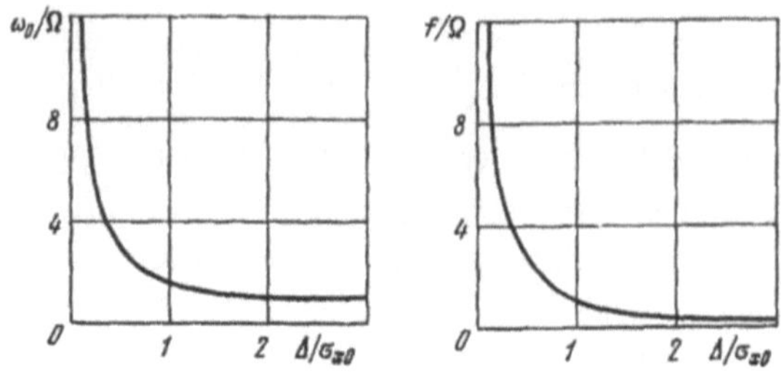

Fig. 13.3

average frequency and frequency of collisions is reduced. It is interesting to note that when $\Delta/\sigma_{x0} > 1$ the influence of increasing gap, as a method of combating impact against the stops, is considerably reduced. This must be taken into account in the design of vibration isolation systems.

Chapter 4 Vibro-impact processes in systems with multiple degrees of freedom

It don't mean a thing if it ain't got that swing.

Duke Ellington.

§14. Vibro-impact interactions of elastic systems

1. We shall now look at vibro-impact interactions in systems with multiple degrees of freedom, implying either lumped or continuous models. As we showed in §6, using dynamic compliance operators makes it possible to develop uniform equations of motion in both cases which differ only in the actual form of the dynamic compliance. The description of analysis methods for problems of existence and stability of periodic motion in the previous chapter can be extended to the class of system described here. A solution for such problems can often be successfully obtained in the general form with the final formula [39].

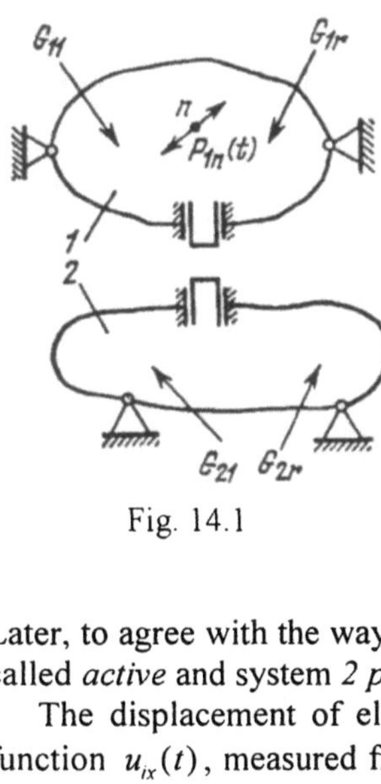

Fig. 14.1

We shall examine two elastic linear systems *1* and *2* (Fig. 14.1) the contact elements of which (assuming particles, plane sections) arranged at distance Δ apart, undergo one-dimensional motion with collisions under the effect of an exciting force $P_{1n}(t)$, applied to element n of system *1*. We shall assume that both systems are loaded by constant forces G_{ir} $(i = 1, 2;\ r = 1, 2, \ldots)$, acting on elements r. Later, to agree with the way vibration energy spreads, system *1* will sometimes be called *active* and system *2 passive*.

The displacement of element x of the i-th system will be described by the function $u_{ix}(t)$, measured from the undeformed position of the element. We shall represents the dynamic compliance operator of the i-th system as $L_{ip}(x,s)$ which connects the movement of element x due to an arbitrary force applied to element p $(p = r, n)$. In this respect, for a system with distributed parameters s implies $\partial/\partial t$. If the contacting element has index zero, we can represent the force properties of the impact interaction in the form

$$F = \Phi(v, sv) \tag{14.1}$$

where $v(t) = u_{10}(t) - u_{20}(t)$

As a result the operator equations for vibration of the system elements can be written in the form

$$u_{1x}(t) = \sum_r L_{1r}(x,0)G_{1r} + L_{1n}(x,s)P_{1n}(t) - L_{10}(x,s)\Phi(v,sv), \tag{14.2}$$

$$u_{2x}(t) = \sum_r L_{2r}(x,0)G_{2r} + L_{20}(x,s)\Phi(v,sv). \tag{14.3}$$

Here the terms under the summation sign describe the static deformation of the elements.

Subtracting (14.3) from (14.2) we obtain when $x = 0$ the equation of relative motion of the colliding elements, which we can write in the following form:

$$v(t) = L_{1n}(0,s)P_{1n}(t) + L(s)\{G - \Phi[v(t), sv(t)]\}, \tag{14.4}$$

where

$$L(s) = L_{10}(0,s) + L_{20}(0,s) \tag{14.5}$$

is the total dynamic compliance of the contacting elements;

$$G = \sum_r [L_{1r}(0,0)G_{1r} - L_{2r}(0,0)G_{2r}]L^{-1}(0) \tag{14.6}$$

is the reduced force of their relative static deformations.

Equation (14.4) is similar to (6.39), and, consequently, operator $L(s)$ produces the filtering effect described in §6. The equivalent linearisation method can effectively be used here for the solution.

We shall assume $P_{1n}(t) = P_{1n}\cos\omega t$. We shall search for an approximate periodic solution of equation (14.4) in the form

$$v(t) \approx m_v + v^0(t), \quad v^0(t) = a_v \cos(\omega t - \varphi_v). \tag{14.7}$$

Applying harmonic linearisation to (14.1)

$$\Phi(v, sv) \approx m_F(m_v, a_v) + k(m_v, a_v)v^0 + \beta(m_v, a_v)sv^0 \tag{14.8}$$

and substituting (14.7),(14.8) into (14.4), we obtain for the constant and periodic components

$$\frac{m_v}{L(0)} = G - m_F(m_v, a_v), \tag{14.9}$$

$$\left[\frac{1}{L(s)} + k(m_v, a_v) + s\beta(m_v, a_v)\right]v^0(t) = \frac{L_{1n}(0,s)}{L(s)}P_{1n}\cos\omega t. \tag{14.10}$$

The right hand side of the operator equation (14.10) contains the reduced force of external excitation

$$\frac{L_{1n}(0,s)}{L(s)}P_{1n}\cos\omega t = a_p \cos(\omega t + \psi), \tag{14.11}$$

where $a_p = \left|L_{1n}(0, j\omega)L^{-1}(j\omega)\right|P_{1n}$; $\psi = \arg L_{1n}(0, j\omega) - \arg L(j\omega)$.

We shall rewrite equation (14.10) in the form

$$W(s, m_v, a_v)v^0(t) = P_{1n}\cos\omega t. \tag{14.12}$$

Here we introduce the definition

$$W(s,m_v,a_v)=\frac{1+[k(m_v,a_v)+s\beta(m_v,a_v)]L(s)}{L_{1n}(0,s)}. \tag{14.13}$$

We will transform equation (14.12) to the equivalent homogeneous form taking account of the form of solution sought (14.7), using the expression

$$P_{1n}\cos\omega t=\frac{P_{1n}}{a_v}\cos[(\omega t-\varphi_v)+\varphi_v]=\frac{P_{1n}}{a_v}\left(\cos\varphi_v+s\frac{\sin\varphi_v}{\omega}\right)v^0(t).$$

Substituting this into (14.12), we obtain

$$\left[W(s,m_v,a_v)-\frac{P_{1n}}{a_v}\left(\cos\varphi_v+s\frac{\sin\varphi_v}{\omega}\right)\right]v^0(t)=\mathcal{D}(s)v^0(t)=0. \tag{14.14}$$

The existence conditions for a periodic solution (14.7) of equation (14.14) occur when the roots $s=\pm j\omega$ of its characteristic equation $\mathcal{D}(s)=0$ exist. Putting $s=j\omega$ into the latter and separating real and imaginary parts we find

$$\sin\varphi_v=a_vV(\omega,m_v,a_v)/P_{1n}, \tag{14.15}$$

$$\cos\varphi_v=a_vU(\omega,m_v,a_v)/P_{1n}. \tag{14.16}$$

Here $U(\omega,m_v,a_v)=\operatorname{Re}W(j\omega,m_v,a_v);V(\omega,m_v,a_v)=\operatorname{Im}W(j\omega,m_v,a_v)$. Eliminating the trigonometric functions from (14.15), (14.16), we obtain

$$a_v=\frac{P_{1n}}{\sqrt{U^2(\omega,m_v,a_v)+V^2(\omega,m_v,a_v)}}=\frac{P_{1n}}{|W(j\omega,m_v,a_v)|}. \tag{14.17}$$

The system of finite equations (14.9), (14.15) to (14.17) define the unknown parameters for the relative motion (14.7) of the colliding elements. For systems with one degree of freedom these are similar to (9.7), (9.11) to (9.13). Their solutions are conveniently found by using the following sequence. Finding from (14.9) the term $m_v(a_v)$, and substituting this into (14.17) we obtain an independent equation to find a_v. Further, we can find the phase φ_v using (14.15), (14.16).

After finding the parameters for the relative motion $v(t)$ and using (14.2), (14.3) we can find the motion of the remaining elements of the system. The approximate periodic solution for equations (14.2), (14.3) will be taken in the form

$$u_{ix}(t)\approx m_{ix}+u_{ix}^0(t),\qquad u_{ix}^0(t)=a_{ix}\cos(\omega t-\varphi_{ix}). \tag{14.18}$$

Substituting (14.18), (14.8) into (14.2) and (14.3) and separating constant and centred components, we obtain, using (14.10)

$$m_{1x}=\sum_r L_{1r}(x,0)G_{1r}-L_{10}(x,0)m_F(m_v,a_v),$$

$$m_{2x}=\sum_r L_{2r}(x,0)G_{2r}-L_{20}(x,0)m_F(m_v,a_v), \tag{14.19}$$

$$u_{ix}^0(t)=P_{1n}\cos\omega t\cdot W_{ix}^{-1}(s,m_v,a_v) \tag{14.20}$$

where

$$W_{1x}(s,m_v,a_v)=\frac{1+[k(m_v,a_v)+s\beta(m_v,a_v)]L(s)}{A}. \tag{14.21}$$

Here

$$A = L_{1n}(x,s)\{1 + k(m_v, a_v) + s\beta(m_v, a_v)]L(s)\} -$$

$$- [k(m_v, a_v) + s\beta(m_v, a_v)]L_{10}(x,s)L_{1n}(0,s), \qquad (14.22)$$

$$W_{2x}(s, m_v, a_v) = \frac{1 + [k(m_v, a_v) + s\beta(m_v, a_v)]L(s)}{[k(m_v, a_v) + s\beta(m_v, a_v)]L_{20}(x,s)L_{1n}(0,s)}.$$

From (14.20) we find the amplitude and phase of the periodic components

$$a_{ix} = P_{1n}\left|W_{ix}(j\omega, m_v, a_v)\right|^{-1}, \qquad (14.23)$$

$$\sin\varphi_{ix} = a_{ix}V_{ix}(\omega, m_v, a_v)/P_{1n}, \qquad (14.24)$$

$$\cos\varphi_{ix} = a_{ix}U_{ix}(\omega, m_v, a_v)/P_{1n}. \qquad (14.25)$$

Here

$$U_{ix}(\omega, m_v, a_v) = \operatorname{Re} W_{ix}(j\omega, m_v, a_v);$$

$$V_{ix}(\omega, m_v, a_v) = \operatorname{Im} W_{ix}(j\omega, m_v, a_v).$$

2. To study the stability of the periodic solutions against small perturbations we can use the approach described in §10. We shall assume that in the vicinity of the stable solution (14.7) relative vibration of the colliding elements takes the form

$$v(t) = \tilde{m}_v(t) + \tilde{a}_v(t)\cos[\omega t - \tilde{\varphi}_v(t)], \qquad (14.26)$$

where $\tilde{m}_v(t), \tilde{a}_v(t)$ and $\tilde{\varphi}_v(t)$ are slowly changing functions of time.

From equation (14.12) we find the work of the dissipative and excitation forces for one period of the steady motion (14.7). The loss of energy E_d is calculated from the equation

$$E_d = \int_0^T \{W(s, m_v, a_v)v^0(t)\}sv^0(t)dt = \pi a_v^2 V(s, m_v, a_v). \qquad (14.27)$$

For the work of the excitation forces E_e, we find

$$E_e = \int_0^T P_{1n}\cos\omega t \cdot sv^0(t)dt = \pi P_{1n}a_v\sin\varphi_v. \qquad (14.28)$$

A specific feature of expression (14.27) in the case of a system with several degrees of freedom is that it can take either a positive or a negative value. The latter situation can arise in the case when the *observed* coordinates, in terms of which the expression has been written (for example $v^0(t)$), does not coincide with the coordinate of the applied force (here $u_n(t)$). As a result, when some forms of vibration occur, an odd number of nodes can be situated between the coordinates and consequently the phase shift between the force and the observed coordinates φ_v is greater than π.

If we require that, when the amplitude $\tilde{a}_v$ diverges from the steady value a_v, the balance of energy is distributed in such a way as to compensate this divergence, we find the energy conditions of stability

$$
\begin{aligned}
&\frac{d}{d\tilde{a}_v}\left[V(\omega,\tilde{m}_v,\tilde{a}_v)-\frac{P_{1n}\sin\tilde{\varphi}_v}{\tilde{a}_v}\right]_{\tilde{a}_v=a_v} > 0,\ \mathit{if}\ \sin\varphi_v > 0,\\
&\frac{d}{d\tilde{a}_v}\left[V(\omega,\tilde{m}_v,\tilde{a}_v)-\frac{P_{1n}\sin\tilde{\varphi}_v}{\tilde{a}_v}\right]_{\tilde{a}_v=a_v} < 0,\ \mathit{if}\ \sin\varphi_v < 0.
\end{aligned}
\tag{14.29}
$$

It is easy to see that the first inequality in (14.29) can be obtained directly from the characteristic equation of the linearised system (see (14.14)) using condition (10.2). Hence the formal application of condition (10.2) to systems with several degrees of freedom requires that the excitation force is reduced first to the coordinate observed and a study of the phase shift between this coordinate and the reduced force. In the given case, for example, during the search for a solution of type (14.7) it is sufficient to set the right hand side of equation (14.10) equal to $a_p \cos\omega t$, while the resulting force $P_{1n}(t)$, using (14.11) can be considered as $P_{1n}\cos(\omega t-\psi)$.

An important detail, however, is that, as will be shown below, both cases of (14.29) lead to the same final expression to find stability, hence this can be taken as the basic case for the phase relationship, for example, in equations of the form (14.12). Really, differentiating the first inequality (14.29) and assuming, using an arbitrary choice of function defined by (14.26), that in the disturbed motion $\tilde{m}_v(t)$ and $\tilde{a}_v$ are related by conditions (14.9) but $\tilde{\varphi}_v$ and $\tilde{a}_v$ by conditions (14.16), we find using (14.15)

$$
\begin{aligned}
&\left(\frac{dV}{d\tilde{a}_v}+\frac{a_p\sin\tilde{\varphi}_v}{\tilde{a}_v^2}-\frac{a_p}{\tilde{a}_v}\cos\tilde{\varphi}_v\frac{d\tilde{\varphi}_v}{d\tilde{a}_v}\right)_{\tilde{a}_v=a_v}=\\
&\qquad=\left(\frac{dV}{d\tilde{a}_v}+\frac{V}{\tilde{a}_v}-U\frac{d\tilde{\varphi}_v}{d\tilde{a}_v}\right)_{\tilde{a}_v=a_v}>0.
\end{aligned}
\tag{14.30}
$$

Differentiating (14.16) as an implicit function $\tilde{\varphi}_v$ of $\tilde{a}_v$, we find using (14.15)

$$
\left(\frac{d\tilde{\varphi}}{d\tilde{a}_v}\right)_{\tilde{a}_v=a_v}=\left(\frac{U+\tilde{a}_v\dfrac{dU}{d\tilde{a}_v}}{P_{1n}\sin\tilde{\varphi}_v}\right)_{\tilde{a}_v=a_v}=\left(\frac{U+\tilde{a}_v\dfrac{dU}{d\tilde{a}_v}}{a_vV}\right)_{\tilde{a}_v=a_v}. \tag{14.31}
$$

Substituting this equation into (14.30), we obtain eventually

$$
\left[U\left(U+\tilde{a}_v\frac{dU}{d\tilde{a}_v}\right)+V\left(V+\tilde{a}_v\frac{dV}{d\tilde{a}_v}\right)\right]_{\tilde{a}_v=a_v}>0. \tag{14.32}
$$

When differentiating the second inequality (14.29) we have the opposite sign in (14.30), however, noting that now $V<0$, after substitution (14.31) and multiplication of both sides of the inequality by the value $\tilde{a}_vV<0$ we obtain (14.32) again.

Later in this chapter we will use condition (14.32) to determine stability, which will allow us to use the general approach for phase reference for all vibrating elements of the system, for example, relative to the applied external excitation at any point.

In the absence of dissipation in the system $V(\omega, m_v, a_v) \equiv 0$ and inequality (14.32) gives the following boundaries for the stability domains:

$$U(\omega, m_v, a_v) = 0,$$
$$\left[U(\omega, \tilde{m}_v, \tilde{a}_v) + \tilde{a}_v \frac{d}{d\tilde{a}_v} U(\omega, \tilde{m}_v, \tilde{a}_v) \right]_{\tilde{a}_v = a_v} = 0. \tag{14.33}$$

In the particular case of a linear oscillator impacting against the stops, $U = \Omega^2 + k - \omega^2)$, $V = (b + \beta)\omega$ and condition (14.32) coincides with equation (10.9) obtained earlier.

The description of the approximate approach, of course, does not completely exhaust the study of the question of stability of the periodic solution for systems with arbitrary degrees of freedom. In the context of applying approximate methods to find periodic solutions, a strict study of this question is often impossible. At the same time it must be borne in mind that, when using approximate methods, as a rule periodic *motion* is assumed *a priori*, mostly for some physical reason. Thus the problem consists only of separating stable and unstable approximate solutions, which can be done using the procedure set out.

Note that a similar approach can be used to study stability of the exact solutions (6.82), (6.83) obtained §6 section 5. Let us assume that the disturbed motion takes the form

$$x(t) = a\cos[\omega t + \tilde{\varphi}(t)] - \tilde{J}(t)\chi(t),$$

where $\tilde{\varphi}(t), \tilde{J}(t)$ are slowly changing functions of time.

Using the first expression (6.83) as the reduced condition for the balance of work between the dissipative and exciting forces, we shall formulate the energy condition for stability in the form

$$\frac{d}{d\tilde{J}} \left\{ \frac{\tilde{J}}{\omega a} \left[\dot{\chi}_-(0) + \frac{1}{(1+R)M} \right] + \sin\tilde{\varphi} \right\}_{\tilde{J}=J} > 0. \tag{14.34}$$

Introducing two arbitrary slow functions in place of one, we will connect them by an expression similar to the second equation in (6.83). Differentiating (14.34) using this connection, we obtain, taking account of (6.83)

$$J > \frac{-\Delta\chi(0)}{\chi^2(0) + \frac{1}{\omega^2}\left[\dot{\chi}_-(0) + \frac{1}{(1+R)M} \right]^2}.$$

Comparing (6.82) with the equation obtained, we can establish that the stability condition is satisfied only by regimes which have the positive sign before the radical.

3. Assuming that the impact interaction between the contacting elements (Fig. 14.1) is purely elastic ($\beta \equiv 0$), we will use the expressions for the harmonic linearisation coefficients in the form (7.65), (7.66) to find the final solution, bearing in mind, that here we assume $m_9 = GL(0)$. Introducing the designation Δ_0 for the reduced value of the gap (interference) taking account of the static loading of the system

$$\Delta_0 = \Delta - GL(0), \tag{14.35}$$

we obtain the equations for the linearisation coefficients in the form

$$m_{I^{\cdot}} = (a_v - \Delta_0)L^{-1}(0), \tag{14.36}$$

$$k = 2(1 - \Delta_0 / a_v)L^{-1}(0), \tag{14.37}$$

by making use of the fact also that, according to (7.63), $m_v + a_v = \Delta$.

We will estimate the behaviour of the system during vibro-impact interactions through the dynamic compliance of its elements at the fundamental. This is adequate, since higher harmonics are localised near the impact pairs and consequently have little influence on the vibration of the elements outside the contact zone (see §15, section 5).

For brevity we will restrict the calculation to the dynamic compliance of the colliding elements. From (14.23) using (14.21), (14.22) we find

$$\frac{a_{10}}{P_{1n}|L_{1n}(0, j\omega)|} = \left|\frac{1 + k(m_v, a_v)L_{20}(0, j\omega)}{1 + k(m_v, a_v)L(j\omega)}\right|, \tag{14.38}$$

$$\frac{a_{20}}{P_{1n}|L_{1n}(0, j\omega)|} = \left|\frac{k(m_v, a_v)L_{20}(0, j\omega)}{1 + k(m_v, a_v)L(j\omega)}\right|. \tag{14.39}$$

For convenience, we take here the relative value of dynamic compliance. From (14.37) it follows that when $a_v >> \Delta_0$ the value $k \to 2/L(0)$. Substituting this value of k in (14.38), (14.39), we obtain for the limiting resonance case, the final equations:

$$\frac{a_{10}}{P_{1n}|L_{1n}(0, j\omega)|} = \left|\frac{L(0) + 2L_{20}(0, j\omega)}{L(0) + 2L(j\omega)}\right|, \tag{14.40}$$

$$\frac{a_{20}}{P_{1n}|L_{1n}(0, j\omega)|} = \left|\frac{2L_{20}(0, j\omega)}{L(0) + 2L(j\omega)}\right|. \tag{14.41}$$

In cases when vibro-impact resonance occurs near to the natural frequency of the impacting linear system, the static compliance in (14.40), (14.41) need not be taken into account compared to the dynamic value of these quantities, and both formulas lead to the identical result

$$\frac{a_{i0}}{P_{1n}|L_{1n}(0, j\omega)|} = \left|\frac{L_{20}(0, j\omega)}{L_{10}(0, j\omega) + 2L(j\omega)}\right| \quad (i = 1, 2). \tag{14.42}$$

From (14.42) it follows that for colliding elements with identical dynamic compliance the latter due to interaction is reduced by half.

To study the basic case we will initially solve equation (14.17) neglecting the dissipation in the system for simplification. In this case the dynamic compliance considered remains real (see §6, section 3). Substituting (14.13), (14.37) into (14.17) and writing using the earlier assumption

$$L_{1n}(0, j\omega) = \mu_{1n}(\omega), \quad L(j\omega) = \mu(\omega) = \mu_{10}(\omega) + \mu_{20}(\omega), \\ \mu(\omega)/\mu(0) = \mu^0(\omega), \tag{14.43}$$

we find after transformation

$$a_v = \frac{\pm P_{1n}\mu_{1n}(\omega) + 2\mu^0(\omega)\Delta_0}{1 + 2\mu^0(\omega)}, \tag{14.44}$$

where

$$U = \frac{1 + 2\mu^0(\omega)(1 - \Delta_0 / a)}{\mu_{1\prime\prime}(\omega)}, \quad V = 0. \tag{14.45}$$

Since a_v is positive, according to (14.44) when satisfying the condition

$$P_{1\prime\prime}\left|\mu_{1\prime\prime}(\omega)\right| > 2\left|\mu^0(\omega)\Delta_0\right| \tag{14.46}$$

a unique solution occurs , defined by the expressions

$$\begin{aligned} &\pm P_{1\prime\prime}\mu_{1\prime\prime}(\omega) > 0 \quad when \quad 1 + 2\mu^0(\omega) > 0, \\ &\pm P_{1\prime\prime}\mu_{1\prime\prime}(\omega) < 0 \quad when \quad 1 + 2\mu^0(\omega) < 0. \end{aligned} \tag{14.47}$$

When the expression (14.46) has opposite sign

$$P_{1\prime\prime}\left|\mu_{1\prime\prime}(\omega)\right| < 2\left|\mu^0(\omega)\Delta_0\right| \tag{14.48}$$

solutions exist in pairs and only when the value of the parameters satisfies the condition

$$\frac{\mu^0(\omega)\Delta_0}{1 + 2\mu^0(\omega)} > 0\,,$$

i.e.

$$\begin{aligned} &\mu^0(\omega)\Delta_0 > 0 \quad \text{when} \quad 1 + 2\mu^0(\omega) > 0, \\ &\mu^0(\omega)\Delta_0 < 0 \quad \text{when} \quad 1 + 2\mu^0(\omega) < 0. \end{aligned} \tag{14.49}$$

Let us look at stability of the solutions obtained. Substituting (14.45) into (14.32) we obtain stability, conditions using (14.44) which coincide with the existence conditions (14.47). Hence, in domains where solutions which satisfy the conditions of uniqueness exist, the existing solutions are always stable. In such cases when solutions exist in pairs, one and only one of them is always stable. This is defined by condition (14.47). Using (14.47), (14.49), (14.44) shows that this solution always corresponds to large amplitude vibration.

Substituting (14.37), (14.43), (14.44) into (14.38), (14.39) and after re-arrangement we finally obtain

$$\frac{a_{10}}{P_{1\prime\prime}\left|\mu_{1\prime\prime}(\omega)\right|} = \left|\frac{1 + \dfrac{2}{\mu(0)}\left[\dfrac{\pm\mu_{10}(\omega)\Delta_0}{P_{1\prime\prime}\mu_{1\prime\prime}(\omega)} + \mu_{20}(\omega)\right]}{1 + 2\mu^0(\omega)}\right|, \tag{14.50}$$

$$\frac{a_{20}}{P_{1\prime\prime}\left|\mu_{1\prime\prime}(\omega)\right|} = \left|\frac{\dfrac{2\mu_{20}(\omega)}{\mu(0)}\left[1 \mp \dfrac{\Delta_0}{P_{1\prime\prime}\mu_{1\prime\prime}(\omega)}\right]}{1 + 2\mu^0(\omega)}\right|. \tag{14.51}$$

Note, that because of nonlinearity of the system examined, the expressions obtained for dynamic compliance depend on the excitation.

Let us now look at the case (Fig. 14.2) when passive system *2* is not fixed and is clamped to active system *1* by a constant force. Then $L_{20}(0,0) = \infty$ and consequently $1/L(0) \equiv 0$. As a result from (14.35) to (14.37) we find

$$m_{r} = G\,, \tag{14.52}$$

$$k = 2G / a_v \tag{14.53}$$

Neglecting dissipative effects in the system and assuming that the passive system is semi-infinite, we find

$$\begin{gathered} L_{1n}(0, j\omega) = \mu_{1n}(\omega), \quad L_{10}(0, j\omega) = \mu_{10}(\omega), \\ L_{20}(0, j\omega) = j\mu_{20}(\omega). \end{gathered} \tag{14.54}$$

Here we take account that wave dissipation of energy occurs in the semi-infinite system.

Substituting (14.13), (14.53), (14.54) into (14.17), we obtain

$$a_v = \frac{\pm P_{1n}\mu_{1n}(\omega)}{\sqrt{\left[1 + \frac{2G}{a_v}\mu_{10}(\omega)\right]^2 + \frac{4G^2}{a_v^2}\mu_{20}^2(\omega)}}, \tag{14.55}$$

for which

$$U = \frac{1 + \frac{2G}{a_v}\mu_{10}(\omega)}{\mu_{1n}(\omega)}, \quad V = \frac{2G\mu_{20}(\omega)}{a_v\mu_{1n}(\omega)}. \tag{14.56}$$

Solving (14.55) for a_v, we find

$$a_v = \pm P_{1n}\mu_{1n}(\omega)\sqrt{1 - \left[\frac{2G\mu_{20}(\omega)}{P_{1n}\mu_{1n}(\omega)}\right]^2} - 2G\mu_{10}(\omega)\,. \tag{14.57}$$

According to (14.57) vibro-impact regimes exist when

$$0 \le 2G\mu_{20}(\omega) \le P_{1n}\left|\mu_{1n}(\omega)\right|, \tag{14.58}$$

whence, if

$$P_{1n}\left|\mu_{1n}(\omega)\right| > \left|2G\sqrt{\mu_{10}^2(\omega) + \mu_{20}^2(\omega)}\right|, \tag{14.59}$$

then a unique regime exists, which, when $\mu_{1n}(\omega) > 0$ corresponds to the positive sign before the radical, but when $\mu_{1n}(\omega) < 0$, to the negative sign. If condition (14.59) is not satisfied, then regimes exist in pairs and only when $\mu_{10}(\omega) < 0$.

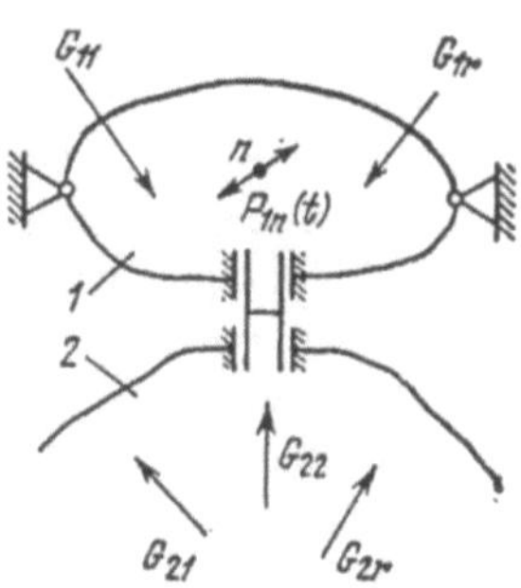

Fig. 14.2

From its physical appearance, inequality (14.59) is similar to that obtained when solving the previous problems for condition (14.46). Both of them determine levels of excitation at which vibro-impact regimes exist for all values of the system parameters.

Let us verify the stability of the solution. Substituting (14.56), (14.57) into (14.32), we obtain after transformation the condition for stability in the form

$$\pm P_{1n}\mu_{1n}(\omega) > 0\,. \tag{14.60}$$

Hence, a unique solution which satisfies the existence conditions is always stable, but when pairs of solutions exist, stability of one and only one of them corresponds to inequality (14.60). According to (14.57) and in the given case, it corresponds to the largest amplitude or vibration.

Let us now find the dynamic compliance of the colliding elements. From (14.38), (14.39) and using equations (14.53), (14.54), (14.57) we obtain after transformation the formula

$$\frac{a_{10}}{P_{1n}|\mu_{1n}(\omega)|}=\sqrt{1+\left[\frac{2G\mu_{10}(\omega)}{P_{1n}\mu_{1n}(\omega)}\right]^2\mp\frac{4G\mu_{10}(\omega)}{P_{1n}\mu_{1n}(\omega)}\sqrt{1-\left[\frac{2G\mu_{20}(\omega)}{P_{1n}\mu_{1n}(\omega)}\right]^2}}, \quad (14.61)$$

$$a_{20}=|2G\mu_{20}(\omega)|. \quad (14.62)$$

In domains defined by the conditions (14.58) where vibro-impact regimes exist, the value of (14.61) changes within the following limits:

$$\frac{a_{10}}{P_{1n}|\mu_{1n}(\omega)|}\in\left\{1\mp\frac{2G\mu_{10}(\omega)}{P_{1n}\mu_{1n}};\sqrt{1+\left[\frac{2G\mu_{10}(\omega)}{P_{1n}\mu_{1n}}\right]^2}\right\}. \quad (14.63)$$

Vibration of system *2* (Fig. 14.2) in a vibro-impact regime appears to be independent of excitation in the active system and its dynamic qualities according to (14.62), and is determined only by the dynamic compliance of the passive system and the value of static clamping. This situation can be used to regulate the transfer of vibrational energy across the joint.

In the following paragraph we will turn to the analysis of the general expressions found and applying them to the problem of vibro-impact processes occurring during collision of rods. Note that the solutions developed can also be used when the active system is excited by the application of several sources of identical frequencies, applied to elements n and having amplitude P_{1n}. In this case we must use $P_{1n}\mu_{1n}(\omega)$ everywhere in place of $\sum_n P_{1n}\mu_{1n}(\omega)$.

4. To consider fundamental and sub-periodic *resonances* of vibro-impact systems of general form, the same approach can be used as was taken in §9 section 4 for a system with one degree of freedom. Basically, such a method to approximate solutions using periodic Green's functions will be realistic only when the contacting elements have mass, hence at the points shown in Fig. 14.3, we will assume that the colliding elements consist of rigid bodies with masses M_1 and M_2 and coefficient of restitution R. In addition for generality we assume that the active system *1* has several synchronised sources of excitation with frequency ω applied, acting on elements n (n = 1, 2,...) with forces P_{1n} (t).

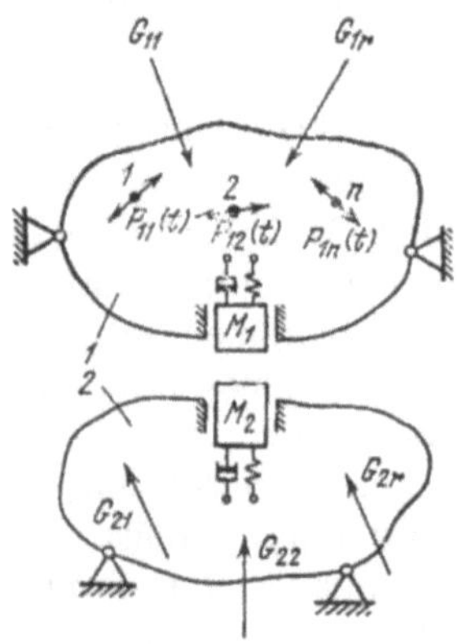

Fig. 14.3

Taking into account that in the system considered, vibro-impact interaction with period

T can be characterised now as a periodic sequence of δ-impulses (see (6.44)), then the equations of vibration of the elements by analogy with (14.2), (14.3) take the following form:

$$u_{1x}(t) = \sum_r L_{1r}(x,0)G_{1r} + \sum_n L_{1n}(x,s)P_{1n}(t) - \\ - L_{10}(x,s)J\sum_{\nu=-\infty}^{\infty}\delta(t-\nu T), \tag{14.64}$$

$$u_{2x}(t) = \sum_r L_{2r}(x,0)G_{2r} + L_{20}(x,s)J\sum_{\nu=-\infty}^{\infty}\delta(t-\nu T). \tag{14.65}$$

Subtracting the second equation from the first, and using (14.5), (14.6), when $x = 0$ we obtain

$$v_{2x}(t) = \sum_n L_{1n}(0,s)P_{1n}(t) + L(0)G - L(s)J\sum_{\nu=-\infty}^{\infty}\delta(t-\nu T). \tag{14.66}$$

The last term on the right hand side of equation (14.66) represents the periodic Green's functions of the first order $\chi_{1J}(t)$ (see §6 section 4) for the systems. Writing

$$\sum_n L_{1n}(0,s)P_{1n}(t) = \vartheta(t), \tag{14.67}$$

where $\vartheta(t)$ as before represents the solution with frequency ω of the corresponding linear problem, obtained by neglecting impact interactions. Measuring the displacement from the static equilibrium condition of the system using the function

$$v_0(t) = v(t) - L(0)G, \tag{14.68}$$

we obtain once again the general solution in the form (6.77)

$$v_0(t) = \vartheta(t) - J\chi_1(t).$$

Time measurement begins at the instant of impact.

The approximate solution for a resonant vibro-impact regime can be written in the form

$$v_0(t) \approx -J\chi_1(t). \tag{14.69}$$

Finding J from the impact conditions $v_0(0) \approx -J\chi_1(0) = v(0) - L(0)G = = \Delta - L(0)G = \Delta_0$, we find $J = -\Delta_0/\chi_1(0)$ and, consequently,

$$v_0(t) \approx \frac{\Delta_0}{\chi_1(0)}\chi_1(t). \tag{14.70}$$

Existence conditions for this solution are found from an energy balance for one period of motion $T = 2\pi l/\omega$ ($l = 1, 2, \ldots$). The energy loss during impact we find similarly to (9.51)

$$E_{d2} = \frac{(1-R)J^2}{2(1+R)M} = \frac{(1-R)\Delta_0^2}{2(1+R)\chi_1^2(0)M} \quad \left(M = \frac{M_1M_2}{M_1+M_2}\right). \tag{14.71}$$

From (14.66) taking account of (14.68), (14.70) we find the work done by dissipative force, acting on linear sub-systems *1*, *2* (Fig. 14.3):

$$E_{d1} = \int_0^T \{L^{-1}(s)v_0(t)\}\dot{v}_0(t)dt. \tag{14.72}$$

Having in mind the appearance of the functions in the integrand, and taking account of the expressions for periodic Green's functions which appear in it in the form of Fourier series (see §6 section4)

$$v_0(t) = \frac{\Delta_0}{\chi_1(0)T} \sum_{m=-\infty}^{\infty} L\left(\frac{mj\omega}{l}\right)\exp\left(\frac{mj\omega}{l}t\right), \tag{14.73}$$

$$\dot{v}_0(t) = \frac{\Delta_0 j\omega}{\chi_1(0)Tl} \sum_{m=-\infty}^{\infty} mL\left(\frac{mj\omega}{l}\right)\exp\left(\frac{mj\omega}{l}t\right) \tag{14.74}$$

we find from (14.72) after integration

$$E_{d1} = \frac{\omega^2\Delta_0^2}{\pi l^2\chi_1^2(0)} \sum_{m=1}^{\infty} m \operatorname{Im} L^{-1}\left(\frac{mj\omega}{l}\right)\left|L\left(\frac{mj\omega}{l}\right)\right|^2. \tag{14.75}$$

The work done by the excitation force is found from (14.66) with the help of the expression

$$E_e = \int_0^T \{L^{-1}(s)\vartheta(t)\}\dot{v}_0(t)dt. \tag{14.76}$$

The value of this integral depends on the phase of the periodic process $\vartheta(t)$ in relation to the instant of impact. Suppose, for example, $\vartheta(t) = a\cos(\omega t + \varphi)$, then

$$L^{-1}(s)\vartheta(t) = a_p \cos(\omega t + \psi), \tag{14.77}$$

where $a_p = \left|L^{-1}(j\omega)\right|a;\ \psi = \varphi - \arg L(j\omega)$. Substituting (14.74), (14.77) into (14.76), we obtain after integration

$$E_e = -\frac{a_p\omega\Delta_0}{\chi_1(0)}\left|L(j\omega)\right|\sin\varphi_0 \quad (\varphi_0 = \psi - \arg L(j\omega)). \tag{14.78}$$

As a result the equation of energy balance

$$E_{d1} + E_{d2} = E_e$$

in the case considered takes the following form:

$$\frac{\Delta_0^2}{\chi_1^2(0)}\left\{\frac{\omega^2}{\pi l^2}\sum_{m=1}^{\infty} mL^{-1}\left(\frac{mj\omega}{l}\right)\left|L\left(\frac{mj\omega}{l}\right)\right|^2 + \frac{1-R}{2(1+R)M}\right\} = -a_p\omega\frac{\Delta_0}{\chi_1(0)}\left|L(j\omega)\right|\sin\varphi_0. \tag{14.79}$$

From this two possible values for the unknown phase for the process φ_0 $(0 \le \varphi_0 \le \pi)$ can be found:

$$\varphi_0 = \arcsin \frac{a_p \omega |L(j\omega)|}{-\dfrac{\Delta_0}{\chi_1(0)}\left\{\dfrac{\omega^2}{\pi l^2}\sum_{m=1}^{\infty} m \operatorname{Im} L^{-1}\left(\dfrac{mj\omega}{l}\right)\left|L\left(\dfrac{mj\omega}{l}\right)\right|^2 + \dfrac{1-R}{2(1+R)M}\right\}},$$
$$\varphi_0 = \pi - \arcsin \frac{a_p \omega |L(j\omega)|}{-\dfrac{\Delta_0}{\chi_1(0)}\left\{\dfrac{\omega^2}{\pi l^2}\sum_{m=1}^{\infty} m \operatorname{Im} L^{-1}\left(\dfrac{mj\omega}{l}\right)\left|L\left(\dfrac{mj\omega}{l}\right)\right|^2 + \dfrac{1-R}{2(1+R)M}\right\}}. \qquad (14.80)$$

Existence conditions for the resonant regimes $|\sin\varphi_0| \le 1$ according to (14.79) can be written as:

$$a_p \omega |L(j\omega)| \ge \frac{\Delta_0}{\chi_1(0)}\left\{\frac{\omega^2}{\pi l^2}\sum_{m=1}^{\infty} m \operatorname{Im} L^{-1}\left(\frac{mj\omega}{l}\right)\left|L\left(\frac{mj\omega}{l}\right)\right|^2 + \frac{1-R}{2(1+R)M}\right\}. \qquad (14.81)$$

In particular, for systems with one degree of freedom, with an operator for the linear portion from (6.72) given by $\operatorname{Im} L^{-1}(mj\omega / l) = bm\omega / l$, we obtain the expression obtained earlier for this case (9.61).

For symmetrical double-sided interaction of the elements of the impacting pair the equation for relative motion of these elements takes the form

$$v_0(t) = \sum_n L_{1n}(0,s) P_{1n}(t) - - L(s) J\left[\sum_{\nu=-\infty}^{\infty} \delta(t - \nu T) - \sum_{\nu=-\infty}^{\infty} \delta\left(t - \nu T - \frac{T}{2}\right)\right]. \qquad (14.82)$$

Writing the general form of solution using periodic Green's functions of second order $\chi_2(t)$:

$$v_0(t) = \vartheta(t) - J\chi_2(t), \qquad (14.83)$$

we arrive at the following approximate solution:

$$v_0(t) \approx \frac{\Delta_0}{\chi_2(0)} \chi_2(t). \qquad (14.84)$$

The existence conditions for this solution, found from the equation for energy balance, now take the following form

$$a_p \omega |L(j\omega)| \ge -\frac{\Delta_0}{\chi_2(0)}\left\{\frac{2\omega^2}{\pi l^2}\sum_{m=1}^{\infty} (2m-1) \operatorname{Im} L^{-1}\left[\frac{(2m-1)j\omega}{l}\right] \times \times \left|L\left[\frac{(2m-1)j\omega}{l}\right]\right|^2 + \frac{1-R}{2(1+R)M}\right\} \qquad (14.85)$$

The fitness of the approximation used as before can be conveniently calculated by using one of the expressions of type (10.66), (10.67).

An important simplification to the approximate solutions (14.70), (14.84) can be obtained by expanding the operator $L(mj\omega / l)$ which appears in the equation, as a series of type (6.61), (6.64) for natural modes of vibration of the linear part of the system with the subsequent use of the basic vibration modes. Suppose, for example, during vibro-impact interaction of identical systems with a finite number of degrees of freedom, only the $\nu-th$ form of vibration appears. Then in agreement with (14.5), (6.61)

$$L\left(\frac{mj\omega}{l}\right)=2L_{10}\left(0,\frac{mj\omega}{l}\right)=\frac{2A_{\nu 0}}{\sqrt{\left[\Omega_\nu^2-\left(\frac{m\omega}{l}\right)^2\right]^2+\frac{4r_\nu\Omega_\nu^2m^2\omega^2}{l^2}}},\tag{14.86}$$

where Ω_ν is the frequency of the $\nu-th$ mode; r_ν is its reduced damping coefficient; $A_{\nu 0}$ is the normalisation coefficient of the structure (see §6 section 3). The frequency response (14.86) is similar to (9.68), hence to find the existence conditions for resonant regimes in this case we may use expressions of types (9.69), (9.71).

Series in m in the equations for the periodic Green's function can be turned into finite expressions (6.75), (6.76) or trimmed for further simplification of the solution by taking the filtering effect of the response (14.86) into account. When $m=1$ we obtain the harmonic approximation for resonant vibration of the vibro-impact system.

Thus the approximate method described allows us to study vibro-impact interactions by examining the equivalent linear problem for the effect of a periodic sequence of impulses and consequently to use the wide opportunities of linear analysis of interacting systems. Note that the form of approximations for the periodic solutions (14.70), (14.84) does not depend on the character of the periodic excitation in the system. The excitation, according to (14.76) influences only the existence conditions of for solutions. This fact reflects the characteristic peculiarity of resonant motion, that its form is determined to considerable extent by the natural characteristics of the system. A convenient detail exists also that, particularly for nonlinear systems, a non-unique solution does not affect the form of the law of motion, but occurs only when finding the value of phase from the existence conditions for the regime. It is clear that for systems with hard nonlinearity, i.e. with a gap ($\Delta>0$), the phase φ_0, corresponding to a stable solution, appears in the range $0\le\varphi_0\le\pi/2$. When $\Delta_0<0$, the corresponding phase range is $\pi/2\le\varphi_0\le\pi$. Using (14.80) we conclude that stable solutions correspond with one and only one of the two phase values which satisfy the existence conditions.

5. The procedure described can be generalised when searching for resonant regimes in systems with many *one-dimensional* impact pairs. Introducing vector $\mathbf{x}(t)$ as the coordinate of the relative disposition of the elements in pairs, we can write an integral equation for it, similar to (6.87):

$$\mathbf{x}(t)=\vartheta(t)-\int_0^T\mathbf{X}(t-\tau)\Phi[\mathbf{x}(\tau),\dot{\mathbf{x}}(\tau)]d\tau .$$

Here $\Phi(\mathbf{x},\dot{\mathbf{x}})$ is the vector for the force characteristics of the impacting pairs; $\vartheta(t)$ is the vector of relative vibration of the elements of the impacting pairs when the impact interaction is neglected; $\mathrm{X}(t)$ is the matrix of periodic Green's functions for the linear part of the system

$$X(t) = \frac{1}{T} \sum_{m=-\infty}^{\infty} \Lambda(mj\omega) \exp(mj\omega t);$$

$\Lambda(j\omega)$ is the matrix of dynamic compliances.

In systems of the type examined it is possible to discover nonlinear (impacting) modes of vibration, characterised by the fact that impact occurs simultaneously for all pairs. The number of impact pairs sharing vibration in various modes can vary. Vibro-impact regimes of the resonant type can occur when such vibration is maintained in the system by external excitation.

For the regimes described we find similarly to (14.69)

$$\mathbf{x}(t) \approx -X(t)\boldsymbol{J};$$

where $\boldsymbol{J}$ is the vector of impact impulses. From the impact conditions $\mathbf{x}(0) = \Delta$, where Δ is the vector of gap (interference) arrangements, we find eventually

$$\mathbf{x}(t) \approx X(t)X^{-1}(0)\Delta. \tag{14.87}$$

The existence conditions for regimes of type (14.87) and the unknown phase of excitation are found from an energy balance of the dissipative and excitation forces in these solutions

$$\int_0^T \{\Lambda^{-1}(s)\vartheta(t)\} \cdot \dot{\mathbf{x}}(t)dt = \int_0^T \{\Lambda^{-1}(s)\mathbf{x}(t)\} \cdot \dot{\mathbf{x}}(t)dt + \sum_i \frac{(1-R_i)J_i^2}{2(1+R_i)M_i},$$

where i is the index of the impact pair under consideration; R_i is the coefficient of restitution for the $i-th$ pair; J_i is their impact impulse; M_i is the reduced mass. The dot indicates a scalar product.

Fig. 14.4 demonstrates the experiment that confirmed the existence of the vibro-impact modes described [246]. A rubber cord with three fixed plastic beads was stretched between the pin of a force transducer and the pushrod of an electrodynamic shaker producing transverse excitation of the cord. The beads had unilateral rigid limitation of motion and it was possible to fix the instant of contact by means of contact pick-ups. The pictures of nonlinear modes (standing waves) were taken by using a stroboscope synchronised with the voltage applied to the exciter.

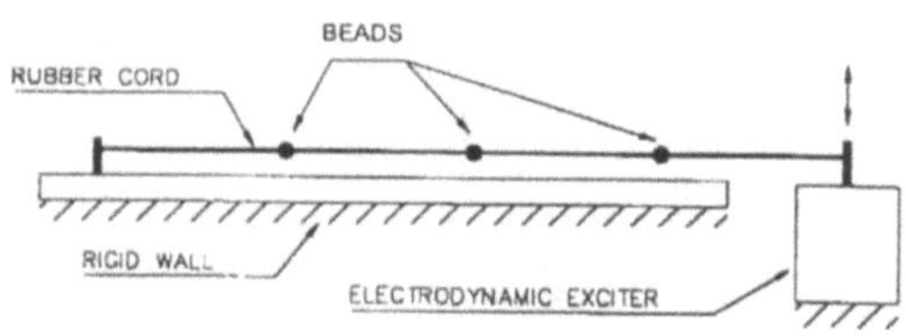

Fig. 14.4

Fig. 14.5, *a* represents the cooperative mode described when the set of beads impacted their limiters simultaneously. Fig. 14.5, *b* confirms this from the records of the transducer signals. Here *1*, *2* and *3* are the signals of the contact pick-ups from the limiters, *4* is the force transducer signal, *5* is the pushrod vibration signal.

The set of beads involved in the cooperative mode of vibration behaved like an impact oscillator with single degree of freedom. These results demonstrate frequency and amplitude pulling, transition from vibro-impact mode to linear mode and back as a result of hard excitation (additional impulse). The frequency band of the cooperative modes is located in the right hand side close to the lowest natural frequency of linear vibration.

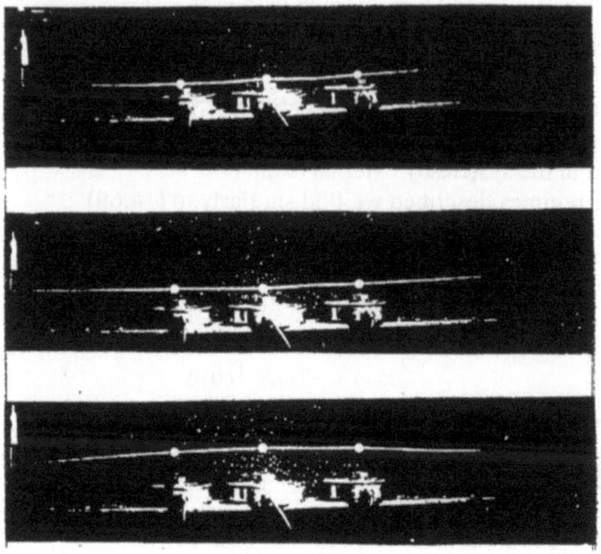

Fig. 14.5, a

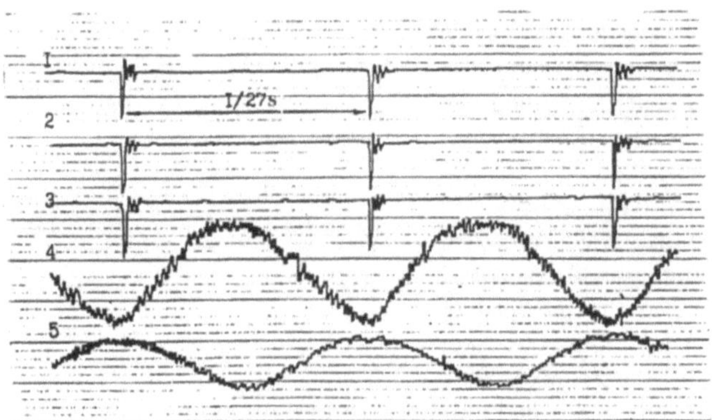

Fig. 14.5, b

In other conditions, the system described demonstrated a localised nonlinear mode of vibration when only one of the beads vibrated with impact against a limiter and others were almost stationary (see Fig. 14.6). The frequency band of these modes was located in the right hand side close to the third natural frequency of the linear system. The impacting particle also demonstrated frequency and amplitude pulling and transitions from vibro-impact to linear mode and back as the result of hard excitation

6. Vibro-impact regimes with large amplitude can exist not only under

conditions of harmonic excitation of the system but also during random action on the system from an exciting force. In the majority of cases such vibration is not stable and after a series of collisions collapses to a regime without impacts. Engineering analysis of similar motion is concerned in principle with finding the conditions under which impact interaction with the limiters will be reduced to a minimum. Using the description of resonant motion of vibro-impact systems in the form (14.70), (14.84) from section 4, allows us to use the method described by M Z Kolovsky in reference [126] to find such estimates.

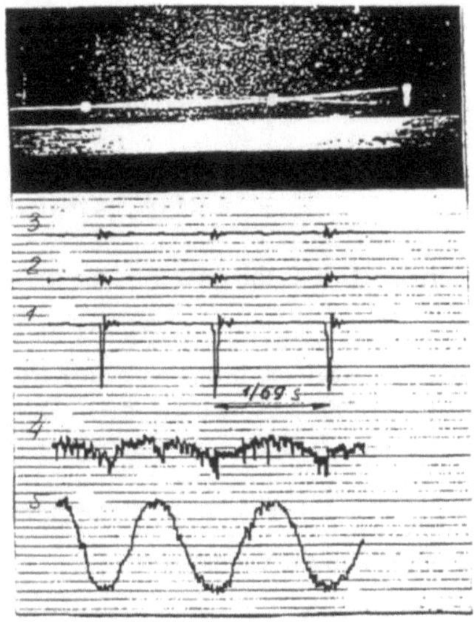

Fig. 14.6

Suppose that in the system described in section 4 (Fig. 14.3), the applied external excitation $P(t) = L^{-1}(s)\vartheta(t)$ is represented by a stationary normal ergodic random process with zero average value. We shall assume that as a result of random thrust in the system a vibro-impact regime develops with frequency ω. For a sufficiently small level of excitation (compared to the level of the inertia and elastic forces) such a regime can only follow a resonant law of motion and, consequently, the corresponding vibration of the relative coordinate $v_0(t)$ of the impacting elements (Fig. 14.3) can be approximated by relationships of type (14.70), (14.84). Let us find the conditions to maintain such a regime with the help of random external excitation $P(t)$.

Representing the power of $P(t)$ for motion $v_0(t)$ by N_e, we find

$$N_e = \frac{1}{T_0}\int_0^{T_0} P(t)\dot{v}(t)dt,$$

where T_0 is the observation time. We find the expected value and mean square value of N_e:

$$\langle N_e \rangle = \left\langle \frac{1}{T_0}\int_0^{T_0} P(t)\dot{v}_0(t)dt \right\rangle = \frac{1}{T_0}\int_0^{T_0} \dot{v}_0(t)\langle P(t)\rangle dt = 0,$$

$$\langle N_e^2 \rangle = \left\langle \frac{1}{T_0^2}\int_0^{T_0} P(t)\dot{v}_0(t)dt \int_0^{T_0} P(t')\dot{v}_0(t')dt' \right\rangle =$$

$$= \frac{1}{T_0^2}\int_0^{T_0}\int_0^{T_0} \dot{v}_0(t)\dot{v}_0(t')R_P(t,t')dtdt'.$$

Here, angle brackets mean statistical averaging; $R_P(t,t')$ is the cross-correlation function of $P(t)$. Taking into account the periodicity of $v_0(t)$ and the stationarity $P(t)$, we find

$$\langle N_e^2 \rangle = \sigma_{N_e}^2 = \frac{1}{T_0^2} \int_0^{T_0}\int_0^{T_0} \dot{v}_0(t)\dot{v}_0(t+\tau)R_p(\tau)dtd\tau . \tag{14.88}$$

The value of N_e follows the normal distribution law $w(\eta)$:

$$w(\eta) = \frac{1}{\sqrt{2\pi}\sigma_{N_e}} \exp\left(-\frac{\eta^2}{2\sigma_{N_e}^2}\right).$$

From this the probability of breakdown of the vibro-impact regime will be as follows:

$$\mathcal{P}\{N_e < N_d(\omega)\} = \int_{-\infty}^{N_d(\omega)} w(\eta)d\eta , \tag{14.89}$$

where $N_d(\omega)$ is the power of the dissipative force:

$$N_d(\omega) = (E_{d1} + E_{d2})/T , \tag{14.90}$$

where $T = 2\pi/\omega$; E_{d1}, E_{d2} are found from formulae of type (14.75), (14.71).

In those cases where $P(t)$ does not contain a periodic component then $\lim_{\tau\to\infty} R_p(\tau) = 0$ and, consequently, as the observation time T_0 increases, the value of $\sigma_{N_e}^2$ decreases. As a result, with the probability prescribed, we can find time T_0 during which the expression

$$N_d(\omega) > N_e \tag{14.91}$$

becomes established.

Suppose that $P(t)$ is white noise with correlation function $R_p(\tau) = S_0\delta(t)$. From (14.88) we find

$$\sigma_{N_e}^2 = \frac{1}{T_0^2} \int_0^{T_0}\int_0^{T_0} \dot{v}_0(t)\dot{v}(t+\tau)S_0\delta(t)dtd\tau = \frac{S_0}{T_0}\langle \dot{v}_0^2(t) \rangle .$$

Assuming, for example, $\mathcal{P}\{N_e < N_d(\omega)\} = 0.98$, we find using the normal distribution law [184]

$$N_d(\omega) > 2\sigma_{N_e} . \tag{14.92}$$

Thus with 98% probability

$$T_0 > \frac{4S_0}{N_d^2(\omega)} \langle \dot{v}_0^2(t) \rangle . \tag{14.93}$$

With the help of Parseval's equality, using (14.74) for the case of a single-sided impact interaction

$$\langle \dot{v}_0^2 \rangle = \frac{2\Delta_0^2\omega^2}{T^2\chi_1^2(0)} \sum_{m=1}^{\infty} m^2 |L(mj\omega)|^2 . \tag{14.94}$$

Substituting (14.90), (14.94) into (14.93) we obtain using (14.71), (14.75)

$$T_0 > \frac{8S_0\omega^2\sum_{m=1}^{\infty} m^2|L(mj\omega)|^2}{\frac{\Delta_0^2}{\chi_1^2(0)}\left[\frac{\omega^2}{\pi}\sum_{m=1}^{\infty} m\,\mathrm{Im}\,L^{-1}(mj\omega)|L(mj\omega|^2 + \frac{(1-R)}{2(1+R)M}\right]^2}. \tag{14.95}$$

For systems with double-sided symmetrical stops we find as the result of a similar calculation

$$T_0 > \frac{8S_0\omega^2\sum_{m=1}^{\infty}(2m-1)^2|L(2m-1)j\omega|^2}{\frac{\Delta_0^2}{\chi_2^2(0)}\left[\frac{2\omega^2}{\pi}\sum_{m=1}^{\infty}(2m-1)\,\mathrm{Im}\,L^{-1}(2m-1)j\omega|L(2m-1)j\omega|^2 + \frac{(1-R)}{2(1+R)M}\right]^2}. \tag{14.96}$$

Let us assume now that $P(t)$ is a harmonic disturbance with random amplitude and phase

$$P(t) = C_1 \sin \nu t + C_2 \cos \nu t .$$

Here C_1 and C_2 have identical mean square values, equal to $\left\langle \tfrac{1}{2} P_0^2(t)\right\rangle$, where $P_0(t)$ is the envelope of the random process [204]. Assuming that the distribution of phase in $P(t)$ is symmetrical we find [204]

$$R_P(\tau) = \left\langle \tfrac{1}{2} P_0^2\right\rangle \cos \nu t . \tag{14.97}$$

Substituting (14.97) into (14.88) we calculate using (14.74) for the case of single-sided collisions

$$\sigma_{N_e}^2 = \frac{\left\langle \tfrac{1}{2} P_0^2\right\rangle \omega^4\Delta_0^2}{\pi^2 i^2 T^2 \chi_1^2(0)} \int_0^{iT}\sum_{m=1}^{\infty} m|L(mj\omega)|\sin(m\omega t)dt \times$$
$$\times \int_0^{iT}\sum_{m=1}^{\infty} m|L(mj\omega)|\sin[m\omega(t+\tau)]\cos \nu\tau d\tau \qquad (i = 1, 2, \ldots). \tag{14.98}$$

When $\nu \neq \omega$ the value $\sigma_{N_e}^2 \to 0$ as i increases and, consequently, a stable regime at frequency ω is not possible. When $\nu = \omega$, from (14.98) we find

$$\sigma_{N_e}^2 = \frac{\left\langle \tfrac{1}{2} P_0^2\right\rangle \omega^2\Delta_0^2}{4\pi^2\chi_1^2(0)}|L(j\omega)|^2 . \tag{14.99}$$

As a result condition (14.92) using (14.90), (14.71), (14.75) takes the form

$$2\left\langle \tfrac{1}{2} P_0^2\right\rangle^{1/2} \omega|L(j\omega)| <$$
$$< -\frac{\Delta_0}{\chi_1(0)}\left\{\frac{\omega^2}{\pi}\sum_{m=1}^{\infty} m\,\mathrm{Im}\,L^{-1}(mj\omega)|L(mj\omega)|^2 + \frac{1-R}{2(1+R)M}\right\}. \tag{14.100}$$

For regimes with symmetrical double-sided collisions we find similarly

$$2\left\langle \tfrac{1}{2} P_0^2\right\rangle^{1/2} \omega|L(j\omega)| <$$
$$< -\frac{\Delta_0}{\chi_2(0)}\left\{\frac{2\omega^2}{\pi}\sum_{m=1}^{\infty}(2m-1)\,\mathrm{Im}\,L^{-1}[(2m-1)j\omega]|L[(2m-1)j\omega]|^2 + \frac{1-R}{2(1+R)M}\right\}. \tag{14.101}$$

Conditions (14.100), (14.101) with probability $\mathcal{P} = 0.98$ guarantee the absence in the system of stable vibro-impact processes.

Sufficient conditions for the satisfactory use of approximation in statistical cases can be framed, by analogy with (9.67), in the form

$$|\Delta_0| > \sigma_{v_0}, \tag{14.102}$$

where σ_{v_0} is the standard deviation of the process calculated when impact disturbance is neglected.

§15. Vibro-impact interaction of visco-elastic bars

1. Using the general approach of §14 for the analysis of vibro-impact interactions in elastic systems, we shall now study the behaviour of straight bars under the effect of periodic lengthwise impacts [19, 20].

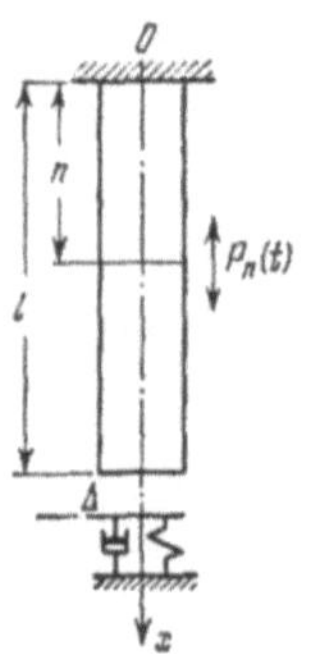

Fig. 15.1

We shall begin by examining a simple system (Fig. 15.1). In a bar of linear visco-elastic material, length l and plane cross-sectional area S, longitudinal waves are excited by a force $P_n(t) = a_p \cos\omega t$, applied at the section with coordinate $x = n$, measured from the fixed end of the bar. During vibration the bottom end of the bar, with coordinate $x = l$, interacts with a limiting stop placed at distance Δ relative to the position of section $x = l$ in the undeformed bar. We shall also write the vibrational displacement of arbitrary section x as a function $u_x(t)$.

By introducing a force characteristic of impact interaction $\Phi(u_l, su_l)$ and dynamic compliance operators $L_n(x,s)$, coupling displacement $u_x(t)$ to the force acting at section $x = n$, we can write the operator equation for the unknown function $u_x(t)$ in the form

$$u_x(t) = L_n(x,s)P_n(t) - L_l(x,s)\Phi(u_l, su_l). \tag{15.1}$$

When $x = l$, equation (15.1) becomes similar to (14.4), hence we can use the solution of the latter.

Suppose, initially, that an exciting force $P_l(t) = a_p \cos\omega t$ is applied to the lower end of the bar which couples with gap (interference) Δ to an elastic stop of stiffness c_2. Under these conditions, the force characteristic of impact interaction takes the form (6.7). For the present situation, to agree with (14.13) we have

$$W(s, m_l, a_l) = L_l^{-1}(l,s) + k(m_l, a_l), \tag{15.2}$$

where $k(m_l, a_l)$ is defined by equation (7.54). Equation (15.2) has a clear physical implication: the dynamic stiffness due to relative motion between the lower end of the bar and the stop is equal to the sum of the dynamic stiffnesses of the elements.

Hence, to complete the solution, it is necessary to find an equation for the operator $L_l(x, j\omega)$, i.e. to establish the reaction at the section with coordinate x on the application of a unit harmonic force (a_p = 1) to the lower end of the bar. We shall solve a suitable linear boundary problem. Using (1.28), we can write the differential equation for the bar in the form

$$s^2 u - (\gamma^2 + bs)\partial^2 u / \partial x^2 = 0\,, \tag{15.3}$$

where $u = u(x,t)$, $s = \partial / \partial t$. The boundary conditions for the fixed and free ends are as follows:

$$u(0,t) = 0, \quad ES\left(1 + \frac{b}{\gamma^2} s\right)\left[\frac{\partial u(x,t)}{\partial x}\right]_{x=l} = \exp(j\omega t) \tag{15.4}$$

where E is Young's Modulus.

Assuming solutions of the form

$$u(x,t) = u_x(t) = L_l(x, j\omega)\exp(j\omega t) \tag{15.5}$$

we obtain, substituting (15.5) into (15.3),15.4):

$$(j\omega)^2 L_l(x, j\omega) - (\gamma^2 + j\omega b)\frac{d^2 L_l(x, j\omega)}{dx^2} = 0 \tag{15.6}$$

$$L_l(0, j\omega) = 0, \quad ES(1 + j\omega\frac{b}{\gamma^2})\left[\frac{dL_l(x, j\omega)}{dx}\right]_{x=l} = 1\,. \tag{15.7}$$

The solution of the linear differential equation (15.6) of second order with constant coefficients satisfied by boundary conditions (15.7) takes the form

$$L_l(x, j\omega) = -\frac{\lambda\gamma^2}{ES\omega^2}\frac{\exp(\lambda x) - \exp(-\lambda x)}{\exp(\lambda l) + \exp(-\lambda l)}\,, \tag{15.8}$$

where

$$\lambda = j\omega / \sqrt{\gamma^2 + bj\omega}\,. \tag{15.9}$$

Since $b = \gamma^2 \varkappa / (\pi\omega)$, then, taking into account the smaller value of the damping coefficient $\varkappa$, the solution obtained can be simplified, by expansion as a series in $\varkappa$ and neglecting terms of higher order. From (15.9), with first degree accuracy we find

$$\lambda = \frac{j\omega}{\gamma} + \frac{\varkappa\omega}{4\pi\gamma}\,. \tag{15.10}$$

Substituting (15.10) into (15.8) and restricting the expansion of the exponent, for small values of $\varkappa$, i.e. linear terms only, we obtain at $x = l$

$$L_l(l, j\omega) = \frac{l}{ES\zeta}\left(\tan\zeta - j\frac{\varkappa}{4\pi}\frac{\zeta + \frac{1}{2}\sin 2\zeta}{\cos^2\zeta}\right), \tag{15.11}$$

where $\zeta = \omega l / \gamma$. In §6 section 3, we obtained a different form for the dynamic compliance operator for a system with distributed parameters in the form of a series for the natural vibration modes of the system. Thus equation (15.11) is equivalent to equation (6.63) by substituting $p = q = l$ into the latter.

Clearly, calculation using equation (15.11) is more convenient, since it does not require a series summation. Moreover, under certain conditions, it is not necessary to calculate the natural frequencies of the system, which is required when using operators of the form (7.63). At the same time the idea of a closed presentation for the dynamic compliance operator is often rather unwieldy and its decomposition into a series for natural vibration, using truncated series including principal terms, can substantially help to simplify the expressions. For a more detailed discussion of the questions mentioned, the reader is recommended to see the book by I M Babakov [24].

Substituting (15.11) into (15.2), we find when $s = j\omega$

$$W(j\omega, m_l, a_l) = \frac{ES}{l}\left[\zeta\cot\zeta + k_0(m_l, a_l) + j\frac{\varkappa\zeta}{4\pi}\frac{\zeta + \frac{1}{2}\sin 2\zeta}{\sin^2\zeta}\right], \qquad (15.12)$$

where $k_0(m_l, a_l) = k(m_l, a_l)l / ES$. Hence,

$$U(\omega, m_l, a_l) = \frac{ES}{l}\left[\zeta\cot\zeta + k_0(m_l, a_l)\right],$$
$$V(\omega) = \frac{ES\varkappa\zeta}{4\pi l}\frac{\zeta + \frac{1}{2}\sin 2\zeta}{\sin^2\zeta}. \qquad (15.13)$$

By analogy with (15.17), we find

$$a_l = \chi\left\{\left[\zeta\cot\zeta + k_0(m_l, a_l)\right]^2 + \left(\frac{\varkappa\zeta}{4\pi}\frac{\zeta + \frac{1}{2}\sin 2\zeta}{\sin^2\zeta}\right)^2\right\}^{-\frac{1}{2}}, \qquad (15.14)$$

where $\chi = a_p l/(ES)$ is the displacement of the lower end of the bar due to the action of static force a_p.

We can write the coefficients of the harmonic linearisation for the static characteristics of the stops (8.53), (8.54) in the following form:

$$h = -a_l c_0\left[\frac{\alpha}{2} - \frac{1}{\pi}(\alpha\arcsin\alpha + \sqrt{1-\alpha^2})\right], \qquad (15.15)$$

$$k_0 = c_0\left[\frac{1}{2} - \frac{1}{\pi}(\arcsin\alpha + \alpha\sqrt{1-\alpha^2})\right]. \qquad (15.16)$$

Here, $h = m_{f^*}l/(ES)$, $c_0 = c_2 l/(ES)$, $\alpha = (\Delta - m_l)/a_l$, when $-1 \le \alpha \le 1$ since only under these conditions that the rod and stop will interact.

To define the basic parameters of motion of the lower end we must add to expressions (15.14) - (15.16) the expressions for the constant components, similar to (15.9)

$$m_l = -L_l(l,0)m_{f^*}(m_l, a_l). \qquad (15.17)$$

Using (15.11), (15.15) this can be transformed to the form

$$a = \Delta\left\{\alpha + c_0\left[\frac{\alpha}{2} - \frac{1}{\pi}(\alpha\arcsin\alpha + \sqrt{1-\alpha^2})\right]\right\}^{-1}. \qquad (15.18)$$

2. Using equations (15.14), (15.16), (15.18), we can construct the amplitude-frequency response for the lower section of the bar. According to (15.14), the expression

$$\zeta_0 \cot \zeta_0 + k_0(m_l, a_l) = 0\,, \tag{15.19}$$

where $\zeta_0 = \omega_0 l / c$ defines all the natural frequencies ω_0 for vibration of the elements of the system examined. Because the system is nonlinear these frequencies depend on the vibration amplitude a_l. Thus, using (15.16) and (15.18),

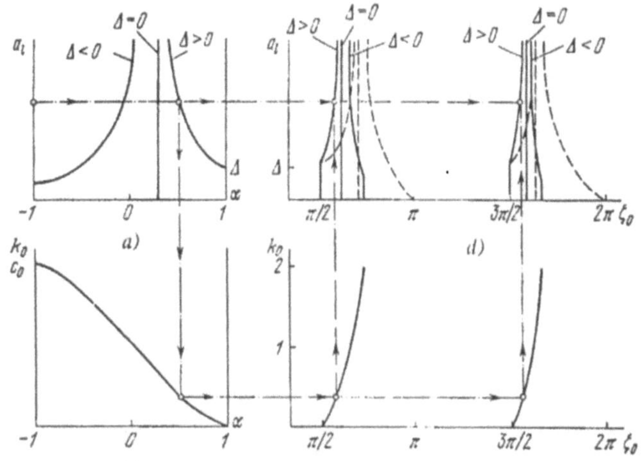

Fig 15.2

equation (15.19) determines the configuration for the backbone curves of the system. To find them, it is convenient to use the graphical method shown in Fig. 15.2. Relationships (15.18), (15.16), (15.19) are shown in Fig. 15.2, *a* - *c* with coordinates shown on the axes. In addition, by choosing various values of a_l, as the ordinate in Fig. 15.2 *a*, and following the construction shown by the arrows, we can determine the corresponding value α on the abscissa and from that the value of k_0 in Fig. 15.2 *b* and ζ_0 in Fig. 15.2 *c*. The resulting backbone curve data for $a_l(\zeta_0)$ is drawn in Fig. 15.2 *d*.

For each value of a_l there are many natural frequencies, ζ_0, corresponding to the different modes of vibration (in Fig. 15.2d the backbone curves for the lower two modes are shown). For a system with gap $(\Delta > 0)$ the backbone curve for each mode has a hard character while for a system with interference $(\Delta < 0)$ the curve is of soft type. In addition, linear vibrations can also occur with both cases. For $(\Delta > 0)$ vibration with amplitude $a_l < \Delta$ occurs; for $(\Delta < 0)$ vibration occurs without loss of contact between the lower end of the bar and the elastic stop. In the latter case, to obtain movement without impact the amplitude of vibration must not exceed the relative deformation of the contacting elements. Taking into account that for an interference Δ the static force of interaction equals $-\Delta ES/l$, while the stiffness of the relative deformation occurring is equal to the sum of the stiffness of the components, we obtain the conditions for non-impact motion in the form

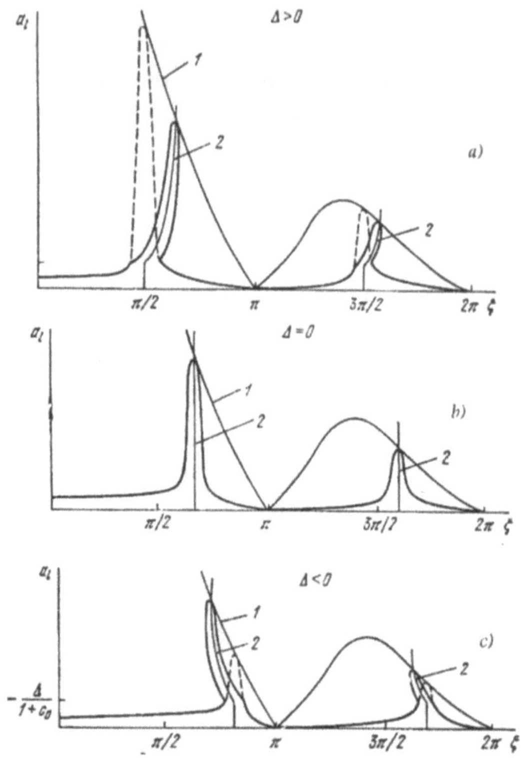

Fig 15.3

$$a_l < \frac{-\Delta ES/l}{ES/l + c_2} = -\frac{\Delta}{1+c_0}.$$

For $(\Delta = 0)$, the system has a spectrum of constant natural frequencies, shifted towards higher values from the corresponding natural frequencies of the bar (Fig. 15.2 *c*).

The degree to which the nonlinear effects manifest themselves depends directly on the increase in rigidity of the stops. In the limit as $c_0 \to \infty$ the equation for k_0 by comparison with (15.37), takes the form

$$k_0 = 2(1 - \Delta / a_l). \tag{15.20}$$

Substituting this value into (15.19) we obtain the equations for the backbone curves

$$a_l = \frac{2\Delta}{\zeta_0 \cot \zeta_0 + 2}. \tag{15.21}$$

Imposing the requirement that $a_l > 0$ in (15.21) then in the case where there is a gap the equation becomes

$$\zeta_0 \cot \zeta_0 + 2 \geq 0, \tag{15.22}$$

The sign of inequality (15.22) changes to the opposite value for interference. From this it follows that the natural frequency of the system lies in the following bands:

$$(2m-1)\pi/2 \le \zeta_0 < \zeta_{0m} \quad \text{when} \quad \Delta > 0,$$

$$\zeta_{0m} < \zeta_0 < \pi m \quad \text{when} \quad \Delta < 0,$$

$$\zeta_0 = \zeta_{0m} \quad \text{when} \quad \Delta = 0 \quad (m = 1,2,\cdots),$$

where ζ_{0m} is the solution of equation (15.22). The backbone curves for the limited system are drawn for comparison in Fig. 15.2 *d* as dotted lines.

Putting $\zeta = \zeta_0$ in (15.14) and taking account of (15.19), we obtain

$$a_l = \frac{4\pi\chi \sin^2 \zeta_0}{\varkappa\zeta_0(\zeta_0 + \frac{1}{2}\sin 2\zeta_0)}. \tag{15.23}$$

This relationship defines the energy boundary of vibration. The intersection of curve *1* (Fig. 15.3), given by equation (15.23), and the backbone line *2,* from (15.14), gives upper values of the amplitude-frequency response. From (15.23) it follows that with increasing mode number and consequently ζ_0, the amplitude of resonance of the relevant forms dies away rapidly. In order to achieve the condition max $a_l > \Delta$ with the initial requirement for $\Delta > 0$ we find from (15.23) that vibration with impact against the stops only occurs for the condition

$$\zeta < \sqrt{4\pi\chi/(\Delta\varkappa)}. \tag{15.24}$$

The relationship (15.24) has a clear physical meaning.

The remaining points on the amplitude curve that correspond to the nonlinear regime are found from the simultaneous solution of the final equations (15.14), (15.15) and (15.18). To achieve this the following method for calculation is recommended. The equation for k_0 (15.15) is substituted into (15.18), and the values subsequently obtained for various values $\zeta = \text{const}$ are plotted as curve *1* (Fig. 15.4); on the relationships (15.18) already used to construct the backbone line (see Fig. 15.2 *a*). The intersection points of the curves shown define corresponding values of the amplitude response a_l and parameter α. Using these values in the expression $\alpha = (\Delta - m_l)/a_l$ the value of m_l can be found. Fig. 15.3 shows examples of amplitude curves for three basic positions of the stop. Only the frequencies of the two lowest resonance zones are shown. The branches of the amplitude curves which correspond to the linear regime of vibration ($a_l < \Delta$ in Fig. 15.3 *a* and $a_l < -\Delta/(1+c_0)$ in Fig. 15.3 *b*) are found from the same relationships by substituting in them the value $\alpha = 1$ when $\Delta > 0$ and $\alpha = -1$ when $\Delta < 0$. Fig. 15.3, also shows dotted the unrealised branches of resonance curves for the linear system, constructed using (15.14) for $k_0 \equiv 0$ when $\Delta > 0$ (Fig. 15.3 *a*) and $k_0 = c_0$ for $\Delta < 0$ (Fig. 15.3, *b*). The first case represents vibration of the bar without stops and the second case represents permanent connection of the lower end of the bar with the stop.

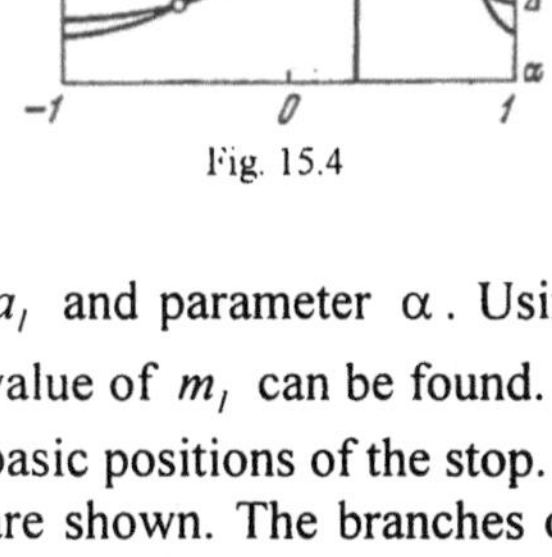

Fig. 15.4

Fig. 15.3 shows that as the result of vibro-impact interaction of the lower end of the bar with the stop a nonlinear distortion occurs for all resonance curves, similar to that observed for systems with one degree of freedom.

For the limiting case as $c_0 \to \infty$ when we substitute (15.20) in (15.14) and put $\varkappa \equiv 0$, we obtain a simple relationship for the resonance curves of nonlinear systems

$$a_l = \frac{2\Delta \pm \chi}{\zeta \cot \zeta + 2}, \tag{15.25}$$

the existence domains of which, according to (15.25), are defined by the

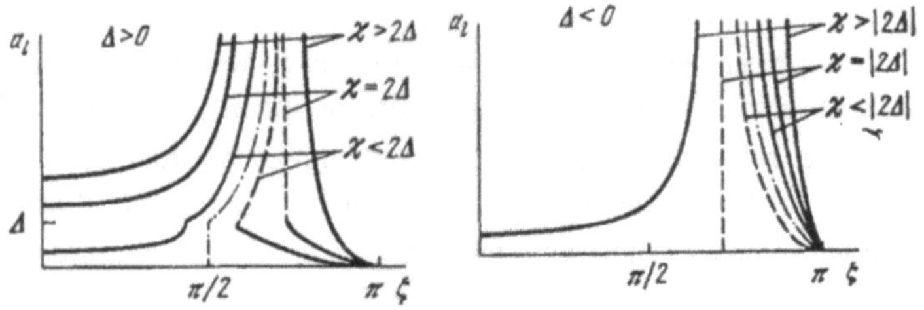

Fig. 15.5

relationships $a_l > 0$ when $\Delta > 0$ and $a_l > 0$ when $\Delta < 0$. Comparison of equations (15.25) and (15.21) shows that as the value of excitation χ increases the amplitude-frequency response, then envelopes the backbone lines (Fig. 15.5) and when $\chi > 2\Delta$ change from double to single values. It is easy to see that this condition represents a particular case similar to the general expression (14.46).

In agreement with (15.25) for $\chi > 2\Delta$ the vibro-impact regime exists for all frequency bands; for values of $\chi < 2\Delta$ the regime exists only for frequencies satisfying the relationships

$$\begin{aligned} \zeta \cot \zeta + 2 > 0 \quad \text{when} \quad \Delta > 0 \\ \zeta \cot \zeta + 2 < 0 \quad \text{when} \quad \Delta < 0 \end{aligned} \tag{15.26}$$

These conditions can also be obtained from the basic relationship (14.49). In accordance with the results of §14 section 3 stability exists for solutions with greater amplitude, i.e. corresponding to the positive sign in (15.25) for $\Delta > 0$ and negative sign for $\Delta < 0$. In Fig. 15.5 unstable conditions are shown dotted.

3. Let us now examine the vibration of a system in which interference occurs as the result of a compressive force (Fig. 15.6 *a*). This consists of a bar, rigidly attached at its upper end to an infinitely large mass whose vibration can be ignored. The bar is excited by a harmonic axial force. The complete system is compressed against the stop by a static force G.

The system shown represents a device for ultrasonic material processing [74, 178]. Fig. 15.6 *b* shows such a device for ultrasonic erosion. The main device of the machine is an acoustic head comprising a body *1*, a piezostrictive or magnetostristive vibrator *2*, with a stepped concentrator *3*, and tool *4*, attached.

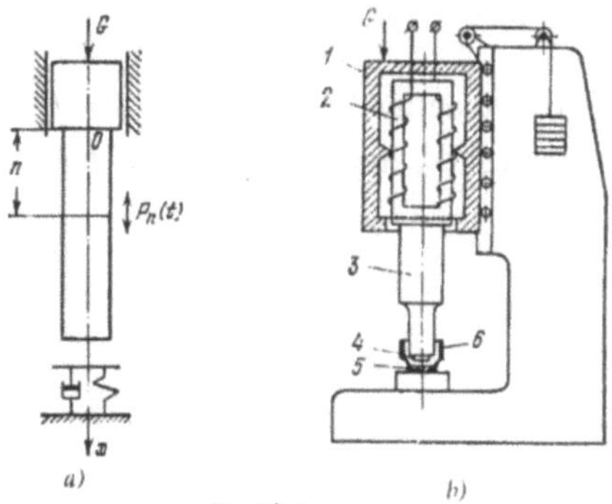

Fig. 15.6

The piezostrictive or magnetostristive device receives impulses from a powerful high frequency generator (at a frequency of $20kHz$). The tool is compressed to workpiece item *5* with a static force *G*. In the erosion zone *6*, an abrasive suspension is placed. Processing takes place as the result of impact of the tool on the abrasive particles which settle on the surface of the workpiece.

The specific feature of the dynamic model described (Fig. 15.6 *a*) consists of the displacement of an upper mass which when vibrating, finds some dynamic equilibrium position. The origin of the displacement function $u_x(t)$ is taken at a prescribed section of the undeformed rod while coordinates of section x are measured from the static position of the top surface of the bar under the action of force G. In the system examined $m_k \equiv G$. Using this, the relative deformation of the bottom section of the bar and the stop of rigidity c_2 under the action of force G, takes the form

$$\frac{Gl}{ES}+\frac{G}{c_2}=h\left(1+\frac{1}{c_0}\right)\quad\left(h=\frac{Gl}{ES}\right). \tag{15.27}$$

Now we can make use of equations for the harmonic linearisation coefficients in the form (15.15), (15.16) using interference values Δ from (15.27). From (15.15) we obtain

$$a_l=\frac{h}{c_0\left[\frac{1}{\pi}(\alpha\arcsin\alpha+\sqrt{1-a^2})-\frac{\alpha}{2}\right]}. \tag{15.28}$$

Assuming as previously that the harmonic excitation is applied to the lower end of the bar we obtain again relationship (15.14). Hence the problem is reduced to the solution of a system of finite equations (15.14), 15.15) and (15.28). This solution is carried out using a technique identical to that explained in section 2.

We shall confine ourselves here to the examination of the limiting case consisting of a rigid stop $(c_2\to\infty)$. Taking (15.53) into account, we find the following equation for resonance curves from (15.14) when $\varkappa\equiv 0$:

$$a_l = \frac{-2h \pm \chi}{\zeta \cot \zeta}, \tag{15.29}$$

and the existence domains are defined from the condition $a_l > 0$.

For $\chi \equiv 0$ this becomes the equation for the backbone curve defined in frequency bands

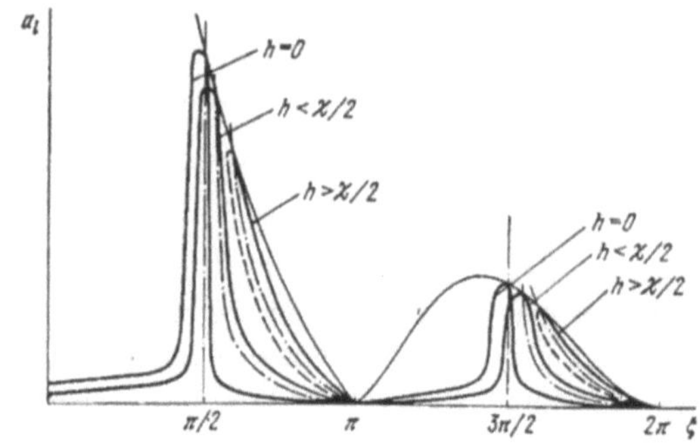

Fig. 15.7

$$\tfrac{1}{2}\pi(2m-1) < \zeta_0 < \pi m \quad (m = 1, 2, \ldots) \tag{15.30}$$

and having fixed asymptotes coinciding with the natural frequencies of the bar $\zeta_0 = \pi(2m-1)/2$ (see (15.11)).

According to (15.29) when $\chi < 2h$ solutions for the parameters exist in pairs in the natural frequency bands of the system (15.30). In accordance with the results of §14 section 3, stable regimes correspond to large amplitude values, i.e. having the negative sign in (15.29).When $\chi > 2h$, the conditions for a unique solution are satisfied, and these are stable for any excitation frequency.

Fig. 15.7 shows the amplitude-frequency response of the system, constructed for the two lower resonance zones with various values of χ and h. Here the chain-dotted line shows the backbone curves, the thin line shows the energy boundaries for vibration (15.23), the thicker line shows branches of the amplitude curve corresponding to stable solutions, while the dotted line shows unstable values. It can be seen that as the compression force increases, so does the frequency of maximum vibration amplitude. When $\chi < 2h$ vibration can occur causing loss of contact at the stops either because of pulling at low frequencies or because of hard excitation.

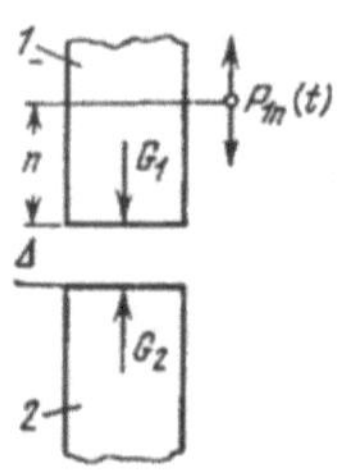

Fig. 15.8

Note that a problem close to the one examined at this point has been solved in reference [154] using a full Fourier series. The results presented agree with results from this reference. The nonlinear phenomena described were observed experimentally in [21].

4 Let us now look at the longitudinal impact interactions between two straight bars with co-linear axes (Fig. 15.8). The two contacting bar ends, whose coordinates are taken as the

origin, are arranged with a gap (interference) Δ between the two and which come into contact under the action of a periodic force $P_{1n}(t) = a_p \cos \omega t$, applied at the section with coordinates $x = n$ in bar *1*. For generality, we will assume that both bars are loaded by static axial forces G_i $(i = 1, 2)$, applied at the origin of each.

As before, assuming that the displacement of sections is $u_{ix}(t)$ from their undeformed position and introducing a characteristic force impact interaction $\Phi(v, sv)$ where $v(t) = u_{10}(t) - u_{20}(t)$, we obtain the following operator equations:

$$u_{1x}(t) = L_{10}(x,0)G_1 + L_{1n}(x,s)P_{1n}(t) - L_{10}(x,s)\Phi(v,sv), \tag{15.31}$$

$$u_{2x}(t) = L_{20}(x,0)G_2 + L_{20}(x,s)\Phi(s,sv) \tag{15.32}$$

where $L_{in}(x, s)$ is the dynamic compliance operator of the *i*-th bar acting at an arbitrary section with coordinates *n* on the section with coordinate *x*.

Subtracting (15.32) from (15.31) when $x = 0$, we obtain the equation for the relative motion of the contact section

$$v(t) = L_{1n}(0,s)P_{1n}(t) + L(s)\{G - \Phi[v(t), sv(t(]\} \tag{15.33}$$

where

$$L(s) = L_{10}(0,s) + L_{20}(0,s);$$

$$G = [L_{10}(0,0)G_1 - L_{20}(0,0)G_2]L^{-1}(0).$$

Equations (15.31) to (15.33) are similar to the general equations (14.2) to (14.4) examined in §14, hence we may use the solutions found in §14 for the problem of interaction between the bars.

To simplify the results, we will assume that the excitation force is applied to the section at the origin of bar *1*, i.e. it takes the form $P_{10}(t) = a_p \cos \omega t$. Then, to obtain solutions relative to the section at the origin, we can solve the problem using only a dynamic compliance factor $L_{i0}(0, j\omega)$ for the contacting section.

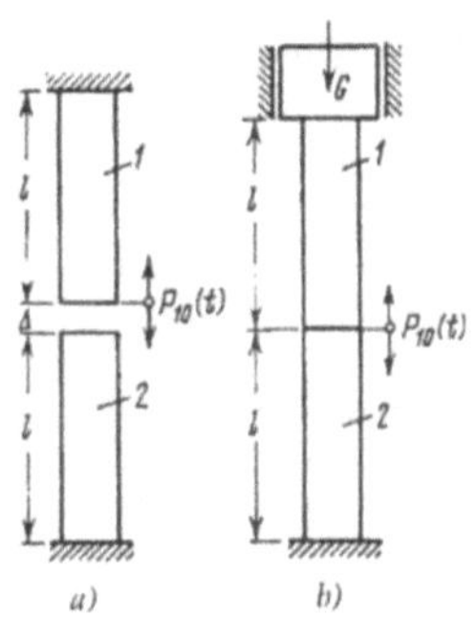

Fig. 15.9

Suppose, as before, that the interacting bars are made from a linear visco-elastic material and are of finite length either built-in at their ends (Fig. 15.9 *a*) or compressed by an axial force *G* (Fig. 15.9b) (in the system shown in Fig. 15.9 *b*, bar *1* is attached to a large mass whose vibration can be ignored), then we may write the equations for the dynamic compliance in the form (15.11) as:

$$L_{i0}(0, j\omega) = \frac{l_i}{E_i S_i \zeta_i}\left(\tan \zeta_i - j\frac{\varkappa_i}{4\pi}\frac{\zeta_i + \frac{1}{2}\sin 2\zeta_i}{\cos^2 \zeta_i}\right). \tag{15.34}$$

The interaction, as in the previous case, will be considered to be perfectly elastic, corresponding to an unlimited rise in the contact stiffness $c_0 \to \infty$.

We can rewrite (15.17), using (15.13) and (15.5) in the following form:

$$a_v = a_p\left|\frac{L_{10}(0, j\omega)}{1 + k(a_v)[L_{10}(0, j\omega) + L_{20}(0, j\omega)]}\right|. \tag{15.35}$$

We shall consider the interaction of identical bars

$$L_{10}(0, j\omega) = L_{20}(0, j\omega), \tag{15.36}$$

Substituting (15.34) and (15.36) into (15.35) we obtain the equation for amplitude a_v, similar to (15.14):

$$a_v = \chi\left\{[\zeta\cot\zeta + 2k_0(a_v)]^2 + \left(\frac{\varkappa\zeta}{4\pi}\frac{\zeta + \frac{1}{2}\sin 2\zeta}{\sin^2\zeta}\right)^2\right\}^{-\frac{1}{2}} \tag{15.37}$$

$$\left(k_0 = \frac{kl}{ES}\right).$$

The term $k(a_v)$ for the case of fixed bars (Fig. 15.9 *a*) to agree with (15.37) and (15.36), takes the form:

$$k(a_v) = \frac{1}{L_{i0}(0,0)}\left(1 - \frac{\Delta_0}{a_v}\right) = \frac{ES}{l}\left(1 - \frac{\Delta_0}{a_v}\right). \tag{15.38}$$

In the case of compressed bars (Fig. 15.9 *b*), equation (15.53) remains valid.

From (15.37) the backbone curves of the amplitude response for relative motion of the system considered become

$$\zeta\cot\zeta + 2k_0(a_v) = 0. \tag{15.39}$$

Substituting in turn from (15.38) and (15.53) into (15.39) we obtain the relationships

$$a_v = \frac{2\Delta_0}{\zeta\cot\zeta + 2}, \tag{15.40}$$

$$a_v = -\frac{4h}{\zeta\cot\zeta} \quad \left(h = \frac{Gl}{ES}\right). \tag{15.41}$$

From (15.37) and taking into account (15.39) we can find the equation for the

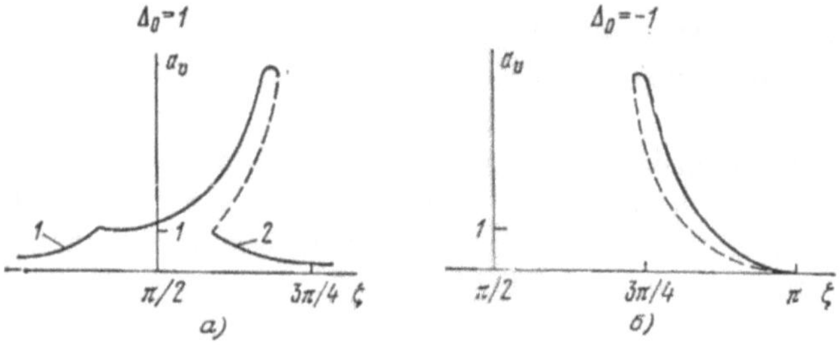

Fig. 15.10

energy boundary of vibration, similar to (15.23)

$$a_v^* = \frac{4\pi\varkappa\sin^2\zeta}{\varkappa\zeta(\zeta + \frac{1}{2}\sin 2\zeta)}, \tag{15.42}$$

According to (15.37), intersections on the (ζ, a_v) plane between the backbone curves and the energy boundary of vibration define the maximum points of the resonance response a_v^*. The remaining points can be found using a simple graphical construction, similar to that shown in section 2. When $\varkappa \equiv 0$, substituting in turn from (15.38) and (15.53) into (15.37), we obtain the explicit expression for the amplitude-frequency response

$$a_v = \frac{2\Delta_0 \pm \chi}{\zeta \cot \zeta + 2}, \tag{15.43}$$

$$a_v = \frac{-4h \pm \chi}{\zeta \cot \zeta}. \tag{15.44}$$

Examples of resonance curves for the relative motion of contacting sections for fixed bars are shown in Fig. 15.10. In addition, the branches *1* and 2 corresponding to linear vibration without impact are also shown. Fig. 15.10 *a* and *b* represent systems with pre-arranged gap or interference. Similar results for compressed bars are shown in Fig. 15.11. Examples for soft and hard excitation of the vibro-impact regime are shown in Fig. 15.11 *a* and *b* respectively. Both here and later in this section, the curves are drawn at the lowest mode of vibration of the bar, with the dotted lines showing branches for unstable solutions.

Let us now examine the absolute vibration of the contacting sections. Substituting (15.34) into (15.38) and (15.39) and re-arranging we obtain

$$a_{10} = \chi \left| \frac{\zeta \cot \zeta + k(a_v) + jB}{\zeta \cot \zeta [\zeta \cot \zeta + 2k(a_v)] + 2jB[\zeta \cot \zeta + k(a_v)]} \right|, \tag{15.45}$$

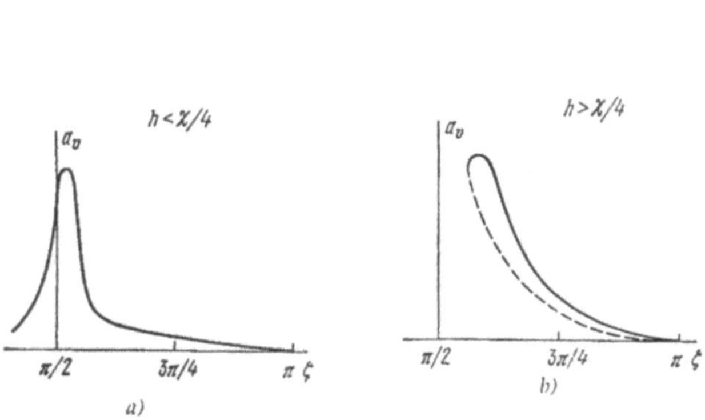

Fig. 15.11

$$a_{20} = \frac{\chi k(a_v)}{|\zeta \cot \zeta [\zeta \cot \zeta + 2k(a_v)] + 2jB[\zeta \cot \zeta + k(a_v)]|} \tag{15.46}$$

Where, for convenience, we make the substitution

$$B = \frac{\varkappa \zeta (\zeta + \frac{1}{2} \sin 2\zeta)}{4\pi \sin^2 \zeta}. \tag{15.47}$$

Neglecting energy dissipation in the system $(\varkappa \equiv 0)$, we can obtain the amplitude equations in an explicit form, which has a clear physical interpretation. For the case of fixed bars (Fig. 15.9 *a*), using (15.38) and (15.43) we find from (15.45) and (15.46) when $B \equiv 0$

$$a_{10} = \left| \frac{\Delta_0 \zeta \cot \zeta \pm \chi(\zeta \cot \zeta + 1)}{\zeta \cot \zeta (\zeta \cot \zeta + 2)} \right|, \tag{15.48}$$

$$a_{20} = \left| \frac{\pm \chi - \Delta_0 \zeta \cot \zeta}{\zeta \cot \zeta (\zeta \cot \zeta + 2)} \right| \tag{15.49}$$

where existence domains for the solution are found using (15.43) and the conditions $a_{i0} > \Delta_0$ for $\Delta_0 > 0$ and $a_{i0} > 0$ for $\Delta_0 < 0$.

When $\chi \equiv 0$ from (15.48) and (15.49) we obtain the equation for the backbone curves for absolute vibration which has half the value of the amplitudes compared to the similar expression (15.43) for relative vibration

$$a_{i0} = \frac{\Delta}{\zeta \cot \zeta + 2} \quad (i = 1, 2). \tag{15.50}$$

At the same time, according to (15.48) and (15.49), in the presence of excitation together with vibration corresponding to the backbone curves (15.50), which we will call *clapping*, the system resonates in absolute motion at the natural frequencies of the bars, satisfying the condition $\cot \zeta = 0$, i.e. at frequencies

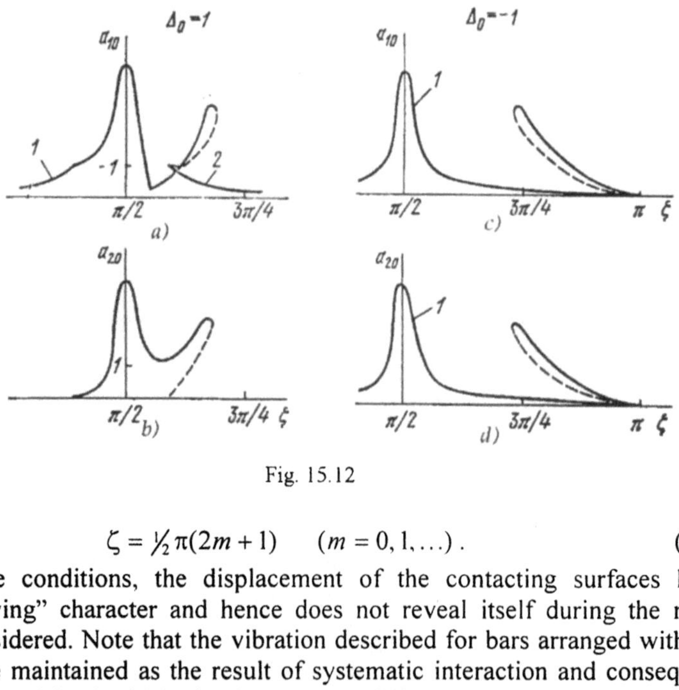

Fig. 15.12

$$\zeta = \tfrac{1}{2}\pi(2m + 1) \quad (m = 0, 1, \ldots). \tag{15.51}$$

Under these conditions, the displacement of the contacting surfaces has an "accompanying" character and hence does not reveal itself during the relative motion considered. Note that the vibration described for bars arranged with a gap will only be maintained as the result of systematic interaction and consequently, can not be explained within the framework of linear behaviour. For this reason they will be called *grazing*.

The maximum vibration amplitude occurs at the intersection of the backbone curves (15.50), (15.51) with the energy boundary for absolute vibration, defined from(15.45) and (15.46). Substituting into these expressions the values from

(15.39) and bearing in mind that according to (15.42), (15.47)

$$B = \chi / a_v^*, \qquad (15.52)$$

we obtain using (15.38)

$$a_{10}^* = a_v^* \sqrt{\frac{(a_v^* - \Delta)^2 + \chi^2}{4(a_v^* - \Delta)^2 + \chi^2}}, \qquad (15.53)$$

$$a_{20}^* = a_v^* \frac{a_v^* - \Delta}{\sqrt{4(a_v^* - \Delta)^2 + \chi^2}}. \qquad (15.54)$$

In these relationships a_v^* is found from equation (15.42). From (15.53) and (15.54) it follows for a sufficiently large value of $a_v^* - \Delta$ compared to χ, corresponding to low internal damping, that

$$a_{10}^* \approx a_{20}^* \approx a_v^* / 2. \qquad (15.55)$$

Fig. 15.12 shows examples of the amplitude-frequency response for absolute

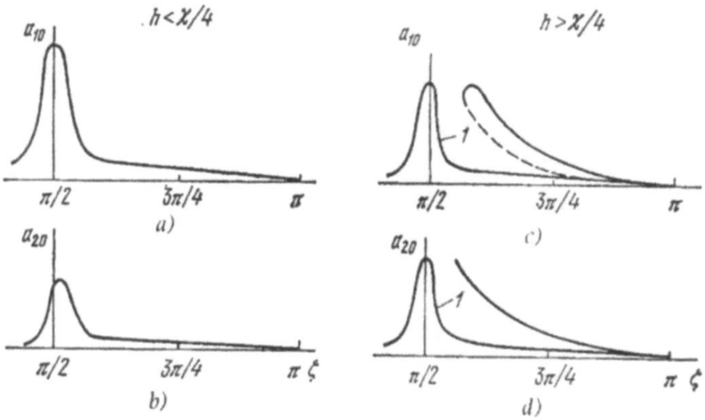

Fig. 15.13

displacements of the contacting sections of fixed bars. Fig. 15.12 *a, b* corresponds to bars installed with a gap while Fig. 12 *c, d* to those with interference. In Fig. 15.12 *a*, apart from branches relative to the vibro-impact regimes, curves *1* and *2* are also shown representing linear vibration of the active bar within the limits of the gap without collisions. In Fig. 15.12 *c, d*, curve 1 shows linear vibration without loss of contact.

For bars with a gap, as the excitation frequency increases at the free end of the active bar, then the free end of the active bar increases its amplitude until interaction with the passive bar occurs. Both bars oscillate in phase, i.e. with identical direction of displacement of similar sections and reach maximum amplitude at a frequency $\zeta = \pi / 2$, equal to their natural linear frequencies.

Once anti-resonance has passed, similar sections move in clapping character, i.e. they move in opposite directions. Under these conditions, impact has the strongest character. As the frequency increases, the amplitude of such motion increases in accordance with the resonance curves for relative vibration (see Fig. 15.10 *a*). When vibration breaks down, as the maximum is passed, the active bar vibrates linearly within the limits of the gap, but the vibration of the passive bar decays. A similar picture occurs, in principle, for all frequency bands corresponding to higher modes of vibration; however, with practical values of internal damping the resonance phenomena at these frequencies is weak.

As the frequency reduces from post-resonance regions (Fig. 15.12 *a, b*) resonance in clapping motion does not happen. The active bar oscillates to the origination of contact, with the result of a decrease in its amplitude after contact. Furthermore, both contact sections pass anti-resonance and consequently resonate with grazing at frequency $\zeta = \pi/2$. Thus, as the frequency of excitation increases, two resonance peaks occur, but with decreasing frequency, only one.

A contrasting picture emerges for bar devices with pre-arranged interference (Fig. 15.12 *c, d*). In this case, as the frequency increases, the bars do not separate and resonate at their own natural frequencies. As the frequency decreases from the post-resonance domain, the bars separate as a result of vibration and they move into a clapping regime until they break apart as they cross the maximum amplitude response, corresponding to the resonance curves of relative vibration (see Fig. 15.10 *b*).

We shall now analyse absolute vibration of contacting sections of compressed bars (Fig. 15.9 *b*). From (15.45) and (15.46) when $\varkappa = 0$ and, consequently $B \equiv 0$, taking into account (15.53) and bearing in mind that a_v in this case is defined by (15.44), we obtain

$$a_{10} = \frac{-2h \pm \chi}{\zeta \cot \zeta}, \tag{15.56}$$

$$a_{20} = \frac{-2h}{\zeta \cot \zeta}. \tag{15.57}$$

When $\chi \equiv 0$ we have the equation for the backbone curves for absolute vibration

$$a_{i0} = \frac{-2h}{\zeta \cot \zeta}, \tag{15.58}$$

which also has twice the minimum amplitude compared to the similar expression (15.44) for relative vibration. For maximum amplitude a_{i0}^* by using (15.53), (15.39) and 15.52) we find from (15.45) and (15.46)

$$a_{10}^* = a_v^* \sqrt{\frac{4h^2 + \chi^2}{16h^2 + \chi^2}}, \tag{15.59}$$

$$a_{20}^* = a_v^* \frac{2h}{\sqrt{16h^2 + \chi^2}}, \tag{15.60}$$

where a_v^* is defined by (15.42). When $\chi << 2h$ we obtain the corresponding amplitude values similar to (15.55).

Fig. 15.13 shows examples of the amplitude-frequency response for absolute motion of the contacting sections of compressed bars. The additional curve *1* corresponds to linear vibration without separation.

Fig. 15.13 *a, b* shows the case of smooth excitation of vibro-impact regimes, occurring with relatively small compressive forces (compare with Fig. 15.11 *a*). As the compressive force increases above the critical value $h = \chi/4$ (see Fig. 15.11 *b*) the configuration of the amplitude-frequency response changes form (Fig. 15.13 *c, d*) and becomes similar to the response of the system with fixed bars, arranged with initial interference (Fig. 15.12 *c, d*), i.e. as the excitation frequency increases linear vibration occurs, but as the frequency decreases through resonance, intense vibro-impact regimes can be established. Such regimes can also occur when the system is excited with an especially large push.

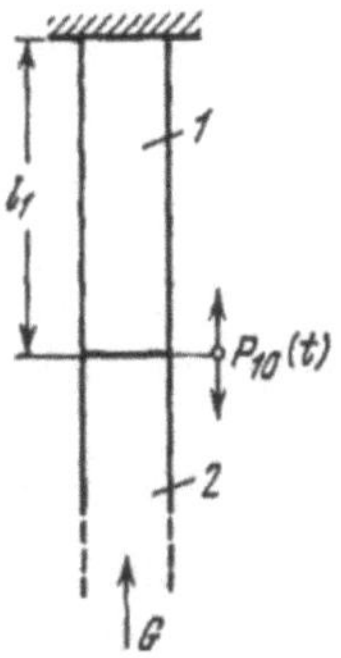

Fig. 15.14

A characteristic feature in the case of compressed bars is that according to (15.57) vibration of the passive bar does not depend on the excitation amplitude of the active bar. This situation, as mentioned in §14, is typical of the interactions examined for systems without damping.

5. Let us now look at the vibration of systems (Fig. 15.14) in which bar *1* length l_1 impacts with a semi-infinite bar *2* under the action of a harmonic force $P_{10}(t) = a_p \cos\omega t$.

Interference is caused by a static force *G*. The system described is a special case of the general problem examined in §14 section 3, hence we will use the results obtained there. To obtain the final expression we must define the dynamic compliance of a semi-infinite bar.

Using equation (15.8), we obtain for a finite bar, measuring coordinates from the free end

$$L_{20}(x, j\omega) = -\frac{\lambda_2\gamma_2^2}{E_2S_2\omega^2}\,\frac{\exp[\lambda_2(l_2 - x)] + \exp[\lambda_2(x - l_2)]}{\exp(\lambda_2 l) + \exp(-\lambda_2 l)}, \tag{15.61}$$

where

$$\lambda_2 = \frac{j\omega}{\sqrt{\gamma_2^2 + b_2 j\omega}} \qquad \left(b_2 = \frac{\gamma_2^2\varkappa_2}{2\pi\omega}\right).$$

Letting $l_2 \to \infty$ in (15.61), we obtain for a semi-infinite bar

$$L_{20}(x, j\omega) = -\frac{\gamma_2^2\lambda_2}{E_2S_2\omega^2}\exp(-\lambda_2 x). \tag{15.62}$$

Linearising this equation in terms of $\varkappa_2$, which occurs in λ_2, we find

$$L_{20}(x, j\omega) = -\frac{\gamma_2}{E_2S_2\omega}\left(j + \frac{\varkappa_2}{4\pi}\right)\exp\left[-\left(j + \frac{\varkappa_2}{4\pi}\right)\frac{\omega x}{\gamma_2}\right]. \tag{15.63}$$

From this it follows that the transmission of dynamic effect to the semi-infinite bar weakens exponentially as distance x from the source of excitation increases. The essential feature of this is that the exponent is proportional to the frequency of excitation. Hence, the high frequencies generated in the contact zone due to impact remain localised to that zone. This once again emphasises the significance of the lower frequency harmonic components of the vibro-impact process. They provide the basis for the transmission of vibro-impact influence in structural elements.

When $x = 0$ we obtain from (15.63) for the contacting section

$$L_{20}(0, j\omega) = -\frac{\gamma_2}{E_2 S_2 \omega}\left(j + \frac{\varkappa_2}{4\pi}\right) = -\frac{1}{N_2 \omega}\left(j + \frac{\varkappa_2}{4\pi}\right), \tag{15.64}$$

where $N_2 = S_2\sqrt{E_2\rho_2}$ is the *specific impedance* of the bar.

When $\varkappa \equiv 0$ the dynamic compliance $L_{20}(0, j\omega)$ becomes imaginary, and this demonstrates the effect of wave dissipation of the vibration energy through the semi-infinite bar.

To determine the amplitude of relative vibration a_v of the contacting sections for the system considered using (15.35) and taking into account the value of the dynamic compliance (15.34) with $i = 1$ and using (15.64), we obtain, neglecting viscosity of the bar materials $(\varkappa_1 = \varkappa_2 \equiv 0)$ and bearing in mind (15.53):

$$a_v = \frac{2h}{\zeta\cot\zeta}\left[\pm\frac{\chi}{2h}\sqrt{1-\left(\frac{2h\cot\zeta}{\chi w}\right)^2} - 1\right]. \tag{15.65}$$

Here, $h = Gl_1/(E_1S_1)$; $\chi = a_p l_1/(E_1S_1)$; $\zeta = \omega l_1/\gamma_1$; $w = N_2/N_1$ is the ratio of the impedances of the impacting bars. This formula represents a special case of the general expression (15.57). According to (15.65) vibro-impact regimes exist when the following condition is satisfied

$$\chi/h > 2|\cot\zeta|/w. \tag{15.66}$$

Fig. 15.15 shows examples of the amplitude-frequency response for relative vibration of the contacting sections for the case $\chi = 1$, $h = 1$, $w = 1$.

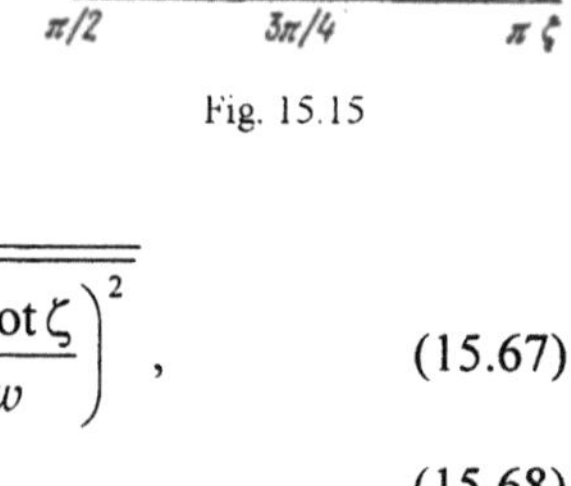

Fig. 15.15

For absolute vibration of these sections we obtain from the general conditions (15.61) and (15.62)

$$a_{10} = \frac{\chi}{|\zeta\cot\zeta|}\sqrt{1+\left(\frac{2h}{\chi}\right)^2 \pm \frac{4h}{\chi}\sqrt{1-\left(\frac{2h\cot\zeta}{\chi w}\right)^2}}, \tag{15.67}$$

$$a_{20} = 2h/(\zeta w). \tag{15.68}$$

These expressions are shown in Fig. 15.15 for the above parameter values. The line *1* in the figure shows amplitude response for vibration without loss of contact. Hence, in the examples considered, the vibro-impact regime can occur only as the result of hard excitation.

Let us find the conditions for smooth opening of the joint. This occurs if the

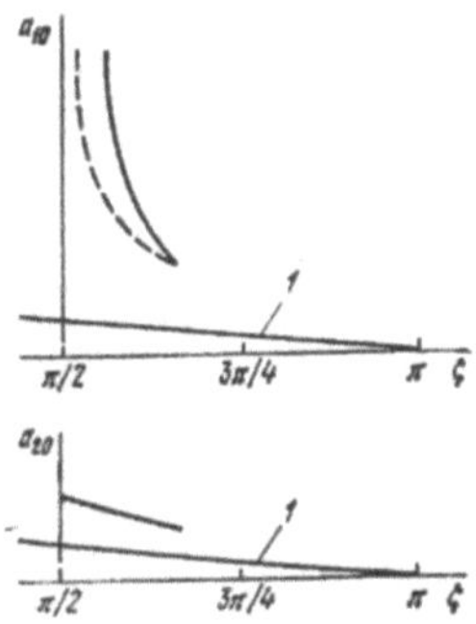

Fig. 15.16

amplitude of the variable force at the joint when moving without loss of contact exceeds compressive force G, i.e.

$$E_2 S_2 (\partial a_{2x} / \partial x)_{x=0} > G\,, \tag{15.69}$$

where a_{2x} is the vibration amplitude of the sections (see (15.18)).

Function $u^0_{2x}(t)$ describing vibration without loss of contact, is found from (15.20) when $\beta \equiv 0,\ k \to \infty$. Bearing in mind that $P_{1n} = a_P$, we find

$$u^0_{2x}(t) = \frac{L_{20}(x,s)}{1 + L_{20}(0,s) L^{-1}_{10}(0,s)} a_p \cos\omega t\,. \tag{15.70}$$

For a dynamic compliance $L_{10}(0, j\omega)$, found from (15.34) when $i = 1$, $L_{20}(x, j\omega)$ and $L_{20}(0, j\omega)$ are given by (15.63) and (15.64), with $\varkappa_1 = \varkappa_2 \equiv 0$, we obtain from (15.70)

$$a_{2x} = \frac{a_P \gamma_2}{E_2 S \omega} \left| \frac{\exp\left[j\left(\dfrac{\pi}{2} - \dfrac{\omega x}{\gamma_2} \right) \right]}{1 - j\dfrac{\cot\zeta}{w}} \right|. \tag{15.71}$$

Using (15.71), condition (15.69) takes the form

$$\frac{\chi}{h} > \sqrt{1 + \left(\frac{\cot\zeta}{w} \right)^2}\,. \tag{15.72}$$

Using (15.66) and (15.72) it is possible to estimate the safety of force-locked rods acted on by dynamic loading.

§16. Vibro-impact regimes of vibrating beams

1. We shall use the approach described above to study bending vibration of a cantilever beam (Fig. 16.1) of length l produced from a linear visco-elastic material and with a point mass M attached to the free end, arranged symmetrically between stops with gap Δ [45]. The coordinate x of a section is measured from the fixed end. The displacement of a section at position x will be denoted by $u_x(t)$, measured from the equilibrium position of the beam. The stops will be considered as elastic with stiffness c_2, later we shall limit the transfer to solutions when $c_2 \to \infty$. Using (6.8) the static characteristics of the impact pairs take the form

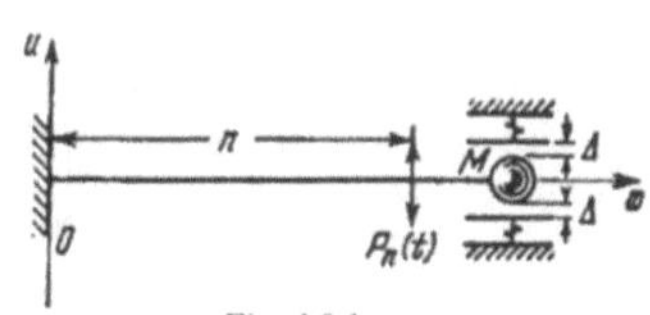

Fig. 16.1

$$\Phi(u_l) = c_2(|u_l| - \Delta)\eta(|u_l| - \Delta)\operatorname{sign} u_l\,. \tag{16.1}$$

The beam undergoes bending vibration under the effect of a transverse force $P_n(t) = a_p \cos\omega t$ applied to the section with coordinate $x = n$. As a result mass M interacts with the stops.

Use of the concept of a dynamic compliance operator for the beam $L_n(x,s)$, establishes the connection between the transverse force in the section with coordinate n and displacement $u_x(t)$, writing the following operator equation for $u_x(t)$:

$$u_x(t) = L_n(x,s)P_n(t) - L_l(x,s)\Phi[u_l(t)] - L_l(x,s)Ms^2 u_l(t)\,. \tag{16.2}$$

The final term on the right hand side of equation (16.2) represents the influence of inertia of the point mass on bending vibration.

The approximate periodic solution of equation (16.2) will be sought in the form

$$u_x(t) = a_x \cos(\omega t - \varphi_x)\,. \tag{16.3}$$

Applying harmonic linearisation to function (16.1), we find

$$\Phi(u_l) \approx k(a_l)u_l\,, \tag{16.4}$$

where according to (7.56) the linearisation coefficient takes the form

$$k(a_l) = \frac{2c_2}{\pi}\left[\arccos\frac{\Delta}{a_l} - \frac{\Delta}{a_l}\sqrt{1 - \left(\frac{\Delta}{a_l}\right)^2}\,\right]. \tag{16.5}$$

Substituting (16.4) into (16.2) we find when $x = l$

$$\{1 + L_l(l,s)[k(a_l + Ms^2]\}u_l(t) = L_n(l,s)P_n(t)\,. \tag{16.6}$$

Introducing the nomenclature

$$W(s,a_l) = \{1 + L_l(l,s)[k(a_l) + Ms^2]\}L_n^{-1}(l,s)\,, \tag{16.7}$$

we obtain once more an equation of type (14.12):

$$W(s,a_l)u_l(t) = a_p \cos\omega t\;. \tag{16.8}$$

Bringing this to the equivalent autonomous form using (16.3), we find

$$\left[W(s,a_l) - \frac{a_p}{a_l}\left(\cos\varphi_l + s\frac{\sin\varphi_l}{w}\right)\right]u_l(t) = \mathcal{D}(s)u_l(t) = 0\,. \tag{16.9}$$

Substituting into the characteristic equation $\mathcal{D}(s) = 0$ the root $s = j\omega$, we obtain, setting the real and imaginary parts to zero:

$$\sin\varphi_l = a_l V(\omega,a_l)/a_p\,, \tag{16.10}$$

$$\cos\varphi_l = a_l U(\omega,a_l)/a_p\,. \tag{16.11}$$

Here, as before, $U = \operatorname{Re}W$, $V = \operatorname{Im}W$. From (16.10), (16.11) we obtain the final equation to find a_l:

$$a_l = a_p[U^2(\omega,a_l) + V^2(\omega,a_l)\}^{-1/2} = a_p|W(j\omega,a_l)|^{-1} \tag{16.12}$$

After finding a_l from equations (16.10), (16.11) we can find the phase φ_l. The periodic motion parameters for the remaining sections of the beam are found from (16.2) using (16.3) to (16.5).

To examine stability of the solutions relative to small perturbations it is possible to use condition (14.32), substituting in it $a_v = a_l$.

2. To simplify the final expressions let us assume that the excitation force is applied at the free end of the beam ($x = l$), where the mass M is attached. Now to apply the formula of section 1 we must put $L_{ll}(x,s) = L_l(x,s)$, and consequently, to solve the problem it is sufficient to find the dynamic compliance $L_l(x,j\omega)$. With this in mind we will turn to the differential equation for bending vibration of a homogeneous beam of constant cross-section, appreciating the dissipation due to the internal damping force with coefficient $\varkappa$ [176]:

$$\rho s^2 u + \left(1 + \frac{\varkappa}{2\pi\omega} s\right) EI \frac{\partial^4 u}{\partial x^4} = 0. \tag{16.13}$$

Here $u = u(x, t)$; ρ is the mass of a unit length of the beam; E is the modulus of elasticity; I is the second moment of area of the beam section. Applying a unit harmonic force to the free end of the beam, we obtain the following boundary conditions for the system under consideration:

$$\begin{gathered} u(0,t) = 0, \quad \frac{\partial^2 u(l,t)}{\partial x^2} = 0, \\ \frac{\partial u(0,t)}{\partial x} = 0, \quad -EI\left(1 + \frac{s\varkappa}{2\pi\omega}\right)\frac{\partial^3 u(l,t)}{\partial x^3} = \exp(j\omega t). \end{gathered} \tag{16.14}$$

Searching for a stable solution for equation (16.13) in the form

$$u(x,t) = u_x(t) = L_l(x, j\omega)\exp(j\omega t), \tag{16.15}$$

we find, substituting (16.15) into (16.13), (16.14) when $s = j\omega$

$$\frac{d^4 L_l(x, j\omega)}{dx^4} + (j\omega)^2 \frac{\rho}{EI\left(1 + j\dfrac{\varkappa}{2\pi}\right)} L_l(x, j\omega) = 0, \tag{16.16}$$

$$\begin{gathered} L_l(0, j\omega) = 0, \quad \frac{d^2 L_l(l, j\omega)}{dx^2} = 0, \\ \frac{dL_l(0, j\omega)}{dx} = 0, \quad \frac{d^3 L_l(l, j\omega)}{dx^3} = -\frac{1}{EI(1 + j\varkappa/(2\pi))}. \end{gathered} \tag{16.17}$$

Searching for a solution of the ordinary linear differential equation (16.16) of the fourth order with constant coefficients in the form

$$L_l(x, j\omega) = C\exp(qx), \tag{16.18}$$

where C is an arbitrary constant, we arrive at the characteristic equation of the fourth degree, the roots of which to first order accuracy in terms of $\varkappa$ take the form

$$\begin{gathered} q_1 = \lambda[1 - j\varkappa/(8\pi)], \quad q_3 = \lambda[-1 + j\varkappa/(8\pi)], \\ q_2 = \lambda[\varkappa/(8\pi) + j], \quad q_4 = -\lambda[\varkappa/(8\pi) + j]. \end{gathered} \tag{16.19}$$

Here $\lambda^4 = \rho\omega^2/(EI)$. As a result, using (16.18), (16.19) the general solution of equation (16.16) can be written as:

$$L_l(x, j\omega) = \sum_{i=1}^{4} C_i(j\omega)\exp(q_i x). \tag{16.20}$$

Applying the boundary conditions (16.17) to (16.20) we obtain a non-homogeneous set of linear equations of the fourth order for the coefficients C_i:

$$\begin{aligned} &C_1 + C_2 + C_3 + C_4 = 0, \\ &C_1 q_1 + C_2 q_2 + C_3 q_3 + C_4 q_4 = 0, \\ &C_1 q_1^2 \exp(q_1 l) + C_2 q_2^2 \exp(q_2 l) + C_3 q_3^2 \exp(q_3 l) + \\ &\qquad + C_4 q_4^2 \exp(q_4 l) = 0, \\ &C_1 q_1^3 \exp(q_1 l) + C_2 q_2^3 \exp(q_2 l) + C_3 q_3^3 \exp(q_3 l) + \\ &\qquad + C_4 q_4^3 \exp(q_4 l) = -\frac{1}{EI(1 + j\varkappa/(2\pi))}. \end{aligned} \tag{16.21}$$

Solving the system of equations (16.21), we find with the help of (16.20) equations for the operators $L_l(x, j\omega)$, which we have not introduced because of the awkwardness of the final expressions.

To construct the amplitude-frequency response for the point mass from equation (16.12) it is satisfactory to use the equation for the local dynamic compliance of the free end of the beam $L_l(l, j\omega)$. Linearising the expression for $L_l(l, j\omega)$ in terms of the small parameter $\varkappa$ using (16.19), we find when $x = l$

$$\begin{aligned} L_l(l, j\omega) = &\frac{l^3(\cosh\zeta\sin\zeta - \sinh\zeta\cos\zeta)}{EI\zeta^3(1 + \cosh\zeta\cos\zeta)} - \\ &\varkappa l^3\left[\frac{\zeta}{2}(\cosh 2\zeta - \cos 2\zeta) + (1 + \cosh\zeta\cos\zeta) \times\right. \\ - j&\frac{\left.\times(\cosh\zeta\sin\zeta - \sinh\zeta\cos\zeta) + 2\zeta\sinh\zeta\sin\zeta\right]}{8\pi EI\zeta^3(1 + \cosh\zeta\cos\zeta)^2}, \end{aligned} \tag{16.22}$$

where $\zeta = \lambda l$. Substituting (16.22) into (16.7) and noting that now $n = l$, we finally obtain

$$\begin{aligned} W(j\omega, a_l) = &\frac{EI}{l^3}\left\{\frac{\zeta^3(1 + \cosh\zeta\cos\zeta)}{\cosh\zeta\sin\zeta - \sinh\zeta\cos\zeta} + k_0(a_l) - M_0\zeta^4\right\} + \\ &\left[\frac{\zeta}{2}(\cosh 2\zeta - \cos 2\zeta) + (1 + \cosh\zeta\cos\zeta) \times\right. \\ + j\frac{EI\varkappa\zeta^3}{8\pi l^3}&\frac{\left.\times(\cosh\zeta\sin\zeta - \sinh\zeta\cos\zeta) + 2\zeta\sinh\zeta\sin\zeta\right]}{(\cosh\zeta\sin\zeta - \sinh\zeta\cos\zeta)^2}. \end{aligned} \tag{16.23}$$

Here $k_0(a_l) = k(a_l)l^3/(EI)$; $M_0 = M/(\rho l)$.

From now on, to eliminate awkward calculation it is convenient to use the frequency functions tabulated in [5] during the solution, which will use the same definitions with a zero index added:

$$\begin{aligned} &U_0(\zeta) = \tfrac{1}{2}(\cosh\zeta - \cos\zeta),\ B_0(\zeta) = \cosh\zeta\sin\zeta - \sinh\zeta\cos\zeta, \\ &S_0(\zeta) = 2\sinh\zeta\sin\zeta,\ E_0(\zeta) = 1 + \cosh\zeta\cos\zeta. \end{aligned} \tag{16.24}$$

From (16.12) using (16.23), (16.24) we obtain the following equation for a_l

$$a_l = \chi\left\{\left[\frac{\zeta^3 E_0(\zeta)}{B_0(\zeta)} + k_0(a_l) + M_0\zeta^4\right] + \right.$$

$$\left. + \left[\frac{\varkappa\zeta^3}{8\pi}\frac{\zeta U_0(2\zeta) + E_0(\zeta)B_0(\zeta) + \zeta S_0(\zeta)}{B_0^2(\zeta)}\right]^2\right\}^{-1/2}, \tag{16.25}$$

where $\chi = a_p l^3/(EI)$.

To find the basic motion parameters of the impacting mass we must add the equation for k_0 to equation (16.25), which, using (16.5) is represented in the form

$$k_0(\alpha) = \frac{2c_2 l^3}{\pi EI}(\arccos\alpha - \alpha\sqrt{1-\alpha^2}), \tag{16.26}$$

where

$$\alpha = \Delta / a_l \qquad (0 < \alpha \leq 1). \tag{16.27}$$

3. Let us construct the amplitude-frequency response for the point mass. According to (16.25), the expression

$$\frac{\zeta_0^3 E_0(\zeta_0)}{B_0(\zeta_0)} + k_0(a_l) - M_0\zeta_0^4 = 0 \qquad \left(\zeta_0 = \sqrt[4]{\frac{\rho\omega_0^2}{EI}}l\right), \tag{16.28}$$

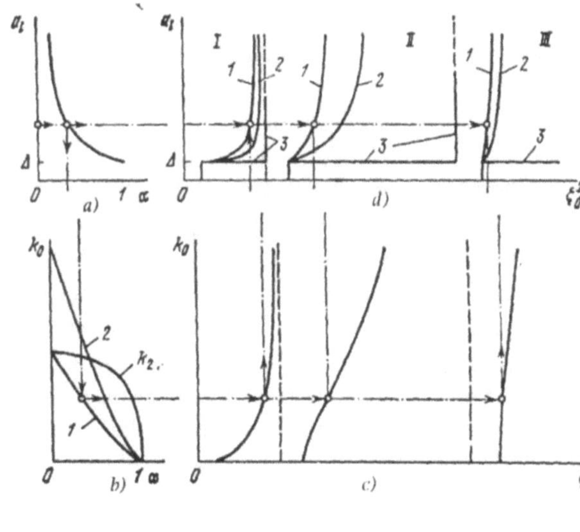

Fig. 16.2

where ω_0 is the frequency of free vibration, defines the configuration of the backbone curves for the system. It is convenient to use the graphical method described in §15, section 2 to show them. Expressions (16.27), (16.26), (16.28) are drawn respectively in Fig. 16.2, *a* to *c* (curves *1* and *2* in Fig. 16.2, *b* represent different stop stiffness c_2, where the larger numbers corresponds to greater stiffness). Choosing a point with value a_l in Fig. 16.2 *a* and making the construction as shown by the chain dotted line, we find in Fig. 16.2, *d* the corresponding point of the backbone curve. Here, for each value a_l there is a countable set of natural frequencies ζ_0. Connecting the points obtained gives a family of backbone curves. Fig. 16.2, *d* shows curves for the three lower modes of vibration. The portions of the backbone curve added correspond to linear vibration of the mass without striking the stops. We will refer to the modes later on by their roman numerals.

Because of the presence of the interacting mass with the elastic stops the frequency of vibration for each mode depends on the amplitude, and this relationship, naturally, has an hard character. As can be seen in Fig. 16.2, *d*, the degree to which the nonlinear effects develop is weakened as the number of modes increases. It can also be seen that the nonlinearity of the system increases as the

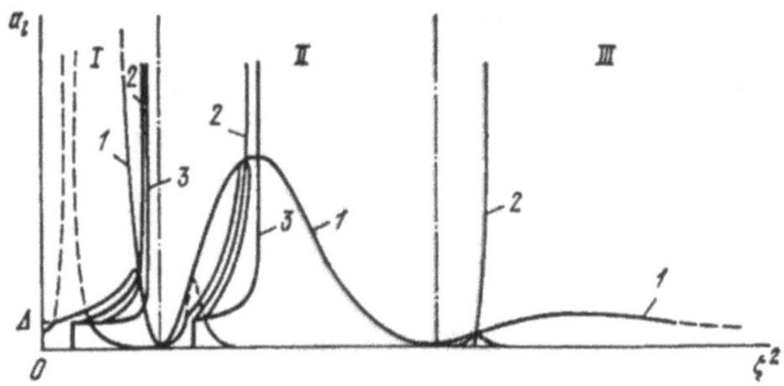

Fig. 16.3

stiffness of the stops increases. The asymptotes of the backbone curve, defined by the sequence of points in expression (16.26) when $\alpha = 0$, converge, when $c_2 \to \infty$, to the asymptotes of expression (16.28). These asymptotes are shown dotted in Fig. 16.2 *c*, *d*. They are situated at the anti-resonance frequency of the linear system. Hence in the limit when $c_2 \to \infty$ for rigid stops we obtain the backbone lines of form *3* (Fig. 16.2, *d*) with the length of the horizontal lines increasing as the number of modes increases and vertical lines situated at the anti-resonance frequency of the linear system. Of course, in the investigation of actual vibration only the horizontal portions of the backbone lines described are of interest.

Putting in (16.25) $\zeta = \zeta_0$, we obtain the energy boundary of vibration, using (16.28)

$$a_l = \frac{8\pi\chi}{\varkappa} \frac{B_0^2(\zeta_0)}{\zeta_0^3 \left|\zeta_0 U_0(2\zeta_0)\right| + E_0(\zeta_0) B_0(\zeta_0) + \zeta_0 S_0(\zeta_0)}. \tag{16.29}$$

According to (16.25), the intersection of the energy boundary of vibration *1* (Fig. 16.3) with the backbone line *2* gives the maximum points of the amplitude-frequency response for the system. It can be shown, that zero points on the energy boundary of vibration correspond to anti-resonance of the linear system arranged between stops. It follows from Fig. 16.3 that the maximum of curve *1* rapidly dies away as the number of the modes increases, hence practical values only occur for some initial modes of vibration.

The remaining points on the amplitude curves, corresponding to nonlinear

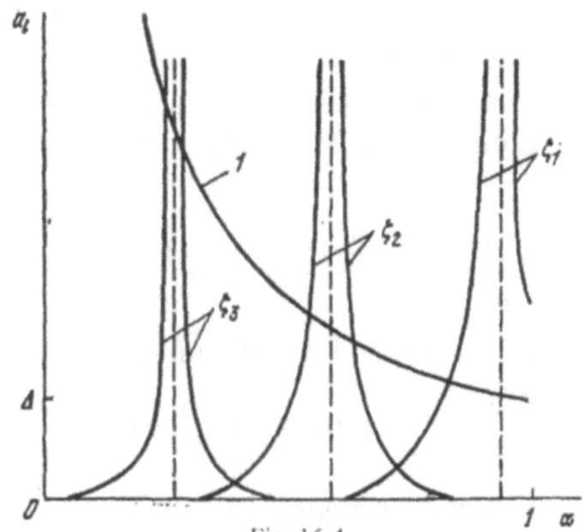

Fig. 16.4

regimes of vibration, are found by jointly solving equations (16.25) to (16.27). For this, the expression (16.26) for $k_0(\alpha)$ is substituted into (16.25), where to simplify calculation it is convenient to substitute $\varkappa \equiv 0$ in (16.25), since as we move away from the peak values the influence of internal damping on the amplitude curves remains insignificant. Subsequently the expression obtain for $a_l = (\alpha, \zeta)$ is drawn on the (α, a_l) plane (Fig. 16.4) for various values of $\zeta = \text{const}$. In Fig. 16.4 this is shown only for the basic modes of vibration for frequencies $\zeta_1 < \zeta_2 < \zeta_3$. The intersection of these curves with $a_l(\alpha)$ (16.27) (curve *1* in Fig. 16.4), which was already used to construct the backbone curves, also produces points for the required values of the expression $a_l(\zeta)$, drawn in Fig. 16.3 for $a_l > \Delta$. When $a_l < \Delta$ a portion of the amplitude-frequency response for linear systems appears. Their continuation to the level $a_l = \Delta$ is shown dotted in Fig. 16.3.

From Fig. 16.3 it follows that in the examples studied, as distinct from the linear case, it is possible to obtain larger amplitudes in the second mode of vibration compared to the first, while they exceed the amplitude of linear vibration substantially in the second mode. It can also be seen that in the third mode only linear regimes of vibration can occur without hitting the stops. Note that the principal qualitative results can also be obtained on the base of the backbone curves and the energy boundary of vibration.

Let us examine the stability of the solutions found. Assuming for simplicity that $\varkappa \equiv 0$ and, consequently $V \equiv 0$, we obtain the following expression for the boundaries of the existence domains from (14.33) using (16.23), (16.24), when $a_v = a_l$:

$$\frac{\zeta^3 E_0(\zeta)}{B_0(\zeta)} + k_0(a_l) - M_0\zeta^4 = 0\,, \tag{16.30}$$

$$\frac{\zeta^3 E_0(\zeta)}{B_0(\zeta)} + k_0(a_l) - M_0\zeta^4 + \frac{a_l l^3}{EI}\frac{dk}{da_l} = 0\,. \tag{16.31}$$

Here, we take into account that $dU/da_l = dk/da_l$. Expression (16.32) coincides with the equation of the backbone curves (16.28), shown by line *2* in Fig. 16.3. To draw (16.31) we will calculate the derivative which appear in it. According to (16.5) we find

$$\frac{dk}{da_l} = \frac{4c_2\Delta\sqrt{1-(\Delta/a_l)^2}}{\pi a_l^2}\,. \tag{16.32}$$

Using (16.32) expression (16.31) can be written in a form similar to (16.30):

$$\frac{\zeta^3 E_0(\zeta)}{B_0(\zeta)} + k_2(\alpha) - M_0\zeta^4 = 0\,, \tag{16.33}$$

where $k_2(\alpha)$ is found from the expression

$$k_2(\alpha) = \frac{2c_2 l^3}{\pi EI}(\arccos\alpha + \alpha\sqrt{1-\alpha^2})\,. \tag{16.34}$$

As a result, using the graphical construction in Fig. 16.2, we can obtain the second boundary branch of the existence domain. The corresponding expressions (16.27), (16.33) are already drawn. It remains to plot expression (16.34) in Fig. 16.2, *b* and to carry out the construction, similar to that which was used when searching for the backbone curves. The limiting branches found in this way, described by expression (16.31) are represented by curve *3* in Fig. 16.3. As a result the regions of instability arranged between curves *2* and *3* enclose vibro-impact regimes, where the lower branches correspond to the double-valued domains of the resonance curves. The upper branches represent stable solutions.

Let us look at the amplitude-frequency response, obtained in the limit as the stops become rigid. As was shown above, when $c_2 \to \infty$, the backbone lines

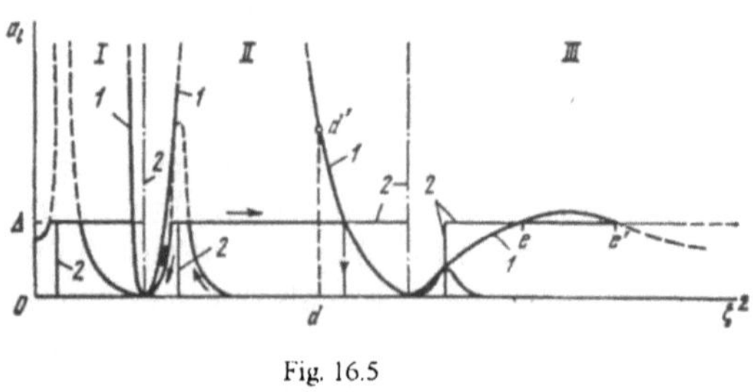

Fig. 16.5

change into line *3* (Fig. 16.2, *d*). During this limited transfer, pairs of nonlinear branches of the amplitude-frequency response (Fig. 16.3) join the horizontal portion of the backbone curves. The resulting characteristics of the system with rigid stops are shown in Fig. 16.5, where the numbering of the curves is similar to that of Fig. 16.3.

In Fig. 16.5, the amplitude curve configuration for the first and second modes of vibration have a similar form and are different from the response of the third mode. From Fig. 16.5 it follows that in the system examined, pulling can occur in vibro-impact regimes in terms of frequency as well as amplitude, similar to that shown for systems with one degree of freedom (see §3). However, in the present system similar effects can arise in several frequency bands, corresponding to different modes of vibration of the beam. Frequency pulling occurs for fixed stops with slowly changing excitation frequency as shown by the arrows, for example, with the second mode of vibration. It can be seen that, depending on the direction in which the frequency changes, the existence band for regimes also changes. In the example shown in Fig. 16.5, pulling in the second mode is of greater significance than in the first. By fixing the excitation frequency, for example at point *d*, it is possible to vary the amplitude of the vibro-impact regime by slowly changing the distance between the stops in range dd'. At point d' the vibration breaks down.

In the third mode of vibration as well as linear resonance, an intensive vibro-impact regime can arise (portion ee'), caused by hard excitation and whose amplitude exceeds the amplitude of linear resonance.

Hence, the presence of various natural modes of vibration for a point mass on elastic supports increases the difficulty of preventing clatter against the stops. This occurs particularly with bending shafts where the destabilising effect of internal damping, occurring at angular velocities greater than the critical, lead to collisions with the bearings.

4. In [296, 298, 333, 334] the term *distributed impact element* was introduced as applied to the study of vibro-impact processes in the tube bundles of heat exchangers and similar extended objects. In such systems a significant variable length of continuous vibrating element (thin-wall tube) may be involved in vibro-impact interaction either with the neighbouring elements or a rigid wall. The object of study was an axially stretched beam located in parallel with a rigid wall that limited the transverse deflections of the beam.

In order to reveal the resonant behaviour of the system, preliminary experiments were carried out. Figure 16.6 represents schematically an experimental rig [333]. A rubber cord was stretched between the pin of a force transducer and the pushrod of an electrodynamic shaker producing transverse excitation of the cord. A rigid wall limited the transverse movement of the cord. The pictures of the standing waves were taken by using a stroboscope synchronised with the voltage applied to the shaker.

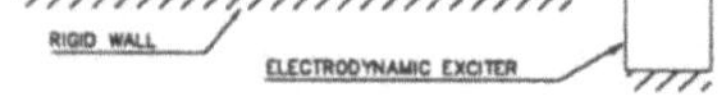

Fig 16.6

As the frequency of excitation exceeded the frequency of the first resonance of the linear system (the cord without motion limiter) the nonlinear mode of cord vibration took a trapezoidal form with simultaneous impact of the entire plane part against the limiter. Fig. 16.7 demonstrates this *slapping mode* at different instants: impact (trapezoidal shape), static equilibrium (straight line) and backward

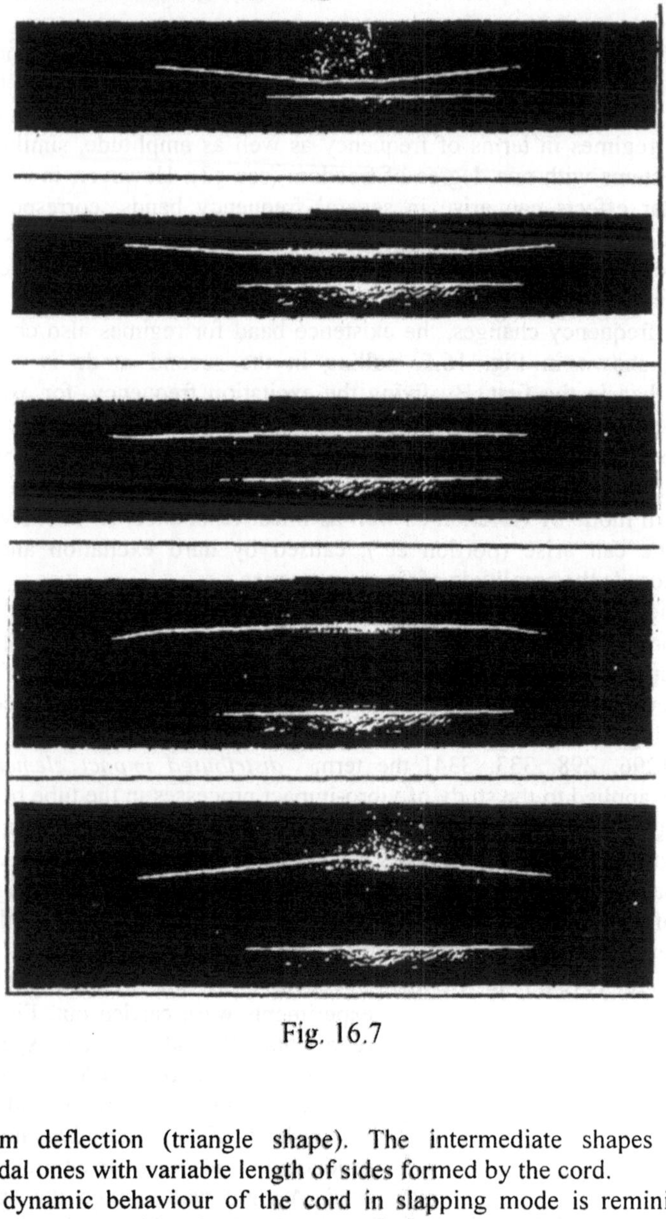

Fig. 16.7

maximum deflection (triangle shape). The intermediate shapes are always trapezoidal ones with variable length of sides formed by the cord.

The dynamic behaviour of the cord in slapping mode is reminiscent of an impact oscillator with single degree of freedom. The mode demonstrates frequency and amplitude pulling, transition from linear to nonlinear mode and back with the help of hard excitation. However, the cord has many nonlinear slapping modes disposed in the right hand side in the vicinity of each natural frequency of the linear system. The nonlinear modes have the same quantity of nodes as a proper linear mode. The frequency bands of nonlinear modes become narrower with increase of mode order.

As it was shown in [333], the method of periodic Green's functions can be applied effectively for regular analysis of slapping modes. This is based on the assumption that the resonant motions are close to free vibration in the corresponding conservative system. The conditions of existence of such modes can be found with the help of an energy balance.

As an example of the approach mentioned, let us consider a continuous system as shown in Fig. 16.8, a. The straight uniform beam of length l is stretched parallel to the rigid limiter with axial force N. The following definitions will be used for description of the system: ρ is the density of the beam material, E, G are elastic and shear modulus, S is cross sectional area, I is area moment of inertia, ε is a small parameter.

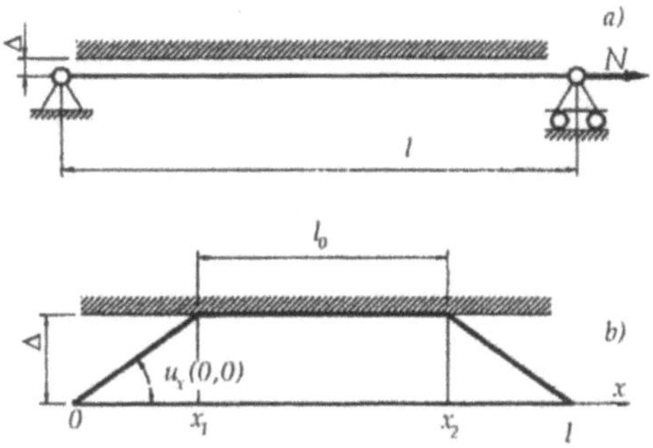

Fig. 16.8

The beam is exposed to the external force with density of distribution $\varepsilon P(x,t)$. It vibrates in the paper plane and can collide with a straight obstacle displaced parallel to the beam with a clearance Δ relative to the position of static equilibrium of the beam.

Following Timoshenko's model of beam bending [207], the following notation will be used for a description of the beam vibration: $u(x,t)$ is the shape of the beam transverse deflection at instant t; $y(x, t)$ is the shape of the beam cross section rotation relative to the neutral line at instant t.

The differential equations of beam oscillations with provision for collisions with the rigid obstacle may be written in a following form:

$$\rho S\frac{\partial^2 u}{\partial t^2}-(\alpha SG+N)\frac{\partial^2 u}{\partial x^2}+\alpha SG\frac{\partial y}{\partial x}+\hat{\Phi}(u)=\varepsilon P(x,t)$$

$$K\frac{\partial^2 y}{\partial x^2}-\alpha SG\left(y-\frac{\partial u}{\partial x}\right)-\frac{I\rho}{S}\frac{\partial^2 y}{\partial t^2}=0 \tag{16.35}$$

with boundary conditions: $u(0,t)=u(l,t)=\dfrac{\partial u(0,t)}{\partial x}=\dfrac{\partial u(l,t)}{\partial x}=0$. Here $K = EI$ is the bending stiffness, α is a form factor, $\hat{\Phi}(u)$ is the impact force density.

Considering T-periodic motions with one impact over the one period of excitation (slapping modes), the impact force density may be presented in the time domain as follows:

$$\hat{\Phi}(u)=\hat{J}(x)\delta^T[t-\varphi(x)] \tag{16.36}$$

where $\hat{J}(x)\geq 0$ and $\varphi(x)$ are the distribution of impact impulse density and the impact phase respectively.

In order to study the beam vibration in a first slapping mode (as described in the experiment) the following initial conditions will be applied (see Fig 16.8,b):

$$u_-(x,0) = u_+(x,0) = u(x,0) = \begin{cases} \dfrac{\Delta}{x_1}x, & x \in [0, x_1] \\ \Delta, & x \in [x_1, l - x_1] \\ \dfrac{\Delta}{x_1}x + \Delta + \dfrac{\Delta}{x_1}(l - x), & x \in [l - x_1, l] \end{cases} \quad (16.37)$$

$$\frac{\partial u_-(x,0)}{\partial t} = \begin{cases} V_0, & x \in [x_1, l - x_1] \\ 0, & x \notin [x_1, l - x_1] \end{cases}, \qquad \frac{\partial u_+(x,0)}{\partial t} = -R\frac{\partial u_-(x,0)}{\partial t}, \quad (16.38)$$

where $R \approx 1 - \varepsilon$ is the coefficient of restitution, plus or minus in the subscript denotes pre-impact or post-impact values.

As a result:

$$\begin{aligned} \hat{J}(x) &= (1 + R)\rho S V_0 = \hat{J} = \text{const}, \quad x \in [x_1, l - x_1] \\ \varphi(x) &= \text{const} = 0, \quad x \notin [x_1, l - x_1] \end{aligned} \quad (16.39)$$

For the description of the slapping mode, let us consider first the conservative case, assuming $\varepsilon = 0$. Using the *T*-periodic Green's function of the distributed conservative system, the following integral presentation may be produced for the *T*-periodic free vibration with single elastic impact over the excitation period:

$$u(x,t) = -\int_0^l \int_0^T \chi(x, z, t - s)\hat{\Phi}[u(z, s)]ds dz, \quad t \in [0, T], \quad x \in [0, l] \quad (16.40)$$

where $\chi(x, z, t)$ is a *T*-periodic Green's function representing a steady state response of the section x on periodic impulses applied to the section z. The periodic Green's function can be expressed by using the transfer dynamic compliance of the linear system $L(x, z, j\omega)$ as follows:

$$\chi(x, z, t) = \frac{2}{T}\sum_{k=1}^{\infty} L(x, z, kj\omega)\cos(k\omega t) \quad (16.41)$$

In a number of cases, such as the thin-wall tubes of heat exchangers, we can neglect the rotary inertia of the cross sections and ignore, therefore, the last term in the second equation of (16.35). Under such assumption, the expression for the dynamic compliance can be presented in the form of the infinite series:

$$L(x, z, j\omega) = -\frac{2}{\rho S l}\sum_{m=1}^{\infty} \frac{\sin\dfrac{\pi m x}{l}\sin\dfrac{\pi m z}{l}}{\left(\omega^2 - \Omega_m^2\right)} \quad (16.42)$$

with the natural frequencies defined in accordance with the following expression

$$\Omega_m = \sqrt{\frac{\left(\dfrac{\pi m}{l}\right)^2\left[\alpha KSG\left(\dfrac{\pi m}{l}\right)^2 + NK\left(\dfrac{\pi m}{l}\right)^2 + \alpha NSG\right]}{\rho S\left[K\left(\dfrac{\pi m}{l}\right)^2 + \alpha NSG\right]}}$$

Substituting (16.42) into the (16.41) and using the finite expression for the Fourier series on the interval of periodicity [191] we find

$$\chi(x,z,t)=\frac{1}{\rho Sl}\sum_{m=1}^{\infty}\frac{\sin\frac{\pi m x}{l}\sin\frac{\pi m z}{l}\cos\Omega_m\left(t-\frac{T}{2}\right)}{\Omega_m\sin\Omega_m\frac{T}{2}},\quad x\in[0,l],\ t\in[0,T]. \tag{16.43}$$

After substituting (16.36), (16.39) and (16.43) into (16.40) and integrating, we have

$$u(x,t)=\frac{2\hat{J}}{\pi\rho S}\sum_{m=1}^{\infty}\frac{(-1)^{m+1}}{\Omega_{2m-1}(2m-1)}\sin\frac{\pi(2m-1)l_0}{2l}\sin\frac{\pi(2m-1)x}{l}\cos\frac{\Omega_{2m-1}\left(t-\frac{T}{2}\right)}{\sin\Omega_{2m-1}\frac{T}{2}}, \tag{16.44}$$

$$x\in[0,T],\quad t\in(0,T)$$

In the expression (16.44) the unknown values $\hat{J}$ and $l_0=2l-x_1$ are functions of period T of nonlinear vibration. For the first nonlinear impact mode (as shown in Fig. 16.8, *b*), from the condition of impact of the middle point of the beam with coordinate $x=l/2$, which must be involved in the impact, we have

$$u(l/2,x)=\frac{2\hat{J}}{\pi\rho S}\sum_{m=1}^{\infty}\frac{1}{\Omega_{2m-1}(2m-1)}\sin\frac{\pi(2m-1)l_0}{2l}\cot\frac{\Omega_{2m-1}T}{2}=\Delta$$

or

$$\hat{J}=\frac{\pi\rho S\Delta}{2\sum_{m=1}^{\infty}\frac{1}{\Omega_{2m-1}(2m-1)}\sin\frac{\pi(2m-1)l_0}{2l}\cot\frac{\Omega_{2m-1}T}{2}}. \tag{16.45}$$

From the simplest geometrical structure (see Fig. 16.8, *b*), it follows that $l_0=l-2x_1=l-\frac{2\Delta}{\tan\frac{\partial u(0,0)}{\partial x}}$. Considering further the case of small vibration $\left(\tan\frac{\partial u(0,0)}{\partial u}\approx\frac{\partial u(0,0)}{\partial u}\right)$, we find finally an equation for defining l_0:

$$l_0=\frac{l\Delta\rho S}{\hat{J}\sum_{m=1}^{\infty}\frac{(-1)^{m+1}}{\Omega_{2m-1}}\sin\frac{\pi(2m-1)l_0}{2l}\cot\frac{\Omega_{2m-1}T}{2}}. \tag{16.46}$$

By substituting (16.45) into (16.46) we have the following transcendental equation for the non-dimensional variable $\lambda=l_0/l$:

$$\lambda=1-\frac{2}{\pi}\frac{\sum_{m=1}^{\infty}\frac{1}{\Omega_{2m-1}(2m-1)}\sin\frac{\pi(2m-1)\lambda}{2}\cot\frac{\pi\Omega_{2m-1}}{\zeta\Omega_1}}{\sum_{m=1}^{\infty}\frac{(-1)^{m+1}}{2m-1}\sin\frac{\pi(2m-1)\lambda}{2}\cot\frac{\pi\Omega_{2m-1}}{\zeta\Omega_1}} \tag{16.47}$$

where $\zeta = \dfrac{2\pi}{T\Omega_1}$ is non-dimensional frequency of vibration.

For the numerical example, round steel tube of total length $l = 2m$ with external diameter $D = 0.016m$ and internal diameter $d = 0.013m$ was considered. The remaining parameters of the model were as follows: $N=2.10^4 N$, $E=2.10^{11}\ N/m^2$, $G=0.75.10^{11}\ N/m^2$, $\rho=8.10^3\ kg/m^3$, $\Delta=0.01m$, $\alpha \approx 1$. After calculations, we have $\rho S \approx 1 kg/m$, $K=360 N/m^2$.

The function $\lambda(\zeta)$ was calculated numerically from (16.47). By substituting $\lambda(\zeta)$ into (16.45) and taking into consideration that $\lambda(\zeta) \in [0,1]$ and $\hat{J}(\zeta) \geq 0$, the frequency band of the first slapping mode was evaluated numerically as $\zeta \in [1,\ 1.16]$. Figure 16.9 shows the full impact impulse $J = \hat{J} l_0$ as function of ζ.

As a result of the study of resonant behaviour, a force of excitation distributed in accordance with the first linear mode $P(x,t) = q \sin\dfrac{\pi x}{l}\cos(\omega t + \varphi)$ was considered, where q is intensity, φ is the phase of excitation relative to impact. The resonant slapping mode $u = (x,t)$ was taken in the form (16.44) with values of $\lambda(\zeta)$ and $\hat{J}(\zeta)$ being evaluated numerically according to (16.45), (16.47). Then the generative solution (16.44) was substituted into the first equation from (16.35), both parts of the equality obtained were multiplied by the function $\dfrac{\partial u(x,t)}{\partial t}$ evaluated analytically from (16.44). The equality prepared in such a manner was integrated along the length of the beam and over the period of vibration $T = 2\pi/\omega$. As a result the equation of energy balance between excitation and dissipation was estimated:

$$\int_0^l\int_0^T \hat{\Phi}(u)\frac{\partial u(x,t)}{\partial t}\,dx\,dt = \varepsilon\int_0^l\int_0^T P(x,t)\frac{\partial u(x,t)}{\partial t}\,dx\,dt\,. \tag{16.48}$$

Because a considerable part of the beam is involved in simultaneous collision, the energy loss can be calculated as the impact of a rigid body with coefficient of restitution R:

$$\int_0^l\int_0^T \hat{\Phi}(u)\frac{\partial u(x,t)}{\partial t}\,dx\,dt = \left(\hat{J} l_0\right)^2 \frac{1-R}{2l_0(1+R)\rho S} \approx \hat{J}^2 l_0 \frac{\varepsilon}{4\rho S}\,.$$

After substituting in the right side of equation (16.48) the excitation and integrating, the equation of energy balance was finally yielded

$$\frac{2q\omega l \sin\varphi \sin\dfrac{\pi l_0}{2l}}{\pi\left(\Omega_1^2 - \omega^2\right)} = \frac{\hat{J} l_0}{4}\,. \tag{16.49}$$

The condition for the existence of the process is $|\sin\varphi| \leq 1$. This condition can be used to define the energy boundary of the slapping mode:

$$q \geq \frac{\pi J\left(1-\zeta^2\right)\Omega_1}{8\zeta l \sin\dfrac{\pi l_0}{2}}. \qquad (16.50)$$

This defines the minimum excitation to support the resonant mode studied.

The procedure described allows the quantitative analysis of the experimentally observed slapping modes. Equation (16.45) describes the phenomena of frequency and amplitude pulling, equation (16.44) presents the trapezoidal shape of the standing wave (slapping mode), and the equality in (16.50) defines the backward jump frequency and through it the frequency band of nonlinear resonance.

§17. Theory of the impact vibration absorber

1. Many references have been devoted to the theoretical and experimental investigation of dynamic systems undergoing impact vibration absorbtion [7, 11, 47, 82, 85, 86, 119, 105, 145, 157 to 159, 205, 226 *et al*]. Theoretical analysis often uses exact methods and is limited to the examination of a double mass dynamic model, showed in Fig. 6.3, by taking the external excitation in the form of a harmonic function.

Using approximate methods described in this book enabled us to obtain several general results, characterising the effect of impact absorbers on linear systems of arbitrary structures, subjected to different forms of excitation [35]. We shall turn to consider these results.

Let us look at the dynamic model shown in Fig. 17.1. Here *1* is an arbitrary linear system, to which a force $P(t)$ is applied at point A. To reduce the vibration of point A an impact vibration absorber *2* of mass M_2 is attached to it. Various arrangements of absorber are possible. We shall look at three variations: "free absorber" (Fig. 17.1, *b*), and "absorber with natural frequency" arranged either with a gap or with interference (Fig. 17.1, *c*), and with double-sided stops (Fig. 17.1, *d*). The model described is a special case of impact interaction of two linear dynamic systems, described in §14. Here system *1* is active while the passive system takes the form of a free mass (Fig. 17.1, *b*) or of a linear oscillator (Fig. 17.1, *c*, *d*). In §14 we used harmonic linearisation by displacement, so here to demonstrate the use of the method of equivalent force linearisation, we will concentrate our principal attention on the analysis of results of calculation and to explaining the basic character of impact vibration absorbtion.

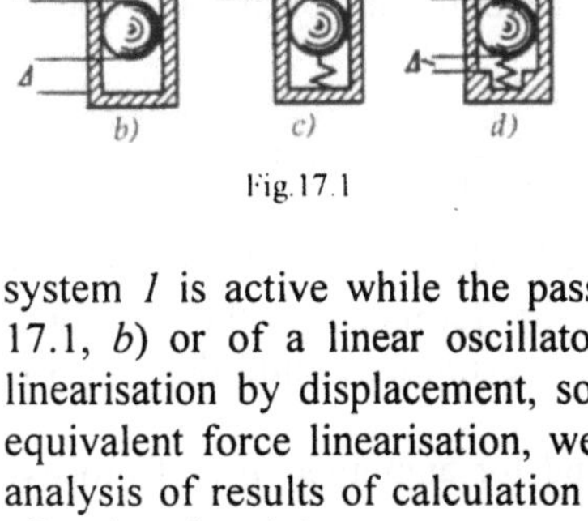

Fig.17.1

The three variations for attaching the damper correspond to the ideas presented in §6 (Fig. 6.2, *b* to *d*) for the static elastic characteristics $\Phi(x)$ relative to the displacement, x, of the mass of the absorber and the damped point I (Fig. 17.1). Then the equation for the system with an absorber considered here can be written in the following form:

$$x_1(t) = L_1(s)\{P(t) + \Phi[x(t)] + \beta sx\}, \tag{17.1}$$

$$M_2 s^2[x_1(t) + x(t)] = -\Phi[x(t)] - \beta sx, \tag{17.2}$$

where x_1 is the absolute displacement of point A, measured from the free position of the system; β is the coefficient of equivalent linearisation for the dissipative losses during impact. We can use here the values (7.61), (7.62) found in §7 when studying harmonic external excitation, since when we introduced the formula, it was stated that for subsequent cases it would only be necessary to adjust the resonance of the "partial frequency" of the impacting element which, as will we show later, is a basic condition for the effectiveness of impact vibration absorbtion. For stationary random excitation we use (8.54) for the value of β, or double for a symmetrical impact pair.

Let us linearise function $\Phi(x)$ using the formula

$$\Phi(x) \approx m_F + k(x - m_x) = m_F + kx^0. \tag{17.3}$$

Substituting (17.3) into (17.2) and noting that the left hand side of (17.2) contains no constant component, we find

$$m_F = 0. \tag{17.4}$$

Furthermore, from (17.1), (17.2) using (17.3), (17.4) we obtain a system of linear equations relative to the generalised coordinates $x_1(t)$ and $x(t)$. Suppose that $P(t) = m_P + P^0(t)$ $(m_P = \text{const})$, we can separate this system into constant and centred components

$$x_1^0(t) = \frac{(M_2 s^2 + \beta s + k)L_1(s)}{[1 + M_2 s^2 L_1(s)](\beta s + k) + M_2 s^2} P^0(t), \tag{17.5}$$

$$x^0(t) = -\frac{M_2 s^2 L_1(s)}{[1 + M_2 s^2 L_1(s)](\beta s + k) + M_2 s^2} P^0(t), \tag{17.6}$$

$$m_{x_1} = L_1(0) m_P. \tag{17.7}$$

Using (17.6), we find the equations for the centred components of process $F(t) = \Phi[x(t)]$:

$$F^0(t) = kx^0(t) = -\frac{M_2 s^2 k L_1(s)}{[1 + M_2 s^2 L_1(s)](\beta s + k) + M_2 s^2} P^0(t). \tag{17.8}$$

2. Let us suppose $P(t) = m_P + a_P \cos(\omega t + \varphi)$. Then for various methods of attaching the absorber we can use corresponding formula for the harmonic linearisation coefficients obtained in §7, section 4, bearing in mind that $k = 1/q$ and substituting condition (17.4) in them directly.

At the same time, according to (17.8) we find for amplitude a_F of process $F^0(t)$

$$a_F = \frac{M_2\omega^2 k(a_F)|L_1(j\omega)|a_p}{\left|[1 - L_1(j\omega)M_2\omega^2][k(a_F) + \beta j\omega] - M_2\omega^2\right|}. \tag{17.9}$$

Hence, in the basic case, equation (17.9) together with one of the equations for k allows us to find the value of a_F, with the help of which we can then find all the remaining parameters for the process in the system examined. When using the formula for k in the forms (7.85), (7.91), (7.95) the equations for a_F can be solved directly.

After finding a_F, the amplitude of vibration of point A in the fundamental is found from (17.5)

$$a_{x_1} = \frac{\left|k(a_F) + \beta j\omega - M_2\omega^2\right||L_1(j\omega)|a_p}{\left|[1 - L_1(j\omega)M_2\omega^2][k(a_F) + \beta j\omega] - M_2\omega^2\right|}$$

or, using (17.9)

$$a_{x_1} = \frac{\left|k(a_F) + \beta j\omega - M_2\omega^2\right|}{M_2\omega^2 k(a_F)} a_F. \tag{17.10}$$

Supposing that the absorber parameters have been chosen so that

$$k(a_F) = M_2\omega^2, \tag{17.11}$$

then for the corresponding frequency ω we find from (17.10)

$$a_{x_1} = \beta a_x / (M_2\omega) \qquad (a_x = a_F / k). \tag{17.12}$$

Hence, similar to the linear dynamic vibration absorber [207], the best tuning of the impact vibration absorber corresponds to the equality of the *partial frequency* of the absorber

$$\omega_0 = \sqrt{k(a_F)/M_2}$$

and to the frequency of excitation ω. The "residual vibration" of the first harmonic is proportional to the loss of energy in the absorber. However, with impact absorbtion the equivalent elasticity k depends not only on the system parameters, but also on the parameters of its own regime, i.e. feedback exists. Thus it is possible to use impact damping to damp vibration with changing frequency.

Substituting the expression for β from (7.61) or (7.62) into (17.12) and bearing in mind that during impact vibration damping $M \approx M_2$, since $M_2 << M_1$ (M_1 is the mass of the protecting system *1* reduced to point A), we obtain

$$a_{x_1} = \rho a_x, \tag{17.13}$$

where $\rho = 0.32(1 - R^2)$ for a double-sided impact pair and $\rho = 0.16(1 - R^2)$ for a single-sided pair. Hence complete damping of the fundamental is possible only when the coefficient of restitution $R = 1$. Taking into account that, according to (17.9) $a_F = a_p$ and substituting this value into (17.1), we can obtain the expression for $a_p(\omega)$, with which complete damping of the fundamental can be obtained in all frequency ranges, which satisfy the existence conditions of the vibro-impact regimes. For example, for the free absorber, using (7.85), (17.4) we find

$$a_P = 4M_2\Delta\omega^2/\pi, \tag{17.14}$$

i.e. it can effectively be applied to damp vibration in regimes with slowly changing frequency due to acceleration or braking of unbalanced masses, since the excitation amplitude here changes according to the law $a_P = \varepsilon\omega^2$. Substituting this expression into (17.14) we obtain the condition for tuning the damper in the given case

$$M_2\Delta = \tfrac{1}{4}\pi\varepsilon \approx 0.78\varepsilon. \tag{17.15}$$

According to (17.15) the mass and gap equally influence the tuning of the free damper. This situation is confirmed by experiments [7].

In engineering practice, the value of the exciting force $P(t)$ is not, as a rule, known in advance, but only the vibrational motion of point A (Fig. 17.1, *a*) without damping. Assuming this is given by $a\cos(\omega t + \psi)$, we find

$$a_P = \frac{a}{|L_1(j\omega)|}.$$

In this case the formula for the best arrangement of the free absorber, using (17.14) takes the form

$$M_2\Delta = \frac{\pi}{4}\frac{a}{|L_1(j\omega)|\omega^2}. \tag{17.16}$$

Let us continue with the study of dynamic systems with free damper (Fig. 17.1, *b*). Using (7.85) we find

$$a_x = qa_F = 4\Delta/\pi. \tag{17.17}$$

Substituting (17.17) in (17.13), we obtain

$$a_{x_i} = 4\rho\Delta/\pi.$$

Hence, for the worst case when $R = 0$ we find $a_{x_i} \approx 0.4\Delta$.

To find the existence conditions of solutions and to construct the amplitude-frequency response for systems with free damper, substitute (17.17) and (7.62) into (17.9). We find

$$\left|[1 - M_2\omega^2 L_1(j\omega)]\left(j\rho M_2\omega^2 + \frac{\pi}{4\Delta}a_F\right) - M_2\omega^2\right| = M_2\omega^2|L_1(j\omega)|\frac{\pi}{4\Delta}a_P. \tag{17.18}$$

Assuming $L_1(j\omega) = \mu_1(\omega) + j\nu_1(\omega)$ and solving equation (17.18) for a_F, we find

$$a_F = \frac{4\Delta M_2\omega^2}{\pi\{[1 - M_2\omega^2\mu_1(\omega)]^2 + M_2^2\omega^4\nu_1^2(\omega)\}}[1 - M_2\omega^2\mu_1(\omega) \pm \sqrt{B}], \tag{17.19}$$

where

$$B = \pi\{[1 - M_2\omega^2\mu_1(\omega)]^2 + M_2^2\omega^4\nu_1^2(\omega)\}\left\{\left(\frac{\pi^2}{16\Delta^2}a_P^2 - \rho^2 M_2^2\omega^4\right)[\mu_1^2(\omega) + \nu_1^2(\omega)] + 2M_2\omega^2\rho[\mu_1(\omega) + \rho\nu_1(\omega)] - \rho^2\right\} - M_2^2\omega^4\nu_1^2(\omega).$$

In reality, of course, only positive values of a_F have a real sense. Analysing

(17.19) it is easy to show that when

$$M_2\omega^2\mu_1(\omega) < 1, \quad B > 0, \tag{17.20}$$

$$1 - \frac{\pi^2}{16\Delta^2} a_P^2[\mu_1^2(\omega) + \nu_1^2(\omega)] + \rho^2\{1 + M_2^2\omega^4[\mu_1^2(\omega) + \nu_1^2(\omega)]\} - \\ - 2M_2\omega^2\rho[\mu_1(\omega) + \rho\nu_1(\omega)] > 0$$

both values of a_F are positive, but when

$$1 - \frac{\pi^2}{16\Delta^2} a_P^2[\mu_1^2(\omega) + \nu_1^2(\omega)] + \rho^2\{1 + M_2^2\omega^4[\mu_1^2(\omega) + \nu_1^2(\omega)]\} - \\ - 2M_2\omega^2\rho[\mu_1(\omega) + \rho\nu_1(\omega)] < 0 \tag{17.21}$$

only one positive value of a_F occurs. If neither of these conditions is not satisfied, then the solution sought does not generally exist. We can assume that in this case vibration at frequency ω can not occur in the system with two impacts per period.

Suppose $\nu_1(\omega) \equiv 0$, i.e. energy dissipation in the damped system is negligible. Then in agreement with (17.19)

$$B = [1 - M_2\omega^2\mu_1(\omega)]^2\left\{\frac{\pi^2}{16\Delta^2} a_P^2\mu_1^2(\omega) - \rho^2[1 - M_2\omega^2\mu_1(\omega)]^2\right\}$$

and, consequently,

$$a_F = \frac{4\Delta M_2\omega^2}{\pi[1 - M_2\omega^2\mu_1(\omega)}\left\{1 \pm \sqrt{\frac{\pi^2}{16\Delta^2} a_P^2\mu_1^2(\omega) - \rho^2[1 - M_2\omega^2\mu_1(\omega)]^2}\right\} \tag{17.22}$$

Conditions (17.20), (17.21) take the corresponding form

$$M_2\omega^2\mu_1(\omega) < 1, \\ \rho\left|1 - M_2\omega^2\mu_1(\omega)\right| < \frac{\pi}{4\Delta} a_P\left|\mu_1(\omega)\right| < \sqrt{1 + \rho^2[1 - M_2\omega^2\mu_1(\omega)]^2}, \tag{17.23}$$

$$\frac{\pi}{4\Delta} a_P\left|\mu_1(\omega)\right| > \sqrt{1 + \rho^2[1 - M_2\omega^2\mu_1(\omega)]^2}\ . \tag{17.24}$$

For the amplitude a_{x_1}, the following expression is obtained:

$$a_{x_1} = \frac{4\Delta}{\pi\left|1 - M_2\omega^2\mu_1(\omega)\right|} \times \\ \times\left[\left\{M_2\omega^2\mu_1(\omega) \pm \sqrt{\frac{\pi^2}{16\Delta^2} a_P^2\mu_1^2(\omega) - \rho^2[1 - M_2\omega^2\mu_1(\omega)]^2}\right\}^2 + \right. \\ \left. + \rho^2[1 - M_2\omega^2\mu_1(\omega)]^2\right]^{1/2} \tag{17.25}$$

The condition for maximum damping of vibration now appear as:

$$M_2\omega^2\mu_1(\omega) \pm \sqrt{\frac{\pi^2}{16\Delta^2} a_P^2\mu_1^2(\omega) - \rho^2[1 - M_2\omega^2\mu_1(\omega)]^2} = 0\ . \tag{17.26}$$

If condition (17.23) is satisfied, then according to (17.26) the maximum damping can only be achieved in one of the possible regimes. In the case when (17.24) is satisfied damping is achieved if $\mu_1(\omega) < 0$, since in this case only the solution corresponding to the positive sign in (17.25) is valid.

Thus the final solution to the behaviour of the system considered can be established only after studying the solution stability. However, before turning to look at stability, we shall, as an example, look at the simpler problem of impact vibration damping of a suspended mass M_1 as an example (Fig. 6.3). In this case when $n_1 \equiv 0$, $L_1(s) = (M_1 s^2 + c_1)^{-1}$. Then $\mu_1(\omega) = (c_1 - M_1\omega^2)^{-1}$, and from (17.25) we obtain

$$a_{x_1} = \frac{4\Delta}{\pi\left|1-(1+\mu)\xi^2\right|} \times \\ \times \sqrt{\{\mu\zeta^2 \pm \sqrt{\chi^2 - \rho^2[1-(1+\mu)\zeta^2]}\}^2 + \rho^2[1-(1+\mu)\zeta^2]^2}, \tag{17.27}$$

where $\mu = M_2/M_1$; $\chi = 4\Delta aP/(\pi c_1)$; $\zeta^2 = M_1\omega^2/c_1$.

According to (17.24), (17.23) when

$$\chi > \sqrt{(1-\zeta^2)^2 + \rho^2[1-(1+\mu)\zeta^2]^2}$$

only one solution is possible, and when

$$\mu\zeta^2/(1-\zeta^2) < 1, \quad \rho\left|1-(1+\mu)\zeta^2\right| < \chi < \sqrt{(1-\zeta^2)^2 + \rho^2[1-(1+\mu)\zeta^2]^2}$$

two solutions are possible. Fig. 17.2 shows curves for the amplitude-frequency response for the damping mass M_1 when $\rho = 0$, $\mu = 0.1$, $\chi = 0.2$. The positive sign in (17.27) corresponds to the full line, the negative sign to the dotted line. The latter can not exist in domains where $\left|1-\zeta^2\right| < \chi$. Later we will show that the solution corresponding to the dotted curves is unstable.

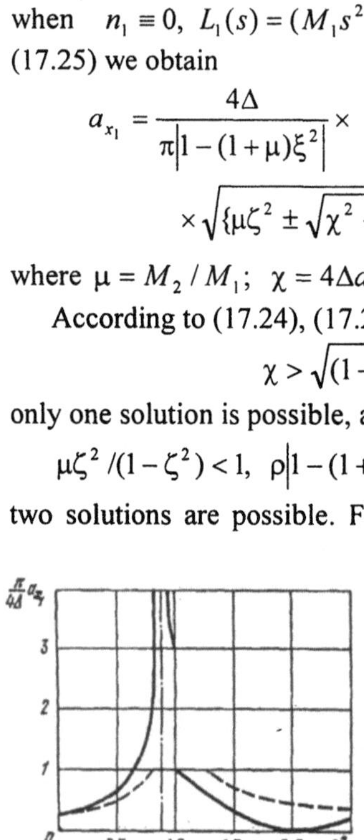

Fig.17.2

3. We shall now turn to look at stability of approximate periodic solutions for the basic model of the system studied (Fig. 17.1, *a*), using for this the results of §14, section 2. In agreement with condition (14.32) and bearing in mind that now $a_v = a_x$, to solve the stability problem we need to find the reduced dynamic stiffness *W*, connecting the external force *P*(*t*) with the "deformation" *x*(*t*) of the linearised impacting element. According to (17.6), this characteristic takes the form

$$W(j\omega, a_F) = \frac{[1 - M_2\omega^2 L_1(j\omega)][\beta j\omega + k(a_F)] - M_2\omega^2}{M_2\omega^2 L_1(j\omega)}. \tag{17.28}$$

When using harmonic linearisation by force the frequency response (17.28) depends on a_F hence to use condition (14.32) it follows that

$$\frac{dU(\omega,\tilde{a}_F)}{d\tilde{a}_x}=\frac{dU(\omega,\tilde{a}_F)}{d\tilde{a}_F}\frac{d\tilde{a}_F}{d\tilde{a}_x}=\frac{dU(\omega,\tilde{a}_F)}{d\tilde{a}_F}k, \tag{17.29}$$

$$\frac{dV(\omega,\tilde{a}_F)}{d\tilde{a}_x}=\frac{dV(\omega,\tilde{a}_F)}{d\tilde{a}_F}\frac{d\tilde{a}_F}{d\tilde{a}_x}=\frac{dV(\omega,\tilde{a}_F)}{d\tilde{a}_F}k \tag{17.30}$$

($U=\operatorname{Re}W$, $V=\operatorname{Im}W$). Substituting (17.29), (17.30) into (14.32) and noting that $\tilde{a}_x k=\tilde{a}_F$, we obtain corresponding conditions for stability of the solution, using the harmonic linearisation by force method

$$\left[U\left(U+\tilde{a}_F\frac{dU}{d\tilde{a}_F}\right)+V\left(V+\tilde{a}_F\frac{dV}{d\tilde{a}_F}\right)\right]_{\tilde{a}_F=a_F}>0. \tag{17.31}$$

When energy dissipation in the system is neglected, $V\equiv 0$ and condition (17.31) is transformed to the form

$$\left[U\left(U+\tilde{a}_F\frac{dU}{d\tilde{a}_F}\right)\right]_{\tilde{a}_F=a_F}>0. \tag{17.32}$$

In (17.28), putting $\beta\equiv 0$, $L_1(j\omega)=\mu_1(\omega)$, we find

$$U=\frac{[1-M_2\omega^2\mu_1(\omega)]k-M_2\omega^2}{M_2\omega^2\mu_1(\omega)}. \tag{17.33}$$

As a result, for example, for systems with free damper ($k=\pi a_F/(4\Delta)$), when we substitute (17.33) into (17.32), we obtain the following necessary conditions for stability:

$$a_F>\frac{4\Delta M_2\omega^2}{\pi[1-M_2\omega^2\mu_1(\omega)]},\quad a_F>\frac{4\Delta M_2\omega^2}{2\pi[1-M_2\omega^2\mu_1(\omega)]}. \tag{17.34}$$

The first of these conditions includes the second.

From (17.22) when $\rho\equiv 0$ we find the value of a_F which corresponds to the regime studied:

$$a_F=\frac{4\Delta M_2\omega^2}{\pi[1-M_2\omega^2\mu_1(\omega)]}\left[1\pm\frac{\pi}{4\Delta}a_P|\mu_1(\omega)|\right]. \tag{17.35}$$

Substituting (17.35) into the first of the conditions (17.34), we obtain eventually

$$\pm\frac{\pi}{4\Delta}a_P|\mu_1(\omega)|>0. \tag{17.36}$$

Hence, regimes which correspond to the negative sign in the solution of the dynamics problem are unstable.

In the application to the example studied (Fig. 17.2) this means that the dotted curves correspond to unstable regimes and, consequently, vibration damping is obtained only when passing through the natural frequency of the damped system. In pre-resonant domains it is possible to get strong oscillation of the system at a frequency $\omega=\sqrt{c_1/(M_1+M_2)}$. These results agree with known results, obtained by using exact methods [84, 85, 86, 157 to 159] and with experimental data [7].

4. Let us now look at the effects of random excitation on systems with an

impact vibration absorber, i.e. we will assume that $P(t) = m_p + P^0(t)$ where $P^0(t)$ is a stationary normal random process with a given spectral density $S_p(\omega)$. Then, for different forms of absorber attachment, we can use the corresponding formula for the statistical linearisation coefficients found in §8 section 4, bearing in mind that $k = 1/q$ and directly substituting in them the condition (17.4).

From (17.8), we obtain also

$$\sigma_F^2 = \frac{1}{2\pi}\int_{-\infty}^{\infty}\frac{M_2^2\omega^4 k|L_1(j\omega)|^2 S_P(\omega)d\omega}{\left|[1-M_2\omega^2 L_1(j\omega)](\beta j\omega+k)-M_2\omega^2\right|^2}. \tag{17.37}$$

The relationship (17.37), together with one of the equations for k allows us to find the value of σ_F, which we can use to find the remaining characteristics of the process in the system studied. In particular, for $\sigma_{x_1}^2$ we find from (17.5)

$$\sigma_{x_1}^2 = \frac{1}{2\pi}\int_{-\infty}^{\infty}\frac{\left|(k+\beta j\omega - M_2\omega^2)L_1(j\omega)\right|^2 S_P(\omega)d\omega}{\left|[1-M_2\omega^2 L_1(j\omega)](\beta j\omega+k)-M_2\omega^2\right|^2}. \tag{17.38}$$

As an example, we shall look at the effects of white noise ($S_P(\omega) = S_0 = \text{const}$) on a single mass damped system, supplied impact absorbtion (Fig. 7.3). In this case, $L_1(s) = (M_1 s^2 + n_1 s + c_1)^{-1}$. As a result, from (17.38), assuming for simplification $R = 1$ $(\beta \equiv 0)$ and using the tabulated values for similar integrals [184] we find

$$\sigma_{x_1}^2 = \frac{1}{2\pi}\int_{-\infty}^{\infty}\frac{(k-M_2\omega^2)^2 S_0 d\omega}{\left|(c_1 - M_1\omega^2 + n_1 j\omega)(k - M_2\omega^2) - M^2\omega^2 k^2\right|^2} = \frac{S_0}{2n_1c_1}, \tag{17.39}$$

i.e. the mean square value of $x_1(t)$ does not depend on k and coincides with expressions for the mean square value of linear single mass systems without absorbtion (see (13.18)). Hence, in the case considered the mean square value of the damped system is not changed with any "elastic" attachment of the absorber.

The reason for this apparent paradoxical effect can be explained by the nature of dynamic damping. During *harmonic* excitation the absorber, thanks to its natural non-isochronous nature, automatically adjusts its own partial frequency to the frequency of excitation. When there is practically no such dominant frequency, for example when excited by white noise, then the influence of the absorber can only dissipate energy when $\beta \neq 0$. Such a method for damping random vibrations would appear to be ineffective, since significant losses in the absorber will be possible only by maintaining intensive vibration of the damping mass.

Chapter 5. Structural synthesis of self-sustaining vibro-impact systems

The scientist describes what is; the engineer creates what never was.

Theodore von Karman

§18. Optimal periodic motions of vibro-impact systems

1. In §12 we have already analysed systems in which the action of a force on a striker appears as a function of its phase coordinates only. Similar methods of realising periodic motion have been widely used in percussion machines which use electrical, pneumatic or hydraulic methods of exciting vibration. Such machines include pneumatic chisels and perforators, hydraulic or pneumatic rotary impact drills, and various types of test machines or test rigs. Applying self-sustaining vibration gives stability to the process compared to the application of forced vibration and often leads to a more simple form of design.

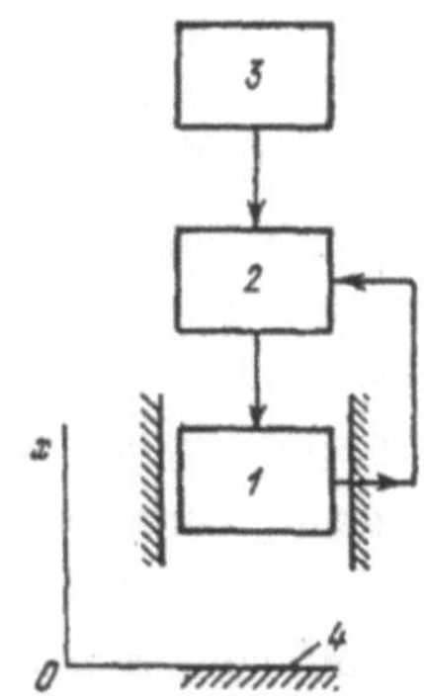

Fig. 18.1

The general schematic for self-sustaining vibro-impact systems can be represented by an impacting mass *1* (Fig. 18.1) with motion $x(t)$, to which is applied a varying force of excitation $u(t) = B[x(t), \dot{x}(t)]$ from an unregulated source of energy *3*, using a controller *2* having characteristic $B(x, \dot{x})$ as the result of which the mass undergoes systematic impacts against an anvil *4*.

The form of the controller characteristic $B(x, \dot{x})$ defines the dynamic behaviour of the system to a considerable degree, making it possible to achieve any type of motion which satisfies the pre-defined technological and energy requirements. Hence, this remaining chapter is devoted to the problem of synthesis of $B(x, \dot{x})$, to produce the required type of periodic system motion under optimal conditions.

In practice, such a problem is not very different from the known problem of the analytic design of controllers [144]. However, using the traditional approach in this case, based on the design of a stabilising controller to compensate the mismatch between the ideal motion required and the real observed motion, does not appear to be applicable for vibro-impact systems because of the great complexity of such control systems.

The specific features of vibro-impact systems provide the opportunity to suggest another approach for synthesis [28, 30 to 32, 29].We seek characteristics for the controller which will enable it to satisfy two basic functions simultaneously: to achieve the required optimal working motion for the system and to compensate external disturbances. Moreover, the required reaction of the system can be pre-programmed also in the structure of the controller within defined limits, as parameters of the system drift from the nominal. These combined effects can be obtained within an autonomous system, which ensures simple and safe engineering arrangements.

The specific feature of vibro-impact systems mentioned is that they should undergo periodic motion, in such a way that the working process (impact) occurs during short time intervals, which are considerably shorter than the period of vibration. This allows us to control the periodic regime of the striker using relay type devices, applying the defined level of force excitation in the given phase plane domains for the striker. In this case, synthesising the form of the controller characteristic $B(x,\dot{x})$ reduces to finding the values of the corresponding control effort U_i and to finding the switching curves $\Pi_i(x,\dot{x})=0$.

Later, we will show that for a wide class of functionals which represent the quality of vibro-impact systems, *optimal control*, $u(t)=B[x(t),\dot{x}(t)]$, prescribed fulfils the required type of periodic motion for the system $x(t)$, is a piecewise-constant function of time with finite number of switches per period (discontinuities). Thus, the search for time control function $u(t)$ and the corresponding system motion $x(t)$ allows us to find a series of points for the switching curve on the phase plane, through which the phase trajectory of the optimal limit cycle for the system passes, and the correct control levels U_i, which support the portion of the trajectory between the switching points (Fig. 18.2). The points indicated have coordinate $\{x(t_i),\dot{x}(t_i)\}$ where t_i is the instant at which $u(t)$ is switched, and completely defines the optimal periodic regime in combination with the given control levels. By adding points defined in this way up to the switching curve, we can guarantee stability of the regime and supply additional useful qualities to the system which characterise its reaction to a variation of the parameters.

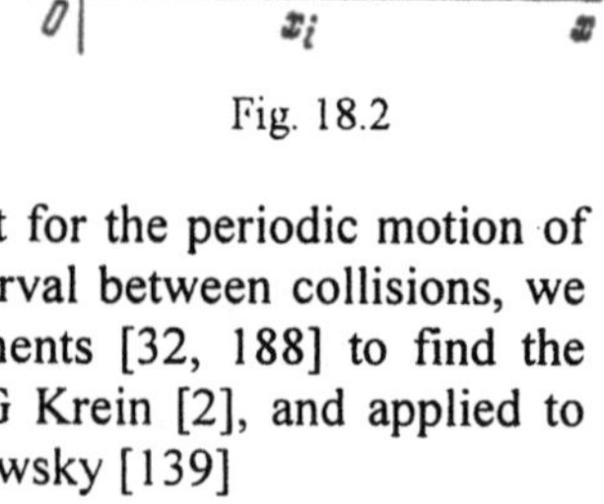

Fig. 18.2

The search for optimal control function $u(t)$ for vibro-impacting systems has specific features [27], since control, which is strong and nonlinear, is sought for the periodic motion of the system. By taking the system as linear in the interval between collisions, we can effectively use the mathematical theory of moments [32, 188] to find the required control function, based on the work of M G Krein [2], and applied to optimal control by N N Krasovsky [137] and R Kulikowsky [139]

2. Let us formulate the mathematical problem of synthesis described above. In the interval between impacts the equations of the system (Fig. 18.1), reduced to unit mass of the striker can be written in the following form:

$$\ddot{x} = B(x,\dot{x}) - g, \tag{18.1}$$

where $B(x,\dot{x})$ is the control characteristic sought; g is the acceleration due to gravity. Impact, originating when $x = 0$ is described by Newton's law

$$\dot{x}_+ = -R\dot{x}_- \qquad R \in [0,1]. \tag{18.2}$$

Taking into account the limitations of the actuator on the value of the forces developed, we naturally assume that

$$|B(x,\dot{x})| \le U = \text{const}. \tag{18.3}$$

As already mentioned, the form of motion for a striker is limited in practice to a periodic function with period T. In addition, during the course of one period a finite number of impacts q of required periodicity and intensity can occur, depending on the technological requirements of the system. Such a situation arises, for example, in actual impact processes where each impact during a cycle represents a separate operation.

Although the problem of producing periodic motion by itself is complex due to the character of such nonlinear systems with ambiguous behaviour, it is convenient to solve the problem of synthesis on the basis of some optimised criteria.

In cases when the working cycle of the proposed system is strictly defined, i.e. given the number of impacts per period, their intensity and time distribution, naturally one tries to obtain such a process using a minimal intensity drive, and choosing as the minimised value of functional I the value

$$I = U. \tag{18.4}$$

For the majority of vibro-impact machines, working process is single impact ($q = 1$) and the regime parameters are not strictly prescribed. In this case we can pose the situation of a maximum intensity impact process with limited effort from the excitation as given by (18.3). The natural intensity of the impacting process is represented by an impacting velocity $\dot{x}_-$ and impact frequency. As a rule, one of the properties is defined from strength or technological consideration, hence the problem of optimisation can be formulated as a maximisation

$$I = \dot{x}_- \tag{18.5}$$

with the given values U and T, or minimising

$$I = T \tag{18.6}$$

with the given values U and $\dot{x}_-$.

Note that with a changing coefficient of restitution R, a more objective characteristic of intensity of the impact process will be the value of the impact impulse J, transmitted to the anvil by the striker, or to the corresponding value of the energy E. Using the linear relationship between impulse and impact velocity $J = (1 + R)\dot{x}_-$, this is preferable in calculations, since later we will also look at the problem of maximisation

$$I = J \tag{18.7}$$

given the values of U and T. The value J/T is a measure of the force and can be used to demonstrate the average efficiency of the machine.

Thus, the problem of synthesis can be put in the following form. *To find the characteristic* $B(x,\dot{x})$ *which satisfies* (19.3), *producing q impact processes with period T in systems described by* (18.1), (18.2), *and optimising the quality functional I, defined by one of the expressions* (18.4), (18.5), (18.6) *or* (18.7).

3. Denoting $B[x(t),\dot{x}(t)] = u(t)$, we can rewrite (18.1) in the form

$$\ddot{x} = u(t) - g \tag{18.8}$$

and we will look at the preliminary problem of finding the optimal control function $u(t)$, realising the periodic motion of the system given in section 2.

We will limit the analysis to single-impact motion ($q = 1$), considered, for generality as a number l sub-periodic regime, i.e. a regime with one impact per l periods of excitation. In this case equation (18.8) acts during the interval $t \in (0, lT)$, where the instant after impact is taken as the starting point for time measurement. It is shown in [225] that for the functionals examined, single impact motion exists as the optimal among all periodic vibro-impact processes. The boundary conditions at the instant of collision with the stationary stop using the conditions of periodicity will be as follows:

$$x(0) = x(lT) = 0, \qquad \dot{x}(0) = \dot{x}(lT)\,. \tag{18.9}$$

Rewriting equation (18.8) in the Cauchy form [142]

$$\begin{aligned} \dot{x}(t) &= \dot{x}_+ - gt + \int_0^t u(\tau)d\tau, \\ x(t) &= x(0) + t\dot{x}_+ - \frac{gt^2}{2} + \int_0^t (t-\tau)u(\tau)d\tau. \end{aligned} \tag{18.10}$$

Applying condition (18.9) to (18.10) we obtain, using (18.2)

$$(1+R)\dot{x}_- + glT = \int_0^{lT} u(\tau)d\tau = l\int_0^T u(\tau)d\tau\,, \tag{18.11}$$

$$RlT\dot{x}_- + \frac{gl^2T^2}{2} = \int_0^{lT} (lT-\tau)u(\tau)d\tau\,. \tag{18.12}$$

Changing the interval of integration in equation (18.12) to the period of control $\tau \in (0,T)$. We find

$$\int_0^{lT} (lT-\tau)u(\tau)d\tau = \sum_{i=0}^{l-1} \int_{iT}^{(i+1)T} (lT-\tau)u(\tau - iT)d\tau\,,$$

due to the periodicity of the control function $u(\tau) = u(\tau - iT)$ $(i = 0,1,\ldots)$. Introducing a new variable $t = \tau - iT$, we obtain

$$\sum_{i=0}^{l-1}\int_{iT}^{(i+1)T}(lT-\tau)u(\tau-iT)d\tau=\sum_{i=0}^{l-1}\int_0^T(lT-iT-t)u(t)dt=$$

$$=\int_0^T\sum_{i=0}^{l-1}(lT-iT-t)u(t)dt=\int_0^T\sum_{i=0}^{l}(iT-t)u(t)dt=$$

$$=\int_0^T\left[\frac{l(l+1)}{2}T-lt\right]u(t)dt.$$

As a result conditions (18.11), (18.12) can be rewritten in the following form:

$$\begin{aligned}(1+R)\dot{x}_-+glT&=\int_0^T lu(t)dt,\\ RlT\dot{x}_-+g\frac{(lT)^2}{2}&=\int_0^T l\left(\frac{l+1}{2}T-t\right)u(t)dt.\end{aligned}\qquad(18.13)$$

Let us discuss from first principles the problem of minimisation of functional (18.4), assuming $\dot{x}_-$ and T have given values. On the left hand side of (18.13) we have a known constant α_l. Regarding the functions $h_1(t)=l$, $h_2(t)=l\left(\frac{l+1}{2}T-t\right)$ and $u(t)$ on the right hand side of equation (18.13) and defined in the sector $t\in[0,T]$ as the vectors of some normalised space, we note that the right hand side of expression (18.13) represents a scalar product of the unknown vector $u(t)$ with the given vector $h_i(t)$ $(i=1,2)$.

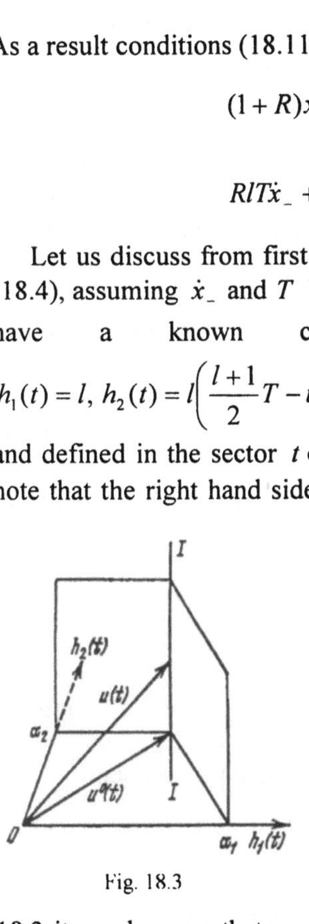

Fig. 18.3

Choosing as a norm ("length") of the vector in $\mathcal{B}^*$ space for the function $u(t)$ with the value

$$U=\max_{t\in[0,T]}|u(t)|,\qquad(18.14)$$

we obtain that finding the optimal control function $u(t)$ from expression (18.13) can be interpreted as the problem of finding vector $u^0(t)\in\mathcal{B}^*$, having the given projection α_i on vectors $h_i(t)$ and possessing the minimum norm (Fig. 18.3). From Fig. 18.3 it can be seen that vector $u^0(t)$ is unique and is found among a number of vectors $u(t)$ the ends of which are arranged along the line of intersection of the planes orthogonal to the given system of vectors $h_i(t)$, which must be linearly independent. In [137] it is shown that in addition, the space $\mathcal{B}^*$ of function $u(t)$ must be "conjugated" to the $\mathcal{B}$ space of function $h_i(t)$ so that choosing the space $\mathcal{B}^*$ in the form (18.14) gives the norm $\mathcal{B}$ in the form

$$\int_0^T|h_i(t)|dt,\qquad(18.15)$$

i.e. the functions $h_i(t)$ must be limited by the norm (18.15), and this is always ensured in problems examined later.

4. The problem described above can be considered as part of a more general mathematical problem. Let a set of m expressions exist

$$\alpha_i = \int_0^T h_i(t)u(t)dt \quad (i = 1, 2, \ldots, m), \tag{18.16}$$

where α_i are known real numbers; $h_i(t)$ are linear independent functions, defined in the sector $t \in [0, T]$, is limited in the sense of (18.15). It is required to find a function $u(t)$, satisfying (18.16) with minimum modulus (18.14).

The problem formulated belongs to that known as a *moment problem*, which has recently received wide application in the theory of optimal control. The complete deduction and proof of the existence of solutions to this problem will emerge in the framework of the present exposition since it is based on the application of functional analysis, with which the reader is not assumed to be familiar. For this reason only a simple explanation is given here, leading to a solution of this particular problem.

Suppose n_i are arbitrary real numbers. Multiplying each of the expressions in (18.16) by n_i and adding these products, we obtain

$$\sum_{i=1}^{m} \alpha_i n_i = \int_0^T \sum_{i=1}^{m} n_i h_i(t)u(t)dt. \tag{18.17}$$

Since the numbers n_i are assumed arbitrary, then they can be chosen, in particular, so that

$$\sum_{i=1}^{m} \alpha_i n_i = 1. \tag{18.18}$$

Then from (18.17) we find

$$1 = \int_0^T \sum_{i=1}^{m} n_i h_i(t)u(t)dt. \tag{18.19}$$

Applying to the right hand side of (18.19) an obvious inequality, we obtain

$$\begin{aligned} 1 = \int_0^T \sum_{i=1}^{m} n_i h_i(t)u(t)dt &\le \int_0^T \left| \sum_{i=1}^{m} n_i h_i(t)u(t) \right| dt \le \\ &\le \max_{t\in[0,T]} |u(t)| \int_0^T \left| \sum_{i=1}^{m} n_i h_i(t) \right| dt. \end{aligned} \tag{18.20}$$

From which we find

$$\max_{t\in[0,T]} |u(t)| \ge \left[\int_0^T \left| \sum_{i=1}^{m} n_i h_i(t) \right| dt \right]^{-1}. \tag{18.21}$$

Inequality (18.21) must be satisfied for any number n_i, satisfying (18.18) and especially for such n_i^0, for which the right hand side of (18.21) takes a maximum value

$$\max_{t\in[0,T]}|u(t)| \geq \left[\min_{n_i}\int_0^T\left|\sum_{i=1}^{m} n_i h_i(t)\right|dt\right]^{-1} \quad \left(\sum_{i=1}^{m} n_i \alpha_i = 1\right). \tag{18.22}$$

From (18.22) it follows that the minimum value of the maximum $|u(t)|$ found among all possible functions $u(t)$, satisfies the condition

$$U_0 = \min_{u}\max_{t\in[0,T]}|u(t)| = \left[\int_0^T\left|\sum_{i=1}^{m} n_i^0 h_i(t)\right|dt\right]^{-1}. \tag{18.23}$$

Now it is necessary to find the function $u(t)$ in such a form that the series of two inequalities (18.20) can be converted to an equality, since this is a necessary condition to satisfy expression (18.16). It is not difficult to prove that this takes place when and only when

$$u(t) = U \mathrm{sign} \sum_{i=1}^{m} n_i h_i(t), \tag{18.24}$$

where U is a constant. Really, the first of the inequalities (18.20) can be converted to an equation, when and only when the function $\sum_{i=1}^{m} n_i h_i(t)$ and $u(t)$ take identical signs at every instant of time, and secondly when and only when $u(t)$ takes the same sign as $\sum_{i=1}^{m} n_i h_i(t)$, and also, a constant value $|u(t)| = U$ for all $t \in [0,T]$. Obviously, if U was not a constant then it can not be factorised out of the integral and the second inequality can not be converted to an equation.

Thus to satisfy (18.16), function $u(t)$ must take the form (18.24) but since we want the value $U = \max_t |u(t)|$ to be a minimum, i.e. equal to U_0, then it must be found from (18.23), where the number n_i^0 is calculated from the solution to the problem: to find

$$\min_{n_i}\int_0^T\left|\sum_{i=1}^{m} n_i h_i(t)\right|dt \tag{18.25}$$

under condition (18.18). Consequently, on a basis of (18.23), (18.24) the solution of the resulting problem takes the form

$$u(t) = \frac{\mathrm{sign}\sum_{i=1}^{m} n_i^0 h_i(t)}{\int_0^T\left|\sum_{i=1}^{\infty} n_i^0 h_i(t)\right|dt}. \tag{18.26}$$

5. We shall apply the results of section 4 to find the optimal control function $u(t)$ for vibro-impact systems. From (18.26), using (18.13) we find

$$u(t)=\frac{sign\left\{n_1^0\left[\frac{l(l+1)}{2}T-lt\right]+n_2^0 l\right\}}{\int_0^T\left|n_1^0\left[\frac{l(l+1)}{2}T-lt\right]+n_2^0 l\right|dt},\tag{18.27}$$

where the numbers n_1^0 and n_2^0 are solutions of the expression (18.25): to find

$$\min_{n_1,n_2}\int_0^T\left|n_1\left[\frac{l(l+1)}{2}T-lt\right]+n_2 l\right|dt=\frac{1}{U_0}\tag{18.28}$$

with the condition

$$n_1\left[RlT\dot{x}_-+g\frac{(lT)^2}{2}\right]+n_2[(1+R)\dot{x}_-+glT]=1.\tag{18.29}$$

In the case under consideration it is not necessary to follow the standard procedure, since the solution can be obtained more simply using only the information from solution of the moment problem that, according to (18.27), the optimal control function is discontinuous with one switching in the interval $t\in(0,T)$, since the 'sign' is applied to a linear function. Thus,

$$u(t)=\begin{cases}+U_0 \text{ when } t\in[0,t_1],\\ -U_0 \text{ when } t\in[t_1,T]\end{cases}\tag{18.30}$$

(the opposite combination of signs is excluded for the obvious mechanical reasons).

Substituting (18.30) into (18.13) we obtain after integrating the system of two finite equations for finding U_0 and t_1:

$$\begin{aligned}(1+R)\dot{x}_-+glT&=(2t_1-T)lU_0,\\ RlT\dot{x}_-+g\frac{(lT)^2}{2}&=\left[\frac{l+1}{2}(2t_1-T)T-t_1^2+\frac{T^2}{2}\right]lU_0.\end{aligned}\tag{18.31}$$

Solving these, we find

$$U_0=\frac{(1+R)\dot{x}_-+glT}{(2t_1-T)l},\tag{18.32}$$

$$t_1=\alpha\pm\sqrt{\alpha^2-\alpha T+\tfrac{1}{2}T^2},\tag{18.33}$$

where

$$\alpha=\frac{l^2T(1+R)\dot{x}_-+lT(1-R)+gl^3T^2}{2[(1+R)\dot{x}_-+glT]}.$$

The sought solution is only satisfied when the values $t_1\in(0,T)$. If both roots $t_1\in(0,T)$, then it is necessary to choose that corresponding to the smaller value of U_0 in (18.32).

The equations (18.31) allow us to find the optimal control function *u*(*t*), giving the maximum for (18.5) for the given values of *U* and *T*. To eliminate $\dot{x}_-$ from (18.31) we find a quadratic equation in terms of t_1:

$$t_1^2 - \left(1 + l\frac{1-R}{1+R}\right)Tt_1 + l\frac{1-R}{1+R}(1+\gamma)T^2 = 0,$$

where $\gamma = g/U_0$.

Solving this, we find

$$t_1 = \frac{T}{2}\left[1 + l\frac{1-R}{1+R} \pm \sqrt{1 - 2\gamma l\frac{1-R}{1+R} + l^2\left(\frac{1-R}{1+R}\right)^2}\right]. \tag{18.34}$$

Limiting the value of γ from the inequality $\gamma < 1$, we find from (18.34) that for a positive sign $t_1 > T$, hence the conditions for the problem are satisfied only if the solution (18.34) agrees with the negative sign. Using this we find from (18.31)

$$|\dot{x}_-| = \left[\gamma + \sqrt{1 - 2\gamma l\frac{1-R}{1+R} + l^2\left(\frac{1-R}{1+R}\right)^2} - l\frac{1-R}{1+R}\right]\frac{lTU_0}{1+R}. \tag{18.35}$$

From (18.35) it can be seen that the maximum $|\dot{x}_-|$ occurs for the maximum possible value of U_0, i.e. when $U_0 = U$.

For the problem of maximising the impact impulse, we find in agreement with (18.35)

$$\max_u J = (1+R)\max_u|\dot{x}_-| =$$
$$= \left[\gamma + \sqrt{1 - 2\gamma\frac{1-R}{1+R} + l^2\left(\frac{1-R}{1+R}\right)^2} - l\frac{1-R}{1+R}\right]lTU. \tag{18.36}$$

In the same way we can find expressions for the problem of optimisation of high speed action, by minimising condition (18.6).

Note that when solving the problem of optimisation the condition $x(t) \geq 0$ was not used, leading, in general, to the infringement of the fact that a unique optimal solution exists. However, depending on the form of the solution obtained, it is easy to show that they always satisfy the required conditions.

Fig. 18.4 shows examples of optimal excitation $u(t)$ and the corresponding motion of the striker when $l = 2$, $R = 0.5$, $\gamma = 0.4$, $U = 1$. Expression (18.36) is drawn in Fig. 18.5 for various values of R, γ, l. It can be seen that increasing the value of l can lead to an increase in the energy of impact.

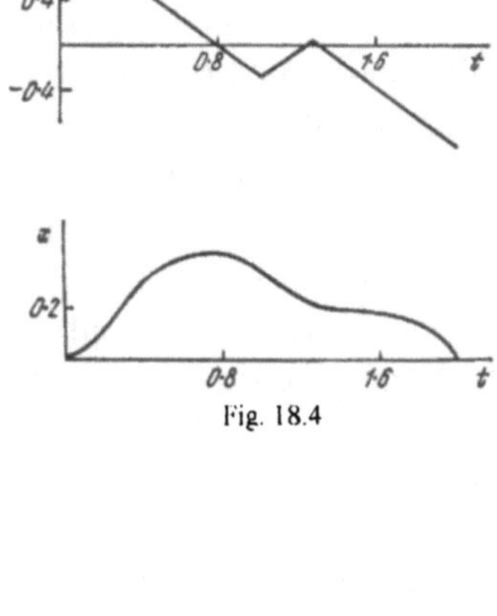

Fig. 18.4

Note that for the case when $l = 1$ reference [188] gives a solution for the problem considered by optimising condition (18.4); in [16] the optimal control

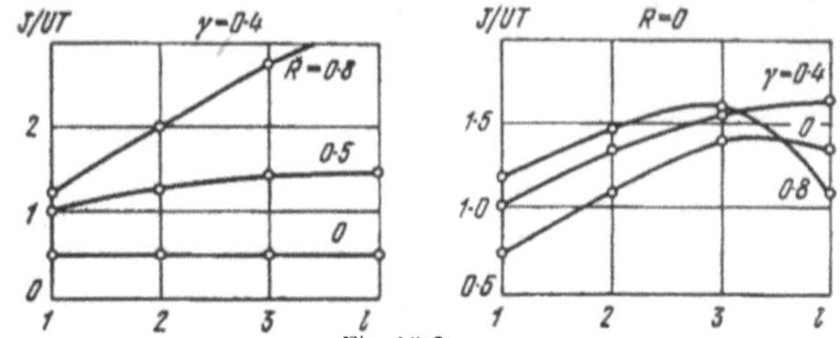

Fig. 18.5

function was found for the maximum speed of systems when $l = 1$, $R = 0$ using special procedures; in [110] the problem is solved for a striker with elastic suspension.

The solutions obtained above for the problem of optimal control will be applied later to the synthesis of regulators, however they are also of separate interest, since they can be used in etalon during the design, adjustment and evaluation of machines with impact effects. Thus, for example, references [16, 16] searched for the best parameters to construct hammer drills with impact effects using similar calculation principles. Here, the optimal laws of motion were used in etalon, and in reference [140] the optimal law was used in a similar manner to synthesise the parameters for various mechanisms with impact effects. Finally, note that frequently the optimal control $u(t)$ can be used in the form of periodic excitation to produce forced vibration of vibro-impact systems. In this case, the question of stability of the optimal motion still remains. The problem of stability of vibro-impact systems excited by arbitrary periodic forces has been studied by the stitching method in references [30, 69].

§19. Synthesis of self-sustaining systems

1. The above solution for optimising striker excitation makes it possible to synthesise self-sustaining systems to give the optimal motion for the striker. This can be achieved by eliminating time from the optimised functions found for $x(t)$, $\dot{x}(t)$ and u(t) (Fig. 19.1). The problem of synthesis now becomes a question of finding a function $B(x,\dot{x})$, which coincides in time with the optimised control function (18.30) $B[x(t),\dot{x}(t)] = u(t)$ as the system moves periodically. This function will be determined completely, given the switching curve $\Pi(x,\dot{x}) = 0$, which divides the phase plane of the system into two domains, corresponding to the different control levels $\pm U$. It is obvious that the switching curve must pass through the point with

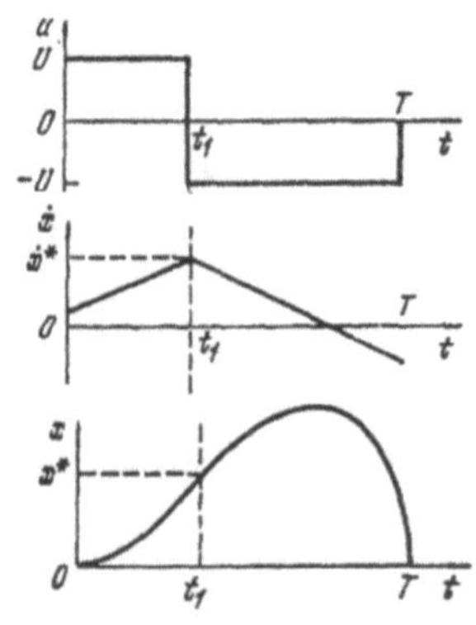

Fig. 19.1

coordinate $\{x(t_1), \dot{x}(t_1)\}$ (Fig. 19.1), which we shall call the *nominal* point, since at this point control switching occurs in periodic regimes of systems whose parameters are nominal.

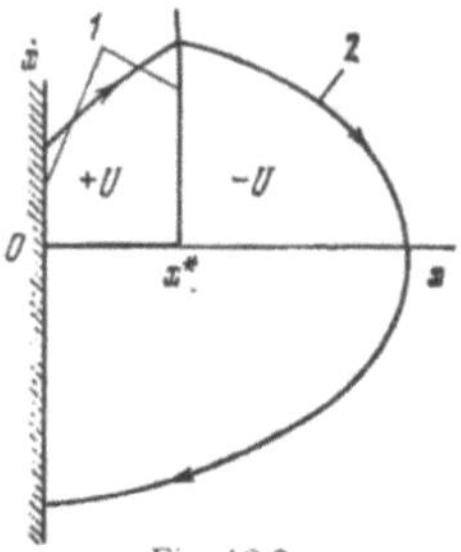

Fig. 19.2

Let us look at the equations for the phase coordinates of the nominal point, obtained from the solution to the problem of finding the control function $u(t)$, optimising the system for various criteria (see §18).

a) maximising J or $\dot{x}_-$, given the values of U and T:

$$x(t_1) = t_1\left[\frac{\dot{x}(t_1)}{UT} - \frac{1-\gamma}{2}\frac{t_1}{T}\right]UT\,, \tag{19.1}$$

$$\dot{x}(t_1) = \left[\frac{R}{1+R}\left(1 - \frac{2t_1}{T} + \gamma\right) + (1-\gamma)\frac{t_1}{T}\right]UT\,, \tag{19.2}$$

$$t_1 = \frac{2 - \sqrt{2(1+R^2) - 2\gamma(1-R^2)}}{2(1+R)}T\,. \tag{19.3}$$

b)`minimising T, given the values of J and U:

$$x(t_1) = \frac{(1-R)J^2}{4(1+R)U}\,, \tag{19.4}$$

$$\dot{x}(t_1) = \sqrt{\frac{1+R^2-\gamma(1-R^2)}{2}}\,\frac{J}{1+R}\,, \tag{19.5}$$

$$t_1 = \frac{J}{U}\frac{1}{(1+R)(1-\gamma)}\sqrt{\frac{1+R^2-\gamma(1-R^2)}{2}} - R\,. \tag{19.6}$$

c) minimising T, given the values of $\dot{x}_-$ and U:

$$x(t_1) = \left(Rt_1 + \frac{1-\gamma}{2}t_1^2\right)\frac{(\dot{x}_-)^2}{U}\,, \tag{19.7}$$

$$\dot{x}(t_1) = [R + (1-\gamma)t_1]|\dot{x}_-|\,, \tag{19.8}$$

$$t_1 = \frac{1}{(1-\gamma)}\left[\sqrt{\frac{1+R^2-\gamma(1-R^2)}{2}} - R\right]\frac{|\dot{x}_-|}{U}\,. \tag{19.9}$$

Note that the nominal point completely defines the optimal limit cycle of the synthesised system. This is typical of the synthesis of systems to produce optimal periodic motion, in contrast to the usual problem of synthesis in the theory of optimal control, meant to produce optimal transient processes [55, 88, 179]. When the problem is restricted to the optimisation of periodic motion, an element of arbitrariness is introduced to the shape of the switching curves, so it is possible to expand nominal switching points to the switching curve in such a way that, apart from the necessary singularity and stability conditions of the nominal limit cycle, the system would ensure the required reaction to drift the parameters from their nominal values, for example, stabilise some parameters of the working regime:

period T, impact velocity $\dot{x}_-$ or impulse J as the coefficient of restitution R changes.

2. Later we will make a detailed study of the additional potential of the synthesis method indicated, but for the moment we will study the switching curve *1* (Fig. 19.2)which is simpler to achieve, and has the form

$$\Pi(x,\dot{x}) = \begin{cases} x - x^* = 0 \text{ when } \dot{x} > 0, \\ x = 0 \text{ when } \dot{x} < 0, \end{cases} \tag{19.10}$$

where $x^* = x(t_1)$ (Fig. 19.1).

As a result the optimised force function (control characteristic) $B(x,\dot{x})$ will be as follows (Fig. 19.2):

$$B(x,\dot{x}) = \begin{cases} +U \text{ when } \dot{x} > 0, & 0 \le x < x^*, \\ -U \text{ when } \dot{x} > 0, & x \ge x^*, \\ \qquad\quad \dot{x} < 0, & x > 0, \end{cases} \tag{19.11}$$

i.e. as the striker move upwards to the switching point x^* it takes the positive value, and is negative for all other conditions.

Let us investigate the state of the vibro-impact system, defined by expressions (18.1), (18.2), during the application of the force characteristic (19.11). We shall take x^* as an arbitrary value.

We will rewrite (18.1) in the form

$$dx/dt = \dot{x}, \quad d\dot{x}/dt = B(x,\dot{x}) - g \tag{19.12}$$

and we will look at the phase trajectory of system *2* (Fig. 19.2). The straight line $x = 0$ represents the position of the stop. The domain of positive values for the force function lies above the abscissa between the lines $x = 0, x = x^*$. The remainder of the phase plane corresponds to negative values of the force function.

At the instant $t = 0$, following the *v-th* collision, the coordinates of the system will be $(0, \dot{x}_{v+})$, $B(x,\dot{x}) = U$. Dividing the second equation in (19.12) by the first,, we obtain

$$d\dot{x}/dx = U(1-\gamma)/\dot{x}\,.$$

Integrating this equation using the initial conditions, we find

$$\dot{x} = \sqrt{(\dot{x}_{v+})^2 + 2U(1-\gamma)x}\,, \tag{19.13}$$

i.e. the system velocity increases continuously and when it reaches the switching point becomes equal to

$$\dot{x}^* = \sqrt{(\dot{x}_{v+})^2 + 2U(1-\gamma)x^*}\,. \tag{19.14}$$

Similarly for the phase trajectory of the system after switching we obtain

$$\dot{x} = \pm\sqrt{(x_{v+})^2 + 2U[2x^* - (1+\gamma)x]}\,, \tag{19.15}$$

where the upper half-space corresponds to the positive sign, and the lower to the negative sign. Assuming $\dot{x} = 0$ in (19.15), we find the maximum displacement of the striker in the *v-th* cycle:

$$x_{\max} = \frac{(\dot{x}_{\nu+})^2 + 4Ux^*}{2U(1+\gamma)}. \tag{19.16}$$

As the representative point reaches the line $x = 0$ the next (ν + 1)-*th* collision occurs. In this situation, in agreement with (19.15)

$$\dot{x}_{(\nu+1)-} = -\sqrt{R^2(\dot{x}_{\nu-})^2 + 4Ux^*}. \tag{19.17}$$

Expression (19.17) defined the point mapping of the line $x = 0$ onto itself. When the regime is periodic, we find from (19.17)

$$\dot{x}_{(\nu+1)-} = \dot{x}_{\nu-} = \dot{x}_- = -2\sqrt{Ux^*/(1-R^2)}. \tag{19.18}$$

Hence, the system has a unique limiting cycle which must be the optimal.

Let us study its stability. From (19.17) we obtain

$$\left|\frac{\partial \dot{x}_{(\nu+1)-}}{\partial \dot{x}_{\nu-}}\right| = \left|\frac{R^2 x_{\nu-}}{\sqrt{R^2(\dot{x}_{\nu-})^2 + 4Ux^*}}\right| < 1$$

for any starting conditions and parameters of the system. Consequently, the periodic regime of the synthesised self-sustaining system is optimal and stable.

The connection between the period T of self-sustaining vibration and the switching coordinate x^* can be obtained conveniently using (18.10). We find

$$x^* = x(t_1) = -R\dot{x}_- t_1 - \frac{gt^2}{2} + \int_0^{t_1}(t_1 - \tau)U d\tau = -t_1 R\dot{x}_- + \frac{t_1^2}{2}U(1-\gamma). \tag{19.19}$$

Substituting here the values for $\dot{x}_-$ and t_1 from (18.34), (18.35), we obtain after transformation

$$T = \frac{1}{1-\gamma^2}\left[\sqrt{2\left(\frac{1+R^2}{1-R^2} - \gamma\right)} + \sqrt{\frac{1+R}{1-R}}\left(\frac{1-R}{1+R} - \gamma\right)\right]\sqrt{\frac{4x^*}{U}}. \tag{19.20}$$

Substituting the optimal value of the impact velocity from (19.18) into (19.16) we find a formula to estimate the length of the system

$$x_{\max} = \frac{2x^*}{(1-\gamma)(1-R^2)}. \tag{19.21}$$

The value of the optimal impact impulse for the system in agreement with (19.18) will be as follows:

$$J = 2\sqrt{\frac{Ux^*(1+R)}{1-R}}. \tag{19.22}$$

Formulae (19.20) to (19.22) show that changing the switching coordinate x^* and force amplitude U is a convenient method to regulate the regime parameters in a synthesised vibro-impact system. Note that by virtue of the unique limit cycle of the synthesised system, the control method shown allows us to obtain optimal regimes for all bands of control.

To estimate the quality of transient processes in systems, using (19.17) we can obtain a general expression to find the sequence of impact velocities for systems during the period when it settles into a steady-state regime. Summing the resulting geometric progression, we find

$$\dot{x}_{\nu-} = -\sqrt{R^{2\nu}(\dot{x}_{0-})^2 + \frac{4Ux^*}{1-R^2}(1-R^{2\nu})} \quad (\nu = 1, 2, \ldots)). \qquad (19.23)$$

where $\dot{x}_{0-}$ is the impact velocity at the start of the transient process.

Let us look at a numerical example of the transient process for a system with the following parameter values: $R = 0.5$, $T = 1s$; $\gamma = 0$, $U = 10m/s^2$. From (19.1) to (19.3) we find the switching point $x^* = 0.45m$; from (19.18) the impact velocity in the steady-state regime $\dot{x}_- = -4.9m/s$. Supposing $\dot{x}_{0-} = 0$, then from (19.23) we find

$$\dot{x}_{1-} = -4.2\,m/s,\ \dot{x}_{2-} = -4.75\,m/s,\ \dot{x}_{3-} = -4.87\,m/s$$

Hence, in practice, the system becomes stable after the third impact.

3. In the system synthesised in section *2* the force function $B(x,\dot{x})$ has two switching points $x = 0$ and $x = x^*$. Accordingly, one switching must occur at the instant of impact. We will look at the influence of delay or advance in switching relative to the instant of impact on the dynamics.

Consider first the case of delayed switching. The phase plane for such a system is drawn in Fig. 19.3. The first switching occurs at some value $x = x_1^*$. As a result the force function $B(x,\dot{x})$ takes the form

$$B(x,\dot{x}) = \begin{cases} +U \ when\ \dot{x} > 0, & x_1^* \le x < x^*, \\ -U \ when\ \dot{x} > 0, & 0 \le x < x_1^*, \\ & x \ge x^*, \\ \dot{x} < 0, & x > 0. \end{cases}$$

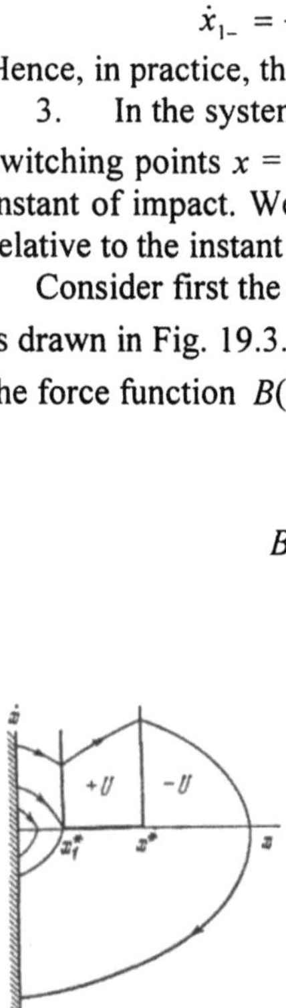

Fig. 19.3

Let us look at the motion of the representative point after the instant of the *v-th* collision (Fig. 19.3). Until the instant of the first switching $B(x,\dot{x}) = -U$, the equation of the phase trajectory takes the form (see section 2)

$$\dot{x} = \sqrt{(\dot{x}_{\nu+})^2 - 2U(1+\gamma)x}\,.$$

According to this law the point will move until such time as it reaches the switching line $x = x_1^*$ or its projection reaches abscissa $x = 0$. The first case will occur if

$$(\dot{x}_{\nu+})^2 > 2U(1+\gamma)x_1^*, \qquad (19.24)$$

the second when the inequality has the opposite sign. Hence, inequality (19.24) defines the value of initial velocity at which self-sustained vibration can possibly occur, since in the second case vibration will be damped out (Fig. 19.3). As a result, the limiting cycle in this case can be achieved with hard excitation.

When condition (19.24) is satisfied after intersecting the line $x = x_1^*$ the equation of the phase trajectory takes the form

$$\tfrac{1}{2}\dot{x}^2 = U(1-\gamma)x + C\,. \qquad (19.25)$$

Finding constant C from the condition $\dot{x} = \sqrt{(\dot{x}_{\nu+})^2 - 2U(1+\gamma)x_1^*}$ when $x = x^*$, we obtain the law of motion for the representative point before the second switching

$$\dot{x} = \sqrt{(\dot{x}_{\nu+})^2 + 2(1-\gamma)Ux - 4Ux_1^*}\ . \tag{19.26}$$

Similarly we find the law of motion for the chosen point after the second switching

$$\dot{x} = \pm\sqrt{(\dot{x}_{\nu+})^2 + 4U(x^* - x_1^*) - (1+\gamma)Ux}\ , \tag{19.27}$$

where the positive sign corresponds to motion in the upper half-plane, and the negative sign to the lower half-plane. Assuming $x = 0$ in (19.27), we obtain the equation for the point mapping of the line $x = 0$ onto itself

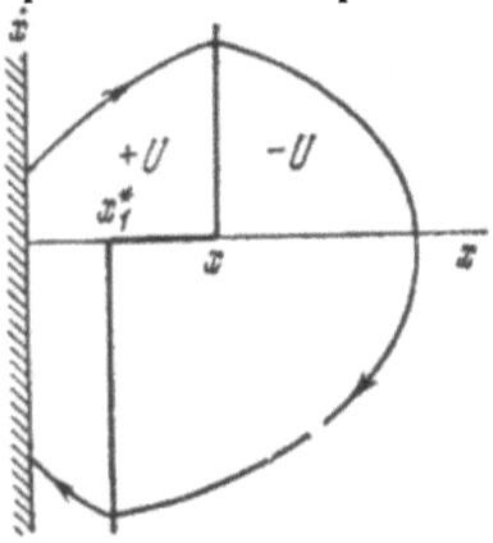

Fig. 19.4

$$\dot{x}_{(\nu+1)} = -\sqrt{R^2(\dot{x}_{\nu-})^2 + 4U(x^* - x_1^*)}\ . \tag{19.28}$$

The unique fixed point of this mapping

$$\dot{x}_{-} = -2\sqrt{\frac{U(x^* - x_1^*)}{1-R^2}} \tag{19.29}$$

determines the stable limit cycle, since

$$\left|\frac{\partial \dot{x}_{(\nu+1)-}}{\partial \dot{x}_{\nu-}}\right| = \left|\frac{R^2\dot{x}_{\nu-}}{\sqrt{R^2(\dot{x}_{\nu-})^2 + 4U(x^* - x_1^*)}}\right| < 1\ . \tag{19.30}$$

However, this limit cycle will not be stable for all initial conditions and parameters, since the initial velocity must satisfy condition (19.24) for it to happen. The latter condition is only a necessary condition for the limit cycle to exist. In reality, if velocity $\dot{x}_+$ corresponding to periodic motion does not satisfy inequality (19.24) then it will be unstable. Hence, from (19.24), (19.29) we can obtain the following condition for stability of the limit cycle:

$$x_1^* < \frac{2R^2/(1-R^2)}{1+\gamma+2R^2/(1-R^2)}x^*\ . \tag{19.31}$$

Let us now look at a system where switching is advanced (Fig. 19.4). Its force function is defined by the expressions

$$B(x,\dot{x}) = \begin{cases} +U \text{ when } \dot{x} > 0,\ 0 \le x < x^*, \\ \qquad\qquad\ \ \dot{x} < 0,\ 0 \le x < x_1^*, \\ -U \text{ when } \dot{x} > 0, \quad x \ge x^*, \\ \qquad\qquad\ \ \dot{x} < 0, \quad x \ge x_1^*. \end{cases} \tag{19.32}$$

The law of motion of the chosen point after impact is described by (19.13). After switching, (19.15) will apply, according to which the point will move until it reaches the line $x = x_1^*$. Thus the motion will be defined by the formula

$$\dot{x} = -\sqrt{(\dot{x}_{\nu+})^2 + 2U(1-\gamma)x + 4U(x^* - x_1^*)}\ . \tag{19.33}$$

If $x^* > x_1^*$, then as it moves according to the (19.33), the chosen point will always reach the line $x = 0$. The corresponding point mapping for this coincides with (19.28) and on the basis of (19.30) we find that the limit cycle in this case is stable for all parameters values and for any initial conditions. From (19.29) for the case considered we find

$$J = 2\sqrt{\frac{1+R}{1-R}U(x^{*} - x_{1}^{*})}\,.$$

4. Let us turn now to explore systems with more general forms of switching curve $\Pi(x,\dot{x}) = 0$. In section 1 we already have shown that widening the class of the system considered allows us to the synthesise the characteristics for regulators, which add *adaptive* properties to systems, for example, the invariance of certain of the regime parameters as external conditions change. The simplest method to construct such a family of switching curves is as follows: the coordinates of nominal points are dealt with as equations for the switching curve in terms of parameter changes, for example *R*, while invariant parameters T or $\dot{x}_{-}$ remain fixed.

Fig. 19.5 shows examples of locus for switching points, constructed using formulae (19.1) to (19.3) for various values of *R* with *T* constant. Using portions of these curves as working parts of the switching curve appropriate to the control of the system over the working zone, it is possible to secure invariance of the period *T* as *R* changes over the interval defined. Both here and later we shall assume that change of *R* occurs slowly by comparison with the time taken for transition of the system process.

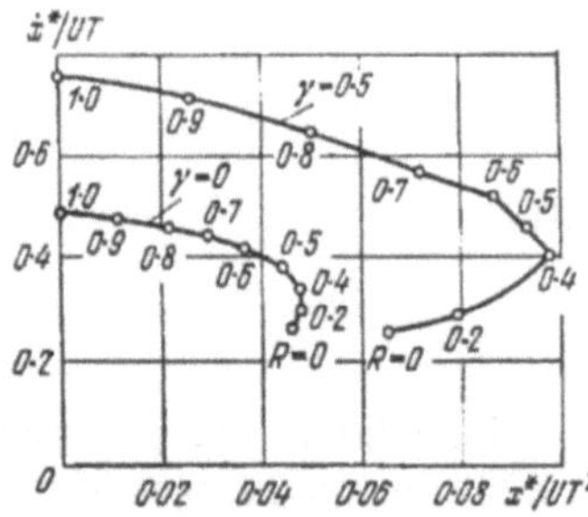

Fig. 19.5

Similarly, we can construct curves for switching which ensure the invariance of J or $\dot{x}_{-}$.

The lengths of the portions obtained are limited by the conditions for the absence of additional intersections with this portion and of the limit cycle for the system. Switching curves which satisfy such conditions will be called *permissible*.

The chosen working portion of the switching curve should supplement the curves which ensure the transition of the system onto this part of the switching curve from the required domains of initial conditions. In the simplest case these curves are produced, as shown, using a portion of the line x = const. In the more general case such portions can be drawn using the required quality of the transfer process.

Let us look at a system with an arbitrary switching curve $\Pi(x,\dot{x}) = 0$. The force function in this case takes the form

$$B(x,\dot{x}) = \begin{cases} U\operatorname{sign}\Pi(x,\dot{x}) & \text{when} \quad \dot{x} > 0, \\ -U & \text{when} \quad \dot{x} < 0. \end{cases} \tag{19.34}$$

After an arbitrary *v-th* collision the representative point will move in agreement with (19.13) until it intersects the switching curve. The coordinates of the switching point $x_{v}^{*}, \dot{x}_{v}^{*}$ are defined by the system of equations

$$\dot{x}_\nu^* = \sqrt{(\dot{x}_{\nu+}^*)^2 + 2U(1-\gamma)x_\nu^*}, \qquad \Pi(x_\nu^*, \dot{x}_\nu^*) = 0. \tag{19.35}$$

After reversal, in agreement with (19.15), the equation for the phase trajectory takes the form

$$\dot{x} = \pm\sqrt{(\dot{x}_{\nu+})^2 + 2U[2x_\nu^* - (1+\gamma)x]}. \tag{19.36}$$

Here the positive sign corresponds with motion in the upper half-plane, and the negative sign with the lower. Substituting $x = 0$ into (19.36), we find the expression for the point mapping

$$\dot{x}_{(\nu+1)-} = -\sqrt{(R\dot{x}_{\nu-})^2 + 4Ux_\nu^*}. \tag{19.37}$$

Assuming $\dot{x}_{(\nu+1)-} = \dot{x}_{\nu-} = \dot{x}_-$, $x_\nu^* = x^*$, $\dot{x}_\nu^* = \dot{x}^*$, we obtain a system of three equations to find the stationary point, corresponding to the periodic regime:

$$\begin{aligned} &(\dot{x}^*)^2 - R^2(\dot{x}_-)^2 = 2U(1-\gamma)x^*, \\ &(1-R^2)(\dot{x}_-)^2 = 4Ux^*, \quad \Pi(x^*, \dot{x}^*) = 0. \end{aligned} \tag{19.38}$$

Eliminating $(\dot{x}_-)^2$ from (19.38) we come to a system of equations which define the switching point

$$(\dot{x}^*)^2 = 2\left(\frac{2R^2}{1-R^2} + 1 - \gamma\right)Ux^*, \tag{19.39}$$

$$\Pi(x^*, \dot{x}^*) = 0. \tag{19.40}$$

In the general case, the system of equations (19.39), (19.40) can not be solved in an explicit form. As a convenient calculation procedure we can take the parabolic approximation of the switching curve in the neighbourhood of the nominal points $x_0^*, \dot{x}_0^*$.

Rearranging (19.40) in the form

$$x^* = a_1(\dot{x}^* - \dot{x}_0^*)^2 + a_2(\dot{x}^* - \dot{x}_0^*) + x_0^*. \tag{19.41}$$

where a_1, a_2 are numerical coefficients, and using conditions (19.39) which can be written in the form $(\dot{x}_0^*)^2 = \alpha x_0^*$, where

$$\alpha = 2\left(\frac{2R^2}{1-R^2} + 1 - \gamma\right)U,$$

we obtain from (19.41) a quadratic equation in $\dot{x}^*$:

$$(1 - \alpha a_1)(\dot{x}^*)^2 + \alpha(2a_1\dot{x}_0^* - a_2)\dot{x}^* - \alpha[a_1(\dot{x}_0^*)^2 - a_2\dot{x}_0^* + x_0^*] = 0. \tag{19.42}$$

Using equation (19.42) it is possible to find the coordinates of the switching curve (19.41) at which switching of control excitation occurs as the system parameters moves away from the nominal. This leads to the selection of the numerical parameters a_1, a_2 to ensure that the required properties arise in systems with slowly changing parameters. Note that the values of T and J in the case of an arbitrary switching curve are defined as before by (19.20), (19.22), however the value of x^* is now found from the set of equations (19.39), (19.40).

5. Let us look at the stability of systems in the region of the nominal limit switching cycle for an arbitrary switching curve. Suppose $x_0^*, \dot{x}_0^*$ are the reversal points corresponding to this cycle; $\dot{x}_{0-}$ is the pre-impact velocity. We shall introduce differences $\delta x_\nu = x_\nu - x$, $\delta\dot{x}_\nu = \dot{x}_\nu - \dot{x}$ $(\nu = 1, 2, \ldots)$ which characterise the variation of the phase trajectory from the nominal cycle. Linearising (19.13) near the nominal cycle, we obtain

$$\delta\dot{x}_\nu - \frac{1}{\dot{x}_{0-}}[R^2\dot{x}_0\delta\dot{x}_\nu + U(1-\gamma)\delta x_\nu]. \tag{19.43}$$

After linearising the switching curve, we find for the switching points in the disturbed motion

$$\delta x_\nu^* = k\delta\dot{x}_\nu^* \tag{19.44}$$

where $k = -\left[\frac{\partial\Pi}{\partial x}(x_0^*, \dot{x}_0^*)\right]^{-1}\frac{\partial\Pi}{\partial\dot{x}}(x_0^*, \dot{x}_0^*)$. From (19.43), (19.44) we find for the instant of switching

$$\delta\dot{x}_\nu^* = -\frac{R^2\dot{x}_-}{\dot{x}_{0-} + kU(1-\gamma)}\delta\dot{x}_\nu. \tag{19.45}$$

For the increment in velocity before the following $(\nu + 1)$-*th* impact, linearising (19.37), we obtain

$$\delta\dot{x}_{(\nu+1)} = \frac{1}{\dot{x}_{0-}}(R^2\dot{x}_{0-}\delta\dot{x}_{\nu-} + 2U\delta x_\nu^*). \tag{19.46}$$

From this using (19.44), (19.45) we find the difference equations for the disturbed motion

$$\delta\dot{x}_{(\nu+1)-} = R^2K\delta\dot{x}_{\nu-}, \tag{19.47}$$

where

$$K = \frac{\dot{x}_{0-} - kU(1+\gamma)}{\dot{x}_{0-} + kU(1-\gamma)}. \tag{19.48}$$

According to (19.47), the stability condition takes the form

$$|R^2K| < 1. \tag{19.49}$$

Note that, in addition to condition (19.49), the value of k must also satisfy another condition by intersecting the switching curve and the limiting cycle at the nominal switching point (Fig. 19.6). The latter takes the form

$$\frac{1}{k} > \frac{d\dot{x}}{dx}(x^*) = \frac{U(1-\gamma)}{\dot{x}^*} \quad \text{when } \frac{1}{k} > 0,$$
$$\left|\frac{1}{k}\right| > \frac{U(1+\gamma)}{\dot{x}^*} \quad \text{when } \frac{1}{k} < 0. \tag{19.50}$$

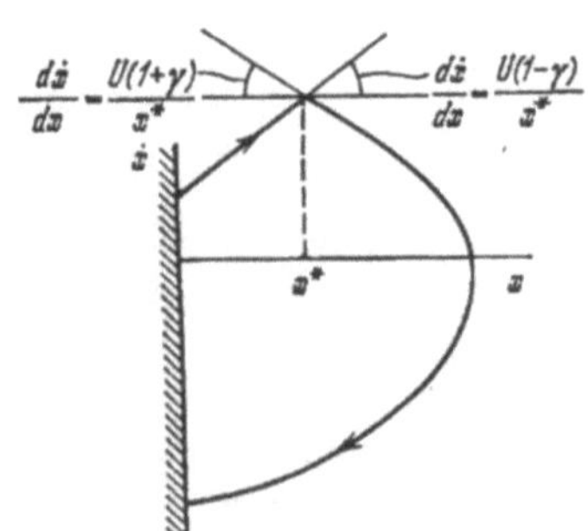

Fig. 19.6

When $\gamma = 0$ both conditions (19.50) can be combined into one:

$$\left|\frac{1}{k}\right| > \frac{U}{\dot{x}^*}. \tag{19.51}$$

The table below shows the limiting values of the term T/k, calculated from (19.50) and corresponding to the problem of maximising the impact impulse for given values of T and U.

R	0	0.2	0.4	0.5	0.6	0.7	0.8	0.9	
$\frac{U(1-\gamma)T}{\dot{x}^*}$	3.34	3.14	2.78	2.63	2.5	2.32	2.22	2.12	$\gamma=0$
	2	1.61	1.22	1.09	0.93	0.86	0.8	0.72	
$\frac{U(1+\gamma)T}{\dot{x}^*}$	6	4.83	3.66	3.27	2.79	2.58	2.4	2.16	$\gamma=0.5$

Numerical calculation shows that for the majority of values of the parameters condition (19.50) is stronger than (19.49). For example, for problem a) (see section 1) when $\gamma=0$, $R=0.2$ we obtain $|k/T|<0.415$ using condition (19.49), while at the same time from the table we find $|k/T|<0.32$.

It follows from (19.47), (19.48), that the corresponding choice of value for k can also influence the speed of the transient process.

6. Let us now look at the problem of synthesising working portions of the switching curve for *small* changes in the coefficient of restitution relative to the nominal value R_0. Suppose that as the result of changing the coefficient of restitution by an amount δR a transient process occurs in the system. Let us linearise the motion of the system near the nominal limit cycle using the change in R. From (19.13) we obtain

$$\delta\dot{x}_{\nu-} = -\frac{1}{\dot{x}_{0-}}[R_0^2\dot{x}_{0-}\delta\dot{x}_{\nu-} + U(1-\gamma)\delta x_\nu] - R_0\dot{x}_{0-}\delta R \tag{19.52}$$

($\dot{x}_{0-}$ is the nominal impact velocity). At the instant of switching, expression (19.44) is satisfied. Using this, we find from (19.52)

$$\delta\dot{x}_\nu^* = -\frac{R_0^2\dot{x}_{0-}}{\dot{x}_{0-}+kU(1-\gamma)}\delta\dot{x}_{\nu-} - \frac{R_0(\dot{x}_{0-})^2}{\dot{x}_{0-}+kU(1-\gamma)}\delta R. \tag{19.53}$$

Linearising (19.37), we find the increment in velocity before impact

$$\delta\dot{x}_{(\nu+1)-} = \frac{1}{\dot{x}_{0-}}(R_0^2\dot{x}_{0-}\delta\dot{x}_{\nu-} + 2U\delta x_\nu^*) + R_0\dot{x}_{0-}\delta R. \tag{19.54}$$

Substituting into (19.54) the expression for δx_ν^* using (19.44), (19.53), we obtain the difference equation for $\delta\dot{x}_-$:

$$\delta\dot{x}_{(\nu+1)-} = R_0^2K\delta\dot{x}_{\nu-} + R_0\dot{x}_{0-}K\delta R. \tag{19.55}$$

From equation (19.47) this changes in the presence of a non-homogeneous term. If the solution of the homogeneous equation (19.47) is stable, i.e. satisfies condition (19.49), then as $\nu\to\infty$, the transient component of the solution of (19.55) tends to zero and only non-homogeneous components remain, which we find by putting $\delta\dot{x}_{(\nu+1)-} = \delta\dot{x}_{\nu-} = \delta\dot{x}_-$ in (19.55):

$$\delta\dot{x}_- = \frac{R_0\dot{x}_{0-}K}{1-R_0^2K}\delta R . \tag{19.56}$$

Expression (19.56) defines the increment in the impact velocity for stable regimes due to a change in the coefficient of restitution. Let us find the corresponding change in the remaining parameters of motion. From (19.53) using (19.56) we find

$$\delta\dot{x}^* = -\frac{(\dot{x}_{0-})^2 R_0}{\dot{x}_{0-}+kU(1-\gamma)}\left(R_0^2+\frac{K}{1-R_0^2K}\right)\delta R .$$

For δJ we obtain directly

$$\delta J = -(1+R_0+\delta R)(\dot{x}_{0-}+\delta\dot{x}_-)-(1+R_0)\dot{x}_{0-} = $$
$$= -\frac{(1+KR_0)\dot{x}_{0-}}{1-KR_0^2}\delta R. \tag{19.57}$$

To find δT we will express T through $\dot{x}_-$ and $\dot{x}^*$. Taking into account that in the interval between impacts and switchings the velocity of the system changes uniformly, we obtain

$$T = \frac{\dot{x}^*+R\dot{x}_-}{U(1-\gamma)}-\frac{\dot{x}_--\dot{x}^*}{U(1+\gamma)} = \frac{2\dot{x}^*+\dot{x}_-[(1+\gamma)R-1+\gamma]}{U(1-\gamma^2)} . \tag{19.58}$$

Linearising (19.58) in terms of δR at the point R_0, we find

$$\delta T = \frac{\{2\delta\dot{x}^*+[(1+\gamma)R_0-1+\gamma]\delta\dot{x}_-\}}{U(1-\gamma)^2}+\frac{\dot{x}_{0-}\delta R}{U(1-\gamma)} . \tag{19.59}$$

(19.56), (19.57) and (19.59) allow us to calculate the sensitivity of the process parameters to changes in the coefficient of restitution. With this aim we can introduce sensitivity characteristics $\delta T/\delta R$, $\delta\dot{x}_-/\delta R$, $\delta J/\delta R$, which depend on the gradient of the switching function k. Choosing this value in the proper manner, it is possible to influence the way in which the parameters of the periodic regime of the system change due to varying the coefficient of restitution from its nominal value. For this the specified regimes will automatically cause the process to function, which gives the adaptive condition required. Fig. 19.7 shows curves for the sensitivity coefficients when

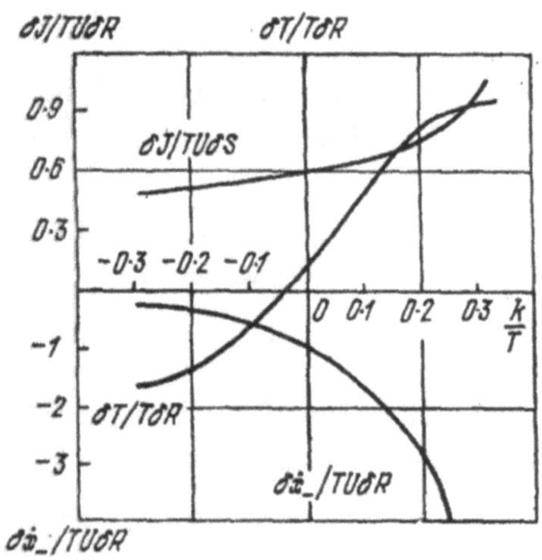

Fig. 19.7

$\gamma = 0$, $R = 0.2$. It was shown above that for these values of the parameters the slope of the switching curve for problem a) is limited to the value $|k/T| \leq 0.32$. In the whole of this range stability of the system is guaranteed for small disturbances.

Reference [29] describes methods for the synthesis for the development of self-sustained vibration in systems causing periodic *multi-impact* regimes with given cycle of operation and intensity of collision.

§20. Synthesis of autoresonant systems

1. We have shown earlier that an effective scheme to excite the striker is to create a driving excitation force on the striker which is a function of its phase coordinate, i.e. using the principle of feedback. The best way to obtain similar excitation of striker *1* (Fig. 20.1) or of the vibro-impact system *2*, including elastic

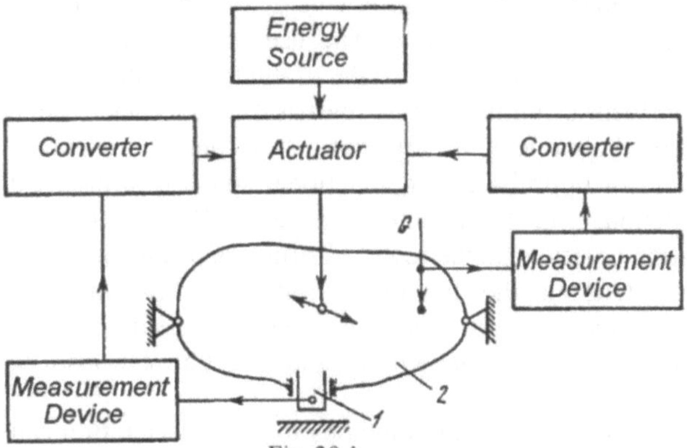

Fig. 20.1

suspension and drive mechanism or wave guide, is to measure and analyse the motion parameters of the striker and the basic loads (for example, the static feed force *G*, with which an impact machine compresses the work piece), to transform and amplify the signal and to generate a driving excitation force to be produced by the actuating arrangement of the drive, based on these signals.

In actual self-sustaining systems, the elements of the control system specified may differ, both in structure and construction. At the same time, as we saw when analysing a pneumatic impacter (see §12 sections 6, 7), the internal dynamics of the system drive elements (such as nonlinearity, hysterisis, the relationship between the output power of the excitation source and the regime in which the system is operating, etc) have a significant influence on system operation. A combination of the internal nonlinear dynamic in vibro-impact systems, with the changing conditions during the operation of the working process (typified by changes in the coefficient of restitution *R*), show that universal control methods, based, for example, on the application of search systems or other adaptive principles, are both expensive and ineffective in such situations.

The best arrangement for systems described would appear to be a self-sustained system tuned to autoresonance, stabilised by nonlinear corrections of perturbation [18] (*rigid adaptation* [89]). With such an approach, the basic emphasis in the construction and adjustment of systems is not to widen the functional potential of the control system, and so increase its complexity by introducing a different analytic structure, but to make a thorough study of systems in order to match all the basic interacting factors. This makes serious demands on the calculation methods to achieve these goals in terms of their universality, flexibility and clarity. The latter quality appears significant, since partially changing, or adding structure to the calculation model by introducing correcting arrangements can often lead directly to the synthesis process from the results of calculations.

Similar to the analysis of vibro-impact systems, the complicated calculation model can give considerable, and often insuperable, difficulties when exact synthesis methods, like those above, are used. The most essential difficulties are related to the increased dimensions of systems and to the estimation of the influence of other nonlinear factors. However, by moving to approximate methods based on equivalent linearisation, the problems of synthesising solutions with the difficulties described can be overcome. The use of equivalent linearisation for nonlinearities makes it possible to reduce the problem of synthesising the required structure to the problem of optimising its linearisation coefficients in terms of the system parameters. Equivalent linearisation is a routine method to reduce a variational problem of optimisation in functional space to a problem of finding *extremum* for functionals in the parameter space. From the values found for the optimal linearisation coefficients the required structure can be produced.

Note that regenerating the functions from their equivalent linearisation coefficients is ambiguous since at this stage of synthesis, additional requirements and limits on characteristics of the synthesised structure can also be taken into account.

The use of harmonic linearisation methods to control vibration can lead to the regulation of amplitude, frequency and phase of the fundamental. In vibro-impact systems of the autoresonant type, the frequency of self-sustaining vibration is defined by the natural characteristics of the vibro-impact system (see §12), hence the problem, as shown, consists of obtaining the maximum intensity of impacts as the tuning of the system changes due to slowly changing the feed G, which influences the system frequency.

A solution to the problem is guaranteed by introducing *feedback* (Fig. 20.2),

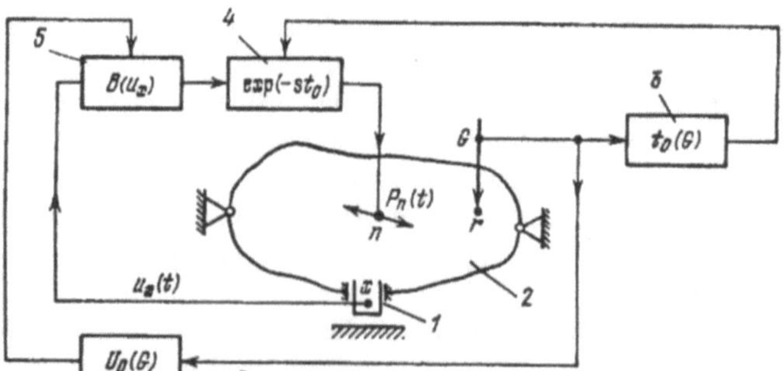

Fig. 20.2

amplifying and transforming the signal $u_x(t)$ proportional to the resulting motion of point x of the striker *1* (the *observed point*) by function $B[u_x(t)]$ in block *5* and attenuated through the delay component *4* at point *n* (the *control point*) of the vibrating system *2*. As a harmonic signal with frequency ω passes through the delay component, a phase lag φ appears between the input and output signals whose value is $\varphi = \omega t_0$, so that, by choosing the delay time t_0 it is possible to guarantee resonant phase at frequency $\omega t_0 = 3\pi/2$ (since $\cos(\omega t - 3\pi/2) = \cos(\omega t + \pi/2)$) on the given autoresonant frequency ω, corresponding to the frequency of resonance with maximum amplitude for the vibro-impact system. By changing the value of feed G applied at point r (the point of disturbance), drift of the natural properties of the vibro-impact system takes place, hence, in order to maintain autoresonance under these conditions it is necessary to control the value of delay t_0 in relation to G to maintain the resonant value of phase $\varphi = -3\pi/2$. Calculating the relationship of ω to G makes it possible to apply such control into practice using a rigid correction coupling using the transformer *3*. By adjusting the transformer 5 using correction component 6, by changing G, it is possible to obtain the required value of autoresonant vibration amplitude in relation to feed G. Hence by arranging the scheme described correctly, autoresonant conditions will be maintained in the system with the prescribed level of vibration whatever the value of G.

2. Let us find the structure of elements 3, 5, 6 for the feedback loop, assuming system *2* is linear and describing the working process executed by the striker by means of a nonlinear reaction $\Phi(u_x, su_x)$ representing the force characteristics of the impact interaction between the striker and the work-piece. We will consider the value of the maximum force, generated by amplifier *5* to be limited to

$$|B(u_x)| \le U_0 \qquad (U_0 = \text{const}). \tag{20.1}$$

Using an operator for the transfer dynamic compliance for system *2* in the form $L_n(x,s)$, connecting the displacement of point x with the force applied at point *n*, we can write the equation for vibration of the striker in the following form:

$$u_x(t) = L_n(x,s)P_n(t) - L_x(x,s)\Phi[u_x(t), su_x(t)] + L_r(x,0)G, \tag{20.2}$$

where, in agreement with the feedback error

$$P_n(t) = B[u_x(t-t_0)] = B[u_x(t)]\exp(-st_0). \tag{20.3}$$

Let us search for an approximate periodic solution to equation (20.2) in the form

$$u_x(t) \approx m_x + u_x^0(t), \qquad u_x^0(t) = a_x \cos\omega t. \tag{20.4}$$

Applying harmonic linearisation to the characteristic $B(u_x)$, we obtain

$$B(u_x) = B(u_x^0) \approx Z(a_x)u_x^0, \tag{20.5}$$

where $Z(a_x)$ is the linearisation coefficient. Here we assume that the feedback filters only the variable components of signal $u_x(t)$.

Taking into account that because of the properties of Fourier series, the limiting level U_0 of the periodic excitation $P_n(t)$ can not have a first harmonic amplitude greater than $4U_0/\pi$ we find from (20.5) using (20.4)

$$a_x Z(a_x) \le 4U_0/\pi\,. \tag{20.6}$$

Linearising the force characteristics of the impact interaction similarly to (14.8), we obtain, substituting (20.4), (20.5) into (20.2) and separating the constant and periodic components:

$$m_x = -L_x(x,0)m_F(m_x,a) + L_r(x,0)G \tag{20.7}$$

$$\begin{gathered}\{1 + L_x(x,s)[k(m_x,a_x) + s\beta(m_x,a_x)] - \\ - L_n(x,s)Z(a_x)\exp(-st_0) = \mathcal{D}(s)u_x^0(t) = 0,\end{gathered} \tag{20.8}$$

where $\mathcal{D}(s)$ is the characteristic equation for the linearised system depending on m_x and a_x as well as the other parameters. Substituting $s = j\omega$ into this and setting the real and imaginary parts to zero we find the existence conditions for self-sustained vibration:

$$Z(a_x)\sin\omega t_0 = -V(\omega,\, m_x,\, a_x), \tag{20.9}$$

$$Z(a_x)\cos\omega t_0 = U(\omega,\, m_x,\, a_x)\,. \tag{20.10}$$

Here $U(\omega,m_x,a_x) = \operatorname{Re} W(j\omega,m_x,a_x)$; $V(\omega,m_x,a_x) = \operatorname{Im} W(j\omega,m_x\, a_x)$; $W(j\omega,m_x,a_x) = [1 + k + j\omega\beta)L_x(x,j\omega)]L^{-1}(x,j\omega)$. .

Eliminating the trigonometric functions from (20.9), (20.10) we obtain

$$Z(a_x) = \sqrt{U^2(\omega,m_x,a_x) + V^2(\omega,m_x,a_x)} = |W(j\omega,m_x,a_x)|\,. \tag{20.11}$$

We will rewrite this expression in the form

$$a_x = a_x Z(a_x)[U^2(\omega,m_x,a_x) + V^2(\omega,m_x,a_x)]^{-1/2}\,. \tag{20.12}$$

Comparing expression (20.12) with (20.6) shows that for all frequencies the maximum amplitude of the striker vibration occurs when the following equation is satisfied

$$a_x Z(a_x) = 4U_0/\pi\,,$$

whence

$$Z(a_x) = 4U_0/(\pi a_x)\,. \tag{20.13}$$

As a result we obtain the value of the harmonic linearisation coefficients for the nonlinear amplifier *5* (Fig. 20.2), which ensure maximum excitation of the striker. It is not difficult to construct the amplifier structural characteristic from it. By analogy with (12.3), (12.20) we observe that expression (20.13) can be obtained, for example, by using a characteristic of saturation for the amplifier $B(u_x)$ in the form (Fig. 20.3):

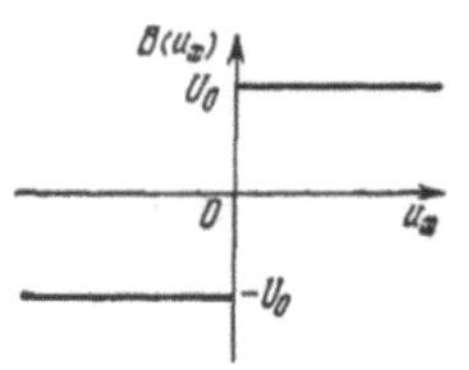

Fig. 20.3

$$B(u_x) = U_0 \operatorname{sign} u_x\,. \tag{20.14}$$

By analogy with (14.17) we can establish that

when an external periodic excitation $P_n(t) = P_n \cos(\omega t + \varphi)$ is applied at point n instead of excitation through feedback, the vibration amplitude of point x takes the form

$$a_x = P_n[U^2(\omega, m_x, a_x) + V^2(\omega, m_x, a_x)]^{-1/2}. \qquad (20.15)$$

Comparing equation (20.12) and (20.15) we can conclude that, taking into account (20.13), when $U_0 = \pi P_n / 4$, the amplitude and frequency of self-sustained vibration of point x, obtained for various delay values t_0, correspond to points on the resonance curves of the striker.

Let us find the delay for which the largest amplitude of self-sustaining vibration occurs (autoresonance). The value $V(\omega, m_x, a_x)$ is proportional to the dissipative coefficients of the vibro-impact system. Assuming that these values are sufficiently small, it is natural to assume that, in agreement with (20.12), the maximum amplitude of vibration a_x^* occurs at the frequency ω^*, which satisfies the condition

$$U(\omega^*, m_x, a_x^*) = 0. \qquad (20.16)$$

Using (20.13), (20.16) we find from (20.12)

$$a_x^* = \frac{4U_0}{\pi V(\omega^*, m_x, a_x^*)}. \qquad (20.17)$$

The system of equations (20.16), (20.17) is similar to expressions (12.11) (12.12) and establishes the expression for the backbone curves of elastic systems and the energy boundary of vibration. Taking into account (20.7), the relationship between m_x and a_x, this system of equations allows us to define the amplitude and frequency of autoresonance as was done in §12. When condition (20.16) is satisfied in agreement with (20.9), (20.10), the delay is found from the expression

$$t_0 = \frac{3\pi}{2\omega^*} + \frac{2\pi i}{\omega^*} \quad (i = 0, 1, \ldots). \qquad (20.18)$$

The system of equations (20.7), (20.16), (20.17) allows us to define the relationship between the autoresonant frequency ω^* and force G by eliminating m_x and a_x. If we define this relationship as $\omega^*(G)$ and substitute it into (20.18), we obtain the characteristic which links the controlling delay 3 (Fig. 20.2) to the feed G. This characteristic depends on the system parameters only and thus the control of delay can be obtained using a rigid nonlinear coupling.

According to (20.17) the amplitude of autoresonance can be easily controlled by varying the saturation value U_0 in amplifier *5*, so that, by eliminating the values of m_x and ω^* from equations (20.7), (20.16), (20.17) we can obtain the relationship $a_x = a_x(G, U_0)$. Given the required value of amplitude a_x which depends on feed G, we can find the corresponding value of U_0 and consequently the characteristic $U_0(G)$ linking the amplitude controller *6* (Fig. 20.2). In particular when it is necessary to stabilise the autoresonant amplitude for all values of G the corresponding characteristic $U_0(G)$ is found from the implicit expression $a_x(G,U_0)=\text{const}$. Since the characteristic $U_0(G)$ is defined in terms of system parameters only, it also can be obtained using a rigid correction coupling.

3. Let us now look at the stability of autoresonance in synthesised systems, using an approach similar to that described in §14 section 2. We will assume that in the neighbourhood of the stable periodic solution (20.4) the vibration of the striker takes the form

$$u_x(t)=\widetilde{m}_x(t)+\widetilde{a}_x(t)\cos\widetilde{\omega}(t)t, \tag{20.19}$$

where the tilde means a slowly changing time function. We shall consider equation (20.9) as a condition for the energy balance between dissipative and excitation forces for the steady-state motion considered, forming the stability conditions for this motion due to the small deviations of amplitude $\widetilde{a}_x$ from the steady-state value a_x in the form

$$\begin{gathered}\frac{d}{d\widetilde{a}}[V(\widetilde{\omega},\widetilde{m}_x,\widetilde{a}_x)+Z(\widetilde{a}_x)\sin\widetilde{\omega}t_0]_{\widetilde{a}_x=a_x}>0 \textit{ when } \sin\omega t_0<0,\\ \frac{d}{d\widetilde{a}}[V(\widetilde{\omega},\widetilde{m}_x,\widetilde{a}_x)+Z(\widetilde{a}_x)\sin\widetilde{\omega}t_0]_{\widetilde{a}_x=a_x}<0 \textit{ when } \sin\omega t_0>0.\end{gathered} \tag{20.20}$$

Noting that as distinct from the case studied in §14, in this system both situations represented by (20.20) are possible when the observation and control points coincide. However, both conditions here lead to identical final forms of the stability condition.

Introducing three unknown functions instead of one in (20.19), we shall connect them through two conditions, assuming that during disturbed motion $\widetilde{m}_x(t)$ and $\widetilde{a}_x(t)$ are connected by equation (20.7) but $\widetilde{\omega}(t)$ and $\widetilde{a}_x(t)$ by equation (20.10). Eliminating $\sin\omega t_0$ from (20.20) using (20.10), we find

$$\begin{gathered}\frac{d}{d\widetilde{a}_x}[V(\widetilde{\omega},\widetilde{m}_x,\widetilde{a}_x)-\sqrt{Z^2(\widetilde{a}_x)-U^2(\widetilde{\omega},\widetilde{m}_x,\widetilde{a}_x)}]_{\widetilde{a}_x=a_x}>0\\ \text{when } \sin\omega t_0<0,\\ \frac{d}{d\widetilde{a}_x}[V(\widetilde{\omega},\widetilde{m}_x,\widetilde{a}_x)-\sqrt{Z^2(\widetilde{a}_x)-U^2(\widetilde{\omega},\widetilde{m}_x,\widetilde{a}_x)}]_{\widetilde{a}_x=a_x}<0\\ \text{when } \sin\omega t_0>0.\end{gathered} \tag{20.21}$$

Differentiating (20.21) and bearing in mind (20.11) we obtain the following conditions for stability for both cases:

$$\left[U(\widetilde{\omega},\widetilde{m}_x,\widetilde{a}_x)\frac{dU(\widetilde{\omega},\widetilde{m}_x,\widetilde{a}_x)}{d\widetilde{a}_x}+V(\widetilde{\omega},\widetilde{m}_x,\widetilde{a}_x)\frac{dV(\widetilde{\omega},\widetilde{m}_x,\widetilde{a}_x)}{d\widetilde{a}_x}-\right.$$
$$\left.-Z(\widetilde{a}_x)\frac{dZ(\widetilde{a}_x)}{d\widetilde{a}_x}\right]_{\widetilde{a}_x=a_x^*}>0. \tag{20.22}$$

Let us apply this to investigate the stability of autoresonance. According to (20.16), (20.11) we find in this case from (20.22)

$$\left[\frac{dV(\widetilde{\omega},\widetilde{m}_x,\widetilde{a}_x)}{d\widetilde{a}_x}-\frac{dZ(\widetilde{a}_x)}{d\widetilde{a}_x}\right]_{\widetilde{a}_x=a_x^*}>0. \tag{20.23}$$

Condition (20.23) generalised expression (12.14) obtained in §12. In accordance with (20.18), (20.9) in the conditions for autoresonance $V(\omega,m_x,a_x)>0$, while this value is linear in terms of coefficient β. As a result, for all forms of the harmonic linearisation coefficients

$$\left[\frac{dV(\widetilde{\omega},\widetilde{m}_x,\widetilde{a}_x)}{d\widetilde{a}_x}\right]_{\widetilde{a}_x=a_x^*}\geq 0.$$

Since according to (20.13)

$$\left[\frac{dZ(\widetilde{a}_x)}{d\widetilde{a}_x}\right]_{\widetilde{a}_x=a_x^*}<0,$$

the stability condition (20.23) is always satisfied for synthesised systems.

4. The method described was used in references [21, 22] to synthesise a system for the automatic adjustment of an ultrasonic machine undergoing an autoresonant regime under various feed forces. The dynamic model of the vibration system for an ultrasonic machine was considered in §15 section 3 where the excitation $P_n(t)$ was produced by an external generator (Fig. 15.6).

Let us now synthesise an autoresonant scheme for excitation of the object described (Fig. 20.4). In this case the excitation is applied in the form of delayed feedback at some section of the bar. Suppose, for example, that a displacement pickup measures the vibration of the bottom section of the bar $(x=l)$. After amplification and transformation of this signal in block *5* it passes through the delay block *4* at the section with coordinate $x=n$. Maintaining the system description of §15, section 3 and assuming for simplicity that $n=l$ we can write the equation of vibration for the bottom section of the bar in the form

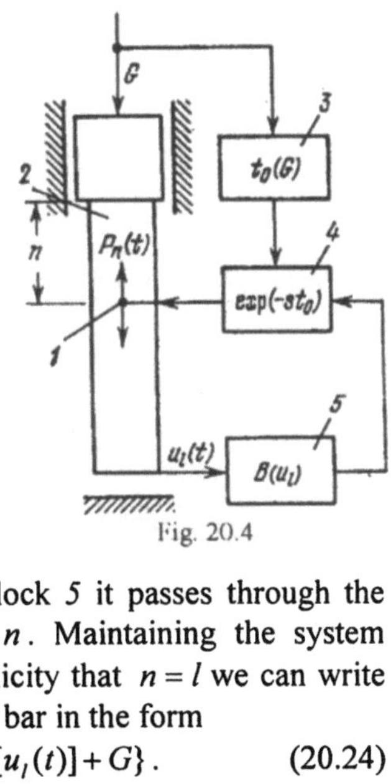

Fig. 20.4

$$u_l(t)=L_l(l,s)\{B[u_l(t)]\exp(-st)-\Phi[u_l(t)]+G\}. \tag{20.24}$$

Here $\Phi(u_l)$ is the static force characteristic of the stop.

Equation (20.24) is a particular case of equation (20.2), hence choosing $B(u_l)$ and t_0 in the forms (20.14), (20.18), we find the required structure for the excitation feedback.

Let us define characteristic $t_0(G)$ as the correction coupling by perturbation of the feed (Fig. 20.4). We shall take advantage of the fact that the equation for the operator $W(j\omega,m_l,a_l)$ coincides in this case with (15.12) (see §15 section 3).

Assuming that the stop is rigid (15.20) we find from (15.29) the following equation for the backbone curves (20.16) when $\chi \equiv 0$:

$$a_l^* = -\frac{2h}{\zeta^* \cot \zeta^*} \qquad \left(\zeta^* = \frac{\omega^* l}{\gamma}\right). \tag{20.25}$$

The equations for the energy boundary of vibration (20.17) coincide here with (15.23) when $\chi = 4U_0 l/(\pi ES)$:

$$a_l^* = \frac{4\pi\chi \sin^2 \zeta^*}{\varkappa\zeta^*(\zeta^* + \frac{1}{2}\sin 2\zeta^*)}. \tag{20.26}$$

Eliminating a_l^* from (20.25),(20.26), we find

$$\frac{G}{U_0} = -\frac{8\sin 2\zeta^*}{\varkappa(2\zeta^* + \sin 2\zeta^*)}. \tag{20.27}$$

In (20.27) expressing ζ^* in terms of t_0 using (20.18), we obtain the relationship sought which defines the characteristic $t_0(G)$ of the delay block regulator. When $i = 0$ it takes the form

$$\frac{G}{U_0} = -\frac{8\sin\dfrac{3\pi}{\tau}}{\varkappa\left(\dfrac{3\pi}{\tau} + \sin\dfrac{3\pi}{\tau}\right)} \qquad \left(\tau = \frac{t_0 l}{\gamma}\right). \tag{20.28}$$

The curve of the characteristic described is drawn in Fig. 20.5 using (20.28). The range of delay change corresponds to autoresonance of the system in the first mode of vibration. Changing the delay outside the band considered can, in principle, induce other modes of vibration, however internal damping in the material of the bar limits such possibilities in practice to only a few lower modes. The monotonic behaviour of the expression obtained for $t_0(G)$ makes it easy to produce in practice.

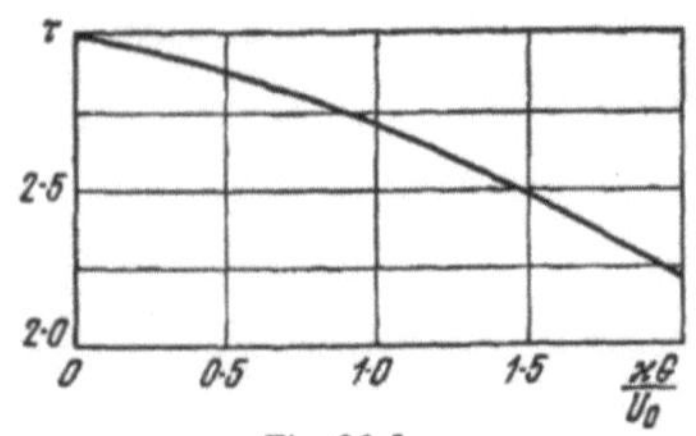

Fig. 20.5

Bibliography

There were giants in the earth in those days.

Genesis, 6: 4

1. Aizerman M A, Smirnova N M, On the application of methods of small parameters in the investigation of periodic regimes in automatic control systems not containing small parameters. Coll. "Tribute tor A A Andronov", 1955. (In Russian).
2. Akhiezer N I, Krein M G, On some questions in the theory of moments. Gonti, Kharkov, 1938. (In Russian).
3. Akynfiev T S, Babitsky V I, Autoresonant regimes in pneumatically driven systems. Machine Studies, 1977, No. 1. (In Russian).
4. Alabuzhev P M, Tsivinsky Yu P, Amplitude-frequency characteristics of vibration for a two-mass system with impact of mass against a stop. Symposium "Mechanisms and machines with impact and rotary impact effects", Novosibirsk, 1963. (In Russian).
5. Aleksandrov E V, Sokolinsky V B, Applied theory and design of impacting systems. Nauka, Moscow, 1969. (In Russian).
6. Ananiev I V Manual for the computation of natural vibration of elastic systems. Gostekhizdat, 1946. (In Russian).
7. Ananiev I V, Timofeev P G, Vibration of elastic systems in aircraft construction and their damping. Mashinostroenie, 1965. (In Russian).
8. Andronov A A, L I, Mandelstam and the theory of nonlinear vibration. Collected works, 1956. (In Russian).
9. Andronov A A, Neimark Yu I. On the motion of an idealised model of a clock with two degrees of freedom. Models of pre-Galilean clocks, 1946, vol. LI, No. 1. (In Russian).
10. Andronov A A, Vitt A A, Khaikyn S E, Theory of oscillations, 1987. Dover Publications Inc (Paperback version of translation pub. Pergamon Press, 1966). (Translation from Russian, Fizmatgiz, 1959)
11. Arnold R N, Response of an impact vibration absorber to forced vibration. Proc IX Cong. Appl Mechs, Brussels, 1956.
12. Arsenin, V Y, Mathematical physics. Basic equations and special functions. Nauka, Moscow, 1966. (In Russian).
13. Artobolevsky I I, Hertz E V, Kobrinsky A E, Raevsky N P, On the dynamics of pneumatic installations. Proceedings of seminars on theory of machines and mechanisms, 1953, No. 51. (In Russian).
14. Arzhanikh I S On novel inequalities in stability. Automation and telemechanics, 1961, No. 4. (In Russian).
15. Ashavsky A M, Sheinbaum V S, On one isoperimetrical problem in the optimisation of the speed of response for a single mass vibro-impact system. Machine Studies, 1970, No. 3. (In Russian).
16. Ashavsky A M, Vladov M V, On the synthesis of optimal vibro-impact machines. Dynamics of machines, Mashinostroenie, 1969. (In Russian).
17. Astashev V K, Babitsky V I, Dolnik E S, About one method of oscillation excitation, Mechanics of Solids, 1972, No. 1. (In Russian).
18. Astashev V K, Babitsky V I, Hertz M E, On the synthesis of autoresonant systems. Vibro-Engineering, Kaunas, 1973, No. 3 (20). (In Russian).
19. Astashev V K, Babitsky V I, Resonant oscillations of a visco-elastic rod with limiter. Mechanics of Solids, 1972, No. 4. (In Russian).
20. Astashev V K, Babitsky V I, Vibro-impact interactions of visco-elastic rods. Machine Studies, 1974, No. 5. (In Russian).
21. Astashev V K, Hertz M E, Excitation and stabilisation of resonant vibration of ultrasonic rod systems. Acoustic Journal, 1976, No. 2. (In Russian).

22. Astashev V K, Hertz M E, Self-sustaining vibration of a visco-elastic rod with limiter under delayed force. Machine Studies, 1973, No. 5. (In Russian).
23. Astashev V K, On the dynamics of an oscillator impacting against a stop. Machine Studies, 1971, No. 2. (In Russian).
24. Babakov I M, Theory of vibrations. Nauka, Moscow, 1968. (In Russian).
25. Babitsky V I, Borovkov B A, Vibro-impact processes in systems with limited excitation. Mechanics of Solids, 1977, No. 1. (In Russian).
26. Babitsky V I, Dzhokhadze Sh R, Kobrinsky A E, Korndorf S F, Polunov Yu L, Pulling effects in vibro-impact systems. Machine Studies, 1968, No. 6. (In Russian).
27. Babitsky V I, Izrailovich M Y, On the optimal motion of vibro-impact systems. Machine Studies, 1968, No. 3. (In Russian).
28. Babitsky V I, Izrailovich M Y, On the synthesis of self-sustaining systems. Machine Studies, 1971, No. 3. (In Russian).
29. Babitsky V I, Izrailovich M Y, Optimal multiple motions of vibro-impact systems. Mechanics of Solids, 1969, No. 3. (In Russian).
30. Babitsky V I, Izrailovich M Y, Some aspects of the optimisation of vibro-impact systems. Machine Studies, 1969, No. 3. (In Russian).
31. Babitsky V I, Izrailovich M Y, Some problems in the synthesis of optimal periodic vibro-impact systems. Mechanics of Machines, 1972, Nos. 33-34. (In Russian).
32. Babitsky V I, Izrailovich M Y, Synthesis of self-sustaining vibro-impact systems with prescribed cycles. Mechanics of Solids, 1970, No. 5. (In Russian).
33. Babitsky V I, Kobrinsky A E, Periodic motion of two-mass vibrating systems in a container. Theory of Machine and Mechanisms, 1964, No. 104. (In Russian).
34. Babitsky V I, Kolovsky M Z, Investigations of linear vibrating systems with limiters by exact and approximate methods. Machine Studies, 1967, No.4. (In Russian).
35. Babitsky V I, Kolovsky M Z, On the dynamics of systems with impact damper. Machine Studies, 1970, No. 2. (In Russian).
36. Babitsky V I, Kolovsky M Z, On the investigation of resonance in vibro-impact systems. Mechanics of Solids, 1976, No. 4. (In Russian).
37. Babitsky V I, Kolovsky M Z, On the theory of vibro-impact systems. Machine Studies, 1970, No.1. (In Russian).
38. Babitsky V I, Kolovsky M Z, Random oscillations of linear systems with limiters. Mechanics of Solids, 1967, No. 3. (In Russian).
39. Babitsky V I, On the analysis of elastic systems with vibro-impact interaction. Mechanics of Solids, 1975, No. 1. (In Russian).
40. Babitsky V I, On the existence of high-frequency oscillations with large amplitude in linear systems with limiters. Machine Studies, 1966, No. 1. (In Russian).
41. Babitsky V I, Oscillations of an inertially suspended pendulum in a vibrating container. Dynamics of machines Mashinostroenie, M., 1966. (In Russian).
42. Babitsky V I, Parametric oscillations of vibro-impact systems. Machine Studies, 1971, No. 1. (In Russian).
43. Babitsky V I, Suvorov G A, Investigation of vibro-impact systems using the universal analogue computer Analak-110, Analogues and analogue-numerical techniques, Mashinstroenie, 1968. (In Russian).
44. Babitsky V I, Tresviatsky A N, About one case of stability loss in a preloaded mechanism, Mechanics of solids, 1975, No.6. (In Russian).
45. Babitsky V I, Tresviatsky A N, Forced oscillations of a visco-elastic beam with fixed particle between limiters, Machine Studies, 1976, No. 2. (In Russian).
46. Babitsky V I, Vibro-impact motion of a pendulum with inertial suspension in a vibrating container. Analysis and synthesis of automatic machines, Nauka, Moscow 1964. (In Russian).
47. Ballo I, Contribution to the investigation of dynamical properties of nonlinear systems with acceleration damper, "Zagadnienia drgan nieliniowych", Warszawa, 1963. (In Polish).
48. Barkan D D, Shekhter O Y, On the theory of forced oscillation of a vibrator with limiter. J. Tech. Phys., 1955, v. 25, No. 13. (In Russian).
49. Batuev G S, Golubkov Yu V, Efremov A K, Fedosov A A, Engineering methods for the study of impact processes, Mashinostroenie, 1969 (In Russian).
50. Bautin N N, Behaviour of dynamic systems near boundary of stability domains. Gostekhizdat, 1949. (In Russian).
51. Bautin N N, Dynamic theory of clock operation without stopping the driving wheel (instantaneous impulse). Engineering, 1955, vol. XXI. (In Russian).

52. Bautin N N, On the Mandelstam problem in the theory of clocks. DAN SSSR, 1949, v. LXV, No. 3. (In Russian).
53. Bautin N N, On the Motion of an ideal model of a clock with two degrees of freedom. Modelling of Galileo-Huygens clocks. DAN SSSR, 1948, v. LXI, No. 1 (In Russian).
54. Beliayev N M, Stability of a prismatic rod subject to variable longitudinal forces. Collection of papers on engineering construction and structural mechanics. Put. Leningrad, 1924. (In Russian).
55. Bellman P, Dynamic programming. Princeton University Press, 1957.
56. Beserkersky V A, Popov E P, Theory of automatic control. Nauka, Moscow, 1966 (In Russian).
57. Bespalova L V, Neimark Yu I, Feigin M I, Dynamics of systems with impact interactions and the theory of nonlinear vibration. Mechanics of solids, 1966, No. 1. (In Russian).
58. Bespalova L V, On the theory of vibro-impact mechanism. OTN, 1957, No.5. (In Russian).
59. Bespechniy V M, Viktorov E D, Propagation of vibration in beams having a gap in the assembly. Mechanics of solids, 1970, No. 4. (In Russian).
60. Biderman V L, Theory of impact. Mashgiz, 1952. (In Russian).
61. Blekhman I I, Dzhanelidze G Yu, Dynamics of the Buass-Sard regulator. OTN, 1955, No. 10. (In Russian).
62. Blekhman I I, Self-synchronisation in some vibration machines. Engineering, 1953, v. XVI. (In Russian).
63. Blekhman I I, Synchronisation of dynamic systems. Nauka, Moscow, 1971. (In Russian).
64. Bogolyubov N N, Mitropolsky Yu A, Asymptotic methods in the theory of nonlinear oscillation. Gordon and Breach, NY. 1961. (Tranlation from Russian, Fizmatgiz, 1958).
65. Bolotin V V, Dynamic stability of elastic systems. Holden Day, San Francisco, 1964.(Translation from Russian, Goztekhizdat, 1956)
66. Booton R C, The analysis of nonlinear control systems with random inputs. Proc. Symp. Nonlinear Circuit Analysis, 1953, v. 2.
67. Boyarinov V S, Neimark Yu I, On vibrations of shafts in ball bearings, Mechanics, 1965, No. 3. (In Russian).
68. Bromberg P V, Matrix methods in the theory of relay and impulse control systems. Nauka, Moscow, 1967. (In Russian).
69. Brunshtein R E, Kobrinsky A E, Periodic motion of systems consisting of a ball in a cavity. OTN, 1959, No. 1. (In Russian).
70. Brunstein R E, Dynamics and stability of oscillating systems occurring due to the effects of arbitrary periodic forces. Mechanics of machines, 1969, Nos. 25-26. (In Russian).
71. Brunstein R E, Kobrinsky A E, On the stability of periodic motion of vibro-impact systems. OTN, 1960, No. 5. (In Russian).
72. Butenin N V, Elements of theory of nonlinear oscillation. Blundell Publishing, NY, 1965. (Translation from Russian, Sudpromgiz, 1962).
73. Cempel Cz, The equations of motion of a system with impacts. Engineering reports, Warsaw, 1969, No.17. (In Polish).
74. Clarc N J, An ultrasonic machine tool. Trans IRE, 1954, No. 10.
75. Colloquium on mechanical impedance methods for mechanical vibrations. Ed . by R Plunkett. ASME, New York, 1958.
76. Crandall S H (Ed.), Random vibration, v. 2. MIT press, Cambridge, Mass., 1963.
77. Crandall S H, Mark W D, Random vibration in mechanical systems. Academic Press, 1963.
78. Deutsch G, Anleitung zum praktischen gebrauch der Laplace-Transformation. R. Oldenburg, Munich, 1956..
79. Ditkyn V A, Prudnikov A P, Integral transformations and operational calculus. Nauka, Moscow, 1974. (In Russian).
80. Dobrogursky S O, On the problem of stresses and forces during impact. In book: "Problems in the design and construction of machine details", Proc. AN SSSR, 1942. (In Russian).
81. Dubowsky S, Freudenstein F, Dynamic analysis of mechanical systems with clearances, parts I, II. J. of eng. for industry, 1971,No. 1.
82. Egle D M, An investigation of an impact vibration absorber. J of Eng. For Industry, 1967, No. 4.
83. Emelianov S V, Systems for automatic control with variable structure. Nauka, Moscow 1967. (In Russian).
84. Feigin M I, Dynamic systems with impact interactions, with the existence of sliding motion. Mechanics of machines, 1972, Nos. 33-34. (In Russian).
85. Feigin M I, On forced vibration of two masses with a gap. Mechanics and mechanical engineering, 1960, No.5. (In Russian).

86. Feigin M I, On the theory of nonlinear damping. Proc. Vuzov, Radio physics, 1959, No.4. (In Russian).
87. Feigin M I, Sliding regimes in dynamic systems with impact interactions. Applied maths. and mechanics, 1967, No. 3. (In Russian).
88. Feldbaum A A, Basic theory of optimal automatic systems. Nauka, Moscow, 1966. (In Russian).
89. Feldbaum A A, Problems in self-adjusting (adaptive) systems. Proc. all-union conf. on self-adjustment, Nauka, Moscow, 1965. (In Russian).
90. Fikhtengolts G M, Course of differential and integral calculus. Nauka, Moscow, 1969, vols. I-III. (In Russian).
91. Frolov K V, Some problems in parametric vibration of machines. Coll. "Vibration and stability of instruments, machines and elements of control systems", Nauka, Moscow, 1968. (In Russian).
92. Fu C C, Dynamic stability of an impact system connected with rock drilling. J. of Appl. Mechanics, 1969, No. 4.
93. Fu C C, Paul B, Stability of motion of impact tools. Int J. of Solids and Structures, 1968, v. 4.
94. Fufaev N A, Theory of electromagnetic contact breakers. Coll. "Memoirs for A A Andronov", 1955. (In Russian).
95. Gantmakher F R, Lectures on analytical mechanics. Fizmatgiz, 1961. (In Russian).
96. Garber E D, Rozenvasser E N, On the study of periodic regimes in systems based on the filter hypothesis. Automation and telemechanics, 1965, No. 2. (In Russian).
97. Gardner M F, Burns J L, Transients in linear systems, vol.1. John Wiley, NY, 1942.
98. Gelfand I M, Shilov G E, Generalised functions and their applications. Fizmatgiz, 1958. (In Russian).
99. Gelfond A O, Calculation using finite differences. Fizmatgiz, 1959. (In Russian).
100. Glatenok I V, On the estimation of domains of occurrence of real periodic solutions found by the harmonic balance method. Automation and telemechanics, 1957, No. 12. (In Russian).
101. Glukharev K K, Frolov K V, Interaction of oscillating systems with two energy sources. Mechanics of solids, 1971, No. 4. (In Russian).
102. Goldfarb L S, On some nonlinearities in control systems. Automation and telemechanics, 1947, No. 5. (In Russian).
103. Goldsmith W, Impact. The theory and physical behaviour of colliding solids. London
104. Gorokhov V A, On periodic motion of systems with impact interactions. Radio physics, 1966, No. 2. (In Russian).
105. Grubin C, On the theory of the acceleration damper, J. of Appl. Mechanics, 1956, No. 3.
106. Hertz E V, Kreinin G V, Theory and design of pneumatically powered systems. Mashgiz, 1960. (In Russian).
107. Hsu J C, Meyer A U, Modern control principles and applications. McGraw-Hill, New York, 1968.
108. Ionov V N, Ogibalov P M, Stresses in bodies due to impulse loading. "Vysshaia Shkola", 1975. (In Russian).
109. Iorish Yu I,The protection of aircraft equipment from vibration. Oborongiz, 1949. (In Russian).
110. Israilovich M Y, The optimal law of motion for single mass oscillating systems. Machine studies, 1969, No.1. (In Russian).
111. Katkovnik V Y, Vibro-impact excitation of high frequency elastic oscillation. Mechanics of solids, 1968, No. 5. (In Russian).
112. Kazakov I E, Approximate methods of statistical study of nonlinear systems. Zhukovsky Aviation Academy, 1954, No. 394. (In Russian).
113. Kazakov I E, Dostupov B G, Statistical dynamics of nonlinear automatic systems.. Fizmatgiz, 1962. (In Russian).
114. Kharkevich A A, Self-sustaining vibration. Gostekhizdat, 1953. (In Russian).
115. Kilchevsky N A, Theory of collisions for solids. "Naukova dumka", Kiev, 1969. (In Russian).
116. Kobrinsky A A, Dynamics and stability of one-dimensional vibro-impact systems. Mechanics of solids, 1970, No. 5. (In Russian).
117. Kobrinsky A E, Dynamics of mechanisms with elastic connections and impact systems. Iliffe Books Ltd., London, 1969. (Translated from Russian, Nauka, Moscow, 1964.)
118. Kobrinsky A E, Kobrinsky A A, Vibro-impact systems. Nauka, Moscow, 1973. (In Russian).
119. Kobrinsky A E, On the theory of impact vibration absorbers. OTN, 1957 No. 5. (In Russian).
120. Kobrinsky A E, Oscillation of two mass systems, moving with periodic collisions. OTN, 1956, No. 5. (In Russian).
121. Kobrinsky A E, Some problems in the theory of vibro-impact systems. In "Collection of problems in the theory of machines and mechanisms", Nauka, Moscow, 1965. (In Russian).

122. Kochenburger R J, Nonlinearity in control systems. Control Engineering, 1956, No. 3.
123. Kolovsky M Z, Automatic control of vibration protection systems. Nauka, Moscow, 1976, (In Russian).
124. Kolovsky M Z, Nonlinear theory of vibration protection systems. Nauka, Moscow, 1966. (In Russian).
125. Kolovsky M Z, On the application of the small parameter method to the calculation of discontinuous periodic solutions. Int. symp. on nonlinear vibration.. AN USSR, Kiev, 1963, v. 1. (In Russian).
126. Kolovsky M Z, On the design of nonlinear vibration isolators with one degree of freedom. OTN, Mechanics and machine engineering, 1960, No. 4. (In Russian).
127. Kolovsky M Z, On the existence conditions for periodic regimes in systems of differential equations with discontinuities, consisting of a small parameter. Appl. maths. and mechanics, 1960, No. 4. (In Russian).
128. Kolovsky M Z, Pervozvansky A A, On linearisation using the distribution function in problems of nonlinear oscillation. OTN, Mechanics and machine engineering, 1962, No. 5. (In Russian).
129. Kolovsky M Z, Pervozvansky A A, On the stability of solutions found using the statistical linearisation method. OTN, Mechanics and machine engineering, 1963, No. 1.
130. Kolovsky M Z, Pervozvansky A A, The application of the distribution function method for poly-harmonic solution of nonlinear equations. Proc. of Vuzov, Mashinostroenie, 1963, No. 4. (In Russian).
131. Kolsky G, Stress waves in solids. Clarendon Press, Oxford, 1953.
132. Kononenko V O, Oscillating systems with restricted excitation. Iliffe, London, 1969. (Translation from Russian, Nauka, Moscow, 1964)
133. Kononenko V O, Problems in the theory of dynamic interaction of machines and sources of energy. Mechanics of solids, 1975, No. 5. (In Russian).
134. Kosterin Yu I, Mechanical self-excited vibration due to dry friction, AN USSR, 1960. (In Russian).
135. Kotelnikov V A, Longitudinal stability of aircraft with autopilots. Trudy LII, M. 1941, No 2. (In Russian).
136. Kozhevnikov S N, Dynamics of machines with elastic components. AN USSR, Kiev, 1961. (In Russian).
137. Krasovsky N N, Theory of motion control." Nauka, Moscow", Moscow, 1968. (In Russian).
138. Krylov N M, Bogolyubov N N, Introduction to nonlinear mechanics. Princeton U P, 1947.(Translation from Russian, Kiev, AN USSR, 1937)
139. Kulikovsky P, Optimal and adaptive processes in automatic control systems. Nauka, Moscow, 1967. (In Russian).
140. Lavendel E E, Synthesis of optimal vibration machines. "Zinatne", Riga, 1970. (In Russian).
141. Lecornu L F A, Regularisation du mouvement dans les machines. Cauthier-Villas et fils, Paris, 1898.
142. Lefshetz S, Differential equations: Geometric theory. Wiley, NY, 1957.
143. Leongard A, Generalised methods of harmonic balance in the analysis of parametric vibration. Proc. II congress IFAK, Nauka, Moscow, 1965, No. 1.
144. Letov A M, Analytic design of controllers, I – IV Automation and telemechanics, 1960, No. 4 – No. 6; 1961, No. 4. (In Russian).
145. Lieber P, Jensen D, An acceleration damper: development, design and some applications. Trans ASME, 1945, v. 67.
146. Lifshits P S, On the problem of forced vibration of a systems impacting against a stop. Tech. Phys., 1952, No. 6. (In Russian).
147. Litovchenko Ts G, Yakovenko Yu P, Analytical and structural description of mechanical transmission systems for automatic controllers with stops and clearances. Automation and telemechanics, 1961, No. 8. (In Russian).
148. Loeb J D, De la mechanique lineaire a la mechanique non-lineaire. Ann. Telecommuns., 1950, No. 2.
149. Loitsiansky L G, Lurie A I, Theoretical mechanics v. II. Gostekhizdat. 1955. (In Russian).
150. Lukomsky S I, Investigation of working regimes of percussion hammers. VNIISTROIDORMASH, 1959, No. 24. (In Russian).
151. Lurie A I, Operator calculus and its application in problems of mechanics. Gostekhizdat, 1951. (In Russian).
152. Lurie A I, Some nonlinear problems in the theory of automatic control. Gostekhizdat, 1951 (In Russian).

153. Lyusternik L A, Sobolev V I, Elements of functional analysis. Nauka, Moscow, 1965. (In Russian).
154. Maezawa S, Steady, forced vibration of unsymmetrical piecewise linear system. Bulletin of JSME, 1961, No. 4.
155. Magnus K, Uber ein Verfahren zur Untersuchung nichtlinearer Schwingungs und Regelungs Systeme. VDI-Forschungshelft, 1955, No. 451.
156. Mandelstam L I, Complete works. AN SSSR, 1955, v. IV. (In Russian).
157. Masri S F, Analytical and experimental studies of multiple-unit impact dampers. J. of the Acoustical Soc. oF Amer., 1969, No. 5.
158. Masri S F, Stability boundaries of the impact damper. J. Appl. Mechanics, 1968, No. 2.
159. Masri, S F, Caughey T K, On the stability of the impact damper. J. Appl. Mechanics, 1966, No. 3.
160. Meiman N I, Some problems on the distribution of zero in polynomials. UMN, 1949, No. 6. (In Russian).
161. Merkin D R, Introduction to the theory of stability of motion. Nauka, Moscow, 1971. (In Russian).
162. Methods for harmonic linearisation in the design of nonlinear systems for automatic control. Ed. by Yu I Topcheev, "Mashinostroenie", M., 1970. (In Russian).
163. Methods for the synthesis of nonlinear systems in automatic control. Ed. By S M Fedorov, "Mashinostroenie", M., 1970. (In Russian).
164. Nagaev R F, Dynamics of a vibro-impact crusher with a pair of self-synchronising vibrators. Mechanics and mechanical engineering, 1963, No. 1 (In Russian).
165. Nagaev R F, General problems of quasi-plastic impact. Mechanics of solids, 1971, No. 3. (In Russian).
166. Nagaev R F, Khodzhaev K Sh, Oscillation of a mechanical system with a periodic structure. "FAN", Tashkent, 1973. (In Russian).
167. Nagaev R F, On the analytical description of quasi-plastic impact. Mechanics of solids, 1970, No. 4. (In Russian).
168. Nagaev R F, Regular impulse motion in one-dimensional systems. Applied maths. and mechanics, 1967, No 2.
169. Naumov B N, Theory of nonlinear automatic systems. Nauka, Moscow, 1972 (In Russian).
170. Neimark Yu I, Point mapping methods in the theory of nonlinear vibration, ch I-III. Radio physics, 1958, No.1, No.2., No. 5 – No. 6. (In Russian).
171. Neimark Yu I, Point mapping methods in the theory of nonlinear vibration., Nauka, Moscow, 1972. (In Russian).
172. Neimark Yu I, Shilnikov L P, On the application of the small parameter method to systems of differential equations with discontinuous right hand sides. Mechanics and mechanical engineering, 1959, No. 6. (In Russian).
173. Nonlinear devices in automatic control systems. Ed. by Yu I Topcheev, "Mashinostroenie", M., 1970. (In Russian).
174. Oppelt W, Method for the frequency locus for controllers with friction. Z. VDI, 1948, 90
175. Paget A L, Vibration in steam turbine buckets and damping by impact. Engineering, 1937, v. 143:305-7.
176. Panovko Y G, Internal damping during oscillation of elastic systems. Fizmatgiz, 1960. (In Russian).
177. Pervozvansky A A, Random processes in nonlinear automatic systems. Fizmatgiz, 1962. (In Russian).
178. Ponomarev S D, Biderman V L, Likharev K S, Makushin V M, Malinin N N, Feodosiev V I, Strength in mechanical engineering, v. III. Mashgiz, 1962. (In Russian).
179. Pontriagin L S, Boltiansky V G, Gamkrelidze R V, Mishenko E F, Mathematical theory of optimal processes. Wiley Interscience, NY, 1962.(Translation from Russian).
180. Popov E P, On the approximate study of self-excited and forced vibration. DAN SSSR, 1954, v. XCV, No.5. (In Russian).
181. Popov E P, Paltov I P, Approximate methods in the study of nonlinear automatic systems. Fizmatgiz, 1960. (In Russian).
182. Poschl T, Der Stoss. Handbuch der Physik. Bd. VI. Springer. 1926.
183. Pugachev V S, Introduction to theory of probability. Nauka, Moscow, 1968. (In Russian).
184. Pugachev V S, Theory of random functions and their application to problems in automatic control. Fizmatgiz, 1960. (In Russian).
185. Ragulskene V L, Vibro-impact systems. Mintis, Vilnius, 1974. (In Russian).

186. Ragulskis K M, Mechanisms on vibration basis. Proc. AN Lit. SSR, Kaunas, 1963. (In Russian).
187. Rakhmatulin Kh A, Demianov Yu A, Stability under intense short time loading. Fizmatgiz, 1961. (In Russian).
188. Raskyn Kh I, 1) Application of the moment method to some problems in vibration transportation. 2) On estimation of deviations of real laws from the ideal. Coll.. "Problems in dynamics and stability", "Zinatne", Riga, 1968, No. 16. (In Russian).
189. Riabov Yu A, On determining fields of application of the small parameter method in problems of nonlinear oscillation. Proc. int. symp. on nonlinear oscillation, Kiev, 1963, v. 1. (In Russian).
190. Rosenberg L D, Kazantsev V F, Makarov L O, Akhimovich L F, Ultrasonic cutting. Proc. AN USSR,1962. (In Russian).
191. Rozenvasser E N, Oscillation of nonlinear systems. Nauka, Moscow, 1969. (In Russian).
192. Rozenvasser E N, Periodic non-stationary control systems. Nauka, Moscow, 1973. (In Russian).
193. Rusakov I G, Kharkevich A A, Forced vibration of a system colliding against a stop. J. Tech. Phys., 1942, Nos. 11 -12. (In Russian).
194. Sadek M M, The behaviou of the impact damper. Proc. Inst. Mech Eng., 1965-1966, v. 180.
195. Samolenko A M, Averaging methods in systems with impulses. In book "Mathematical Physics", "Naukova dumka", Kiev, 1971, No. 9. (In Russian).
196. Schwartz L, Méthodes mathématiques pour les sciences physiques. Paris, 1961 (In French).
197. Sears J E, On the longitudinal impact of metal rods with rounded ends. Trans. Camb. Phil. Soc., 1912, v. 21(2).
198. Shilov G E, Mathematical analysis. Second special course. Nauka, Moscow, 1965. (In Russian).
199. Shock and vibration handbook. Ed. By C M Harris and C E Crede. McGraw-Hill Co., New York, 1975.
200. Sikarskye D L, Paul B, Periodic motions of a two-body system subjected to repititive impact. J. of Eng, for Ind., 1969, No. 4.
201. Sneddon I N, Berry D S, Classical theory of elasticity. Handbuch der Physik, v.6, Ed. S Flugge, Springer Verlag, Berlin, 1958.
202. Staerman I Y, Contact problems in the theory of elasticity. Gostekhizdat, 1949. (In Russian).
203. Stoker J, Nonlinear vibration in mechanical and electrical systems. Interscience Pub. NY, 1950.
204. Sveshnikov A A, Applied methods in the theory of random functions. Nauka, Moscow, 1968. (In Russian).
205. Sysoev V I, Pendulum vibration absorber installation of tower type. Coll. "Study of dynamic installations", Gosstroiizdat, 1957. (In Russian).
206. Takeda, On the coefficient of restitution. Mekhanika, 1956, No. 1282. (In Russian).
207. Timoshenko S P, Vibration problems in engineering. D Van Nostrand Co. Ltd., 1955.
208. Timoshenko S P, Zur Frage nach der Wirkung eines Stosses auf einen Balken. Zeits. Math Phys., Bd. 62, 1913.
209. Topcheev Yu I, Analysis of stability and quality of nonlinear systems of automatic control by the logarithmic frequency characteristic method. In book Theory of automatic control. Ed. by. V V Solodovnikov, "Mashinostroenie", M., 1969, v. 3, part 1. (In Russian).
210. Tsuda K, Studies of oscillating systems, transmitting energy in the presence of angular gaps. Coll. translations, "Mekhanika", M., 1961, No. 1. (In Russian).
211. Tsypkyn Ya Z, On the connection between the equivalent amplification factor of a nonlinear element and its characteristics. Automation and telemechanics, 1956, No.4. (In Russian).
212. Tsypkyn Ya Z, Relay automatic systems. Nauka, Moscow, 1974. (In Russian).
213. Tsypkyn Ya Z, Sampling systems and their application. Pergamon, Oxford, 1962. (Translation from Russian).
214. Vavilov A A, Periodic mehods in the study of nonlinear systems. Energia, Leningrad,. 1970. (In Russian).
215. Veits V L, Dynamics of machine units. Mashinostroenie, Leningrad, 1969. (In Russian).
216. Veits V L, Kochura A E, The effect of limited excitation in machines with internal combustion engines, ch. I, II. Vibrotekhnika, 1973, No. 2 (19). (In Russian).
217. Veluswami M A, Crossley F R E, Horvay G, Multiple impacts of a ball between two plates, p II, Mathematical Modelling. J. of Eng. For Industry, 1975, No. 3.
218. Ventsel E S, Theory of probability. Nauka, Moscow1969.
219. Voloshenko-Klimovitsky Yu Y, Dynamic limits of yield. Nauka, Moscow 1965. (In Russian).
220. Vulfson I I, Dynamic design of cyclic mechanisms. Mashinostroenie, Leningrad, 1976. (In Russian).
221. Vulfson I I, Kolovsky M Z, Nonlinear problems in machine dynamics. Mashinostroenie. Leningrad, 1968. (In Russian).

222. Zadeh L A, On the identification problem. Trans IRE, 1956, No. 4.
223. Zaretsky L B, Synchronisation of centrifugal excitation in systems with discontinuous characteristics. Mechanics of solids, 1968, No. 1. (In Russian).
224. Zevin A A, On the stability of periodic motion due to the collision of masses. Machine studies, 1968, No. 4. (In Russian).
225. Zevin A A, Optimal regimes in vibro-impact systems having one degree of freedom. Mechanics of solids, 1976, No. 2. (In Russian)
226. Zevin A A, Periodic regimes in systems with impact vibration absorbers. Mechanics of machines, 1969, No. 17-18. (In Russian).
227. Zhivkov V S, The influence of electromagnetic inertia of motors on the stability of oscillation of mechanical systems with centrifugal excitation. Machine Studies, 1971, No.4. (In Russian).

Additional bibliography

This list compiles the publications that arrived after the first edition of the book was published. We have tried here to present both development of the ideas of the book and achievements in other directions. Preference was given to publications where analysis is supplemented with physical experiments or engineering applications.

228. Astashev V K, About the influence of high frequency vibration on the processes of plastic deformation. Machine Studies, 1983, No. 2. (in Russian).
229. Astashev V K, Effect of ultrasonic vibration of a single-point tool on the process of cutting. J. of Machinery Manufacture and Reliability, 1992, No. 3. (in Russian).
230. Astashev V K, Babitsky V I, Dynamics of resonant machines. In book "Dynamics and control of machines". Ed. Kreinin G V, Machinostroenie, Moscow, 1988, (in Russian).
231. Astashev V K, Babitsky V I, Efficiency of resonant tuning in ultrasonic machines. Machine Studies, 1981, No. 6. (in Russian).
232. Astashev V K, Babitsky V I, Methods of efficiency improvement of ultrasonic machines. Stanki i Instrument (Machine Tools and Tools) 1982, No. 3. (in Russian).
233. Astashev V K, Babitsky V I, Methods of resonant machine tuning. Machine Studies, 1982, No. 1. (in Russian).
234. Astashev V K, Babitsky V I, Ultrasonic cutting as a nonlinear (vibro-impact) process. Ultrasonics, 1998, v. 36, No. 1-5.
235. Astashev V K, Babitsky V I, Hertz M E, Adaptive control system of ultrasonic machine. Instruments & Control systems, 1977, No. 2. (in Russian).
236. Astashev V K, Krupenin V L, Tresviatsky A N, Experimental investigation of synchronisation of impacts in continuous systems with variable quantity of impact pairs. J. of Machinery Manufacture and Reliability, 1996, No. 2. (in Russian).
237. Babitsky V I, Autoresonant mechatronic systems. Mechatronics, 1995, v. 5 No. 5.
238. Babitsky V I, Hand-held percussion machine as discrete non-linear converter. J. of Sound and Vibration, 1998, v. 214, No. 1.
239. Babitsky V I, Some applications of autoresonant strongly nonlinear vibratory systems. Machine Vibration, 1992, v. 1, No.2.
240. Babitsky V I, Borovkov B A, Stabilisation of resonance in vibro-impact machines. Machine Studies, 1979, No.1. (in Russian).
241. Babitsky V I, Kovaleva A S, Optimal control of non symmetric vibro-impact systems. Mechanics of Solids, 1988, No. 1. (in Russian).
242. Babitsky V I, Kovaleva A S, Krupenin V L, Investigation of quasi-conservative vibro-impact systems by averaging method. Mechanics of Solids, 1982, No. 1. (in Russian).
243. Babitsky V I , Krupenin V L, Oscillations in strongly nonlinear systems. Nauka, Moscow, 1985. (in Russian).
244. Babitsky V I, Krupenin V L, To the theory of parametric resonances in vibro-impact systems. Mechanics of Solids, 1978, No. 4. (in Russian).
245. Babitsky V I, Vedenova E G, Goihman L V, Manevich L I, Oscillating localisation in nonlinear beam system with periodic structure. Proc. USSR Academy of Sciences, 1985, v. 281, No. 2. (in Russian).

246. Babitsky V I, Veprik A M, Krupenin V L, Vibro-impact phenomena due to limited oscillations of one-dimensional elasto-connected particles. Proc. USSR Academy of Sciences, 1988, v. 300, No. 3. (in Russian).
247. Babitsky V I, Veprik A M, Universal bumpered vibration isolator for severe environment. J. of Sound and Vibration, 1988 (in print).
248. Babitsky V I, Vozniuk P D, Perminov S M, Excitation of wide band vibration using auto-stochasticity. Machine Studies, 1986, No. 4. (in Russian).
249. Bansevicius R, Ragulskis K, Vibromotors. Mokslas, Vilnius, 1981. (in Russian).
250. Bapat C N, The general motion of an inclined impact damper with friction. J. of Sound and Vibration, 1995, v. 184, No. 3.
251. Bapat C N, Sancar S, Popplewell N, Repeated impacts on a sinusoidally vibrating table reappraised. J. of Sound and Vibration, 1986, v. 108, No.1.
252. Bapat C N, Popplewell N, Several similar vibro-impact systems. J. of Sound and Vibration, 1987, v. 113, No.1.
253. Bapat C N, Popplewell N, MacLachlan K, Stable periodic motion of an impact pair. J. of Sound and Vibration, 1983, v. 87, No.1.
254. Bapat C N, Sancar S, Single unit impact damper in free and force vibration. J. of Sound and Vibration, 1985, v. 99, No.1.
255. Bishop S R, Impact oscillators. Phil. Trans. of the Royal Society, London, 1994, 347A.
256. Brach R M, Mechanical impact dynamics. Rigid body collisions. John Wiley & Sons, New York, 1991.
257. Brogliato B, Nonsmooth impact mechanics. Models, Dynamics and Control. Lectures notes in control and information sciences 220, Springer-Verlag, London, 1996.
258. Budd C, Dux F, Chattering and related behaviour in impact oscillator. Phil. Trans. Royal Society, London, 1994, A347.
259. Budd C, Dux F, Cliffe A. The effect of frequency and clearance variations on single-degree-of-freedom impact oscillator. J. of Sound and Vibration, 1995, v.184, No. 3.
260. Cabannes H. Cordes virantes avec obstacles. Acustica, 1984, v. 55.
261. Cabannes H, Haraux A, Almost periodic motion of a string against fixed obstacle. Nonlinear Analysis: Theory, Methods and Applications, 1983, No. 7.
262. Chatterjee S, Mallik A K, Ghosh A, Impact dampers for controlling self-excited oscillation. J. of Sound and Vibration, 1996, v. 193, No. 5.
263. Chu F, Zhang Z, Bifurcation and chaos in a rub-impact Jeffcott rotor system. J. of Sound and Vibration, 1998, v. 210, No. 1.
264. Comparin R J, Singh R, Nonlinear frequency response characteristics of an impact pair. J. of Sound and Vibration, 1989, v. 134, No. 2.
265. Dimentberg M, Statistical dynamics of nonlinear and time-varying systems. Research Studies Press, Taunton, 1988 (Translation from Russian, Nauka, Moscow, 1980).
266. Feigin M I, Forced oscillation in systems with discontinuous nonlinearities. Nauka, Moscow, 1994 (in Russian).
267. Foale S, Bishop S, Bifurcations in impact oscillator. Nonlinear Dynamics, 1994, v. 6.
268. Haug E J, Wu S C, Yang S M, Dynamics and mechanical systems with Coulomb friction, stiction, impact and constraint addition-deletion I. Mechanisms and Machine Theory, 1986, v. 21, No. 5.
269. Heiman M S, Sherman P J, Bajaj A K, On the dynamics and stability of an inclined impact pair. J. of Sound and Vibration, 1987, v. 114, No. 3.
270. Hertz M E, Makarov N S, Tresviatsky A M, Experimental investigation of autoresonant system with electro-dynamic excitation. Machine Studies, 1989, No. 3. (in Russian).
271. Hogan S J, On the motion of a rigid block, tethered at one corner, under harmonic forcing. Proc. Royal Society, 1992, London A 439.
272. Hogan S J, Heteroclinic bifurcations in damped rigid block oscillations. Proc. Royal Society, 1992, London A 439.
273. Hogan S J, Slender rigid block motion. J. of Engineering Mechanics, 1994, 120.
274. Hogan S J, On the dynamics of rigid block motion under harmonic forcing. Proc. Royal Soc, 1989, London A 425.
275. Holmes P J, The dynamics of repeated impacts with sinusoidally vibrating table. J. of Sound and Vibration, 1982, v. 84, No. 2.
276. Hongler M O, Streit L, On the origin of chaos in gearbox models. Physica, 29D, 1988.

277. Ivanov A P, Dynamics of systems with mechanic collisions. Int. Programme of Education, Moscow, 1997. (in Russian).
278. Ivanov A P, Bifurcations in impact systems. Chaos, Solitons and Fractals, v. 7, No. 10.
279. Ivanov A P, Bifurcations associated with grazing collisions. In Nonlinear dynamics: New theoretical and applied results. Academic Verlag, Berlin, 1995.
280. Ivanov A P, Impact oscillations: linear theory stability and bifurcations. J. of Sound and Vibration, 1994, v. 178, No. 3.
281. Jean M, Moreau J J, Dynamics in the presence of unilateral contacts and dry friction; a numerical approach. In "Unilateral problems in structural analysis 2", Ed. G. Del Piero and F. Maceri, 1987, CISM courses and lectures 304, Springer-Verlag.
282. Kahraman A, Singh R, Nonlinear dynamics of a spur gear train. J. of Sound and Vibration, 1990, v. 142, No. 1.
283. Kahraman A, Singh R, Nonlinear dynamics of a gear rotor system with multiple clearances. J. of Sound and Vibration, 1990, v. 144, No. 3.
284. Kahraman A, Singh R, Dynamics of an oscillator with both clearance and continuous nonlinearities. 1992, v. 153, No. 1.
285. Karagiannis K, Pfeiffer F, Theoretical and experimental investigations of gear rattling. Nonlinear Dynamics, 1991, v. 2.
286. Karyeaclis M P, Caughey T K, Stability of a semi-active impact damper: part 1- Global behaviour. ASME J. of Applied Mechanics 1989, v. 56.
287. Karyeaclis M P, Caughey T K, Stability of a semi-active impact damper: part 2- Periodic solutions. ASME J. of Applied Mechanics 1989, v. 56.
288. Keller J B, Impact with friction. ASME J. of Applied Mechanics, 1986, v.53.
289. Keller J B, Impact with an impulsive frictional moment. ASME J. of Applied Mechanics, 1987, v. 54.
290. Khulief Y A, Shabana A A, A continuous force model for the impact analysis of flexible multibody systems. Mechanisms and Machine Theory, 1987, v. 22.
291. Kobrinsky A A, Kobrinsky A E, Two-dimensional vibro-impact systems. Nauka, Moscow, 1981. (in Russian).
292. Kovaleva A S, Control of vibratory and vibro-impact systems. Nauka, Moscow, (in Russian).
293. Kovaleva A S, Optimal control of periodic motions in vibro-impact systems. Mechanics of Solids, 1982, No. 6. (in Russian).
294. Kovaleva A S, Optimisation of periodic motions of some vibro-impact systems. Mechanics of Solids, 1983, No. 3. (in Russian).
295. Kovaleva A S, Investigation of vibro-impact resonances in systems with random parametric excitation. Machine Studies, 1983, No. 2. (in Russian).
296. Krupenin V L, About oscillation of vibro-impact systems with distributed impact elements. Machine Studies, 1985, No. 4. (in Russian)
297. Krupenin V L, To the calculation of resonant oscillation of thread interacting with point limiter. J of Machinery Manufacture and Reliability, 1992, No. 2. (in Russian).
298. Krupenin V L, To the theory of vibro-impact systems with distributed impact elements. Mechanics of Solids, 1986, No. 1. (in Russian).
299. Krupenin V L, Transformation of modes in a string interacting with two lengthy limiters. Proc. USSR Academy of Sciences, 1990, v. 313, No. 6. (in Russian).
300. Kumabe D, Vibratory cutting. Dzikke Sjuppan, Tokio, 1979. (in Japanese).
301. Kurilo R, Ragulskene V, Two-dimensional vibration drives. Mokslas, Vilnius, 1986. (in Russian).
302. Lancarani H M, Nikravesh P E, Continuous contact force models for impact analysis in multibody systems, Nonlinear Dynamics, 1994, No. 5.
303. Landa P S, Stratonovich R L, Calculation of stationary distribution function for one of the simplest strange attractors. Proc. USSR Academy of Sciences, 1982, v. 267, No. 4. (in Russian).
304. Landa P S, Stratonovich R L, Probability characteristics of stochastic oscillations for controlled pendulum. Mechanics of Solids, 1984, No. 4. (in Russian).
305. Luo G.-W, Xie J.-H, Hopf bifurcation of a two-degree-of freedom vibro-impact system. J. of Sound and Vibration, 1998, v. 213, No. 3.
306. Manevish L I, Michlin J V, Pilipchuk V N, Method of normal forms for essentially nonlinear systems, Nauka, Moscow, 1989. (in Russian).
307. Moore D B, Shaw S W, The experimental response of an impacting pendulum system., Int. J. Non-linear Mechanics, 1990, v. 25.

308. Moreau J J, Standard inelastic shocks and the dynamics of unilateral constraints. CISM courses and lectures, 1985, No 288, Springer-Verlag.
309. Moreau J J, Some numerical methods in multibody dynamics: applications to granular materials. European J. of Mechanics A/Solids, 1994, v.13, No. 4.
310. Moreau J J, Panagiotopoulos P D, (Eds) Nonsmooth mechanics and applications. CISM courses and lectures, 1988, No 302, Springer-Verlag.
311. Muszunska A, Rotor-to-stationary element rub-related vibration phenomena in rotating machinery - literature survey. The Shock and Vibration Digest, v. 21, No. 9.
312. Nagaev R F, Mechanical processes with repeated attenuated impacts. Nauka, Moscow, 1985. (in Russian).
313. Natsiavas S, Stability and bifurcation analysis for oscillators with motion limiting constraints. J. of Sound and Vibration, 1990, v. 141, No. 1.
314. Natsiavas S, Dynamics of multiple-degree-of freedom oscillators with colliding components. J. of Sound and Vibration, 1993, v. 165, No. 3.
315. Neimark J I, Landa P S, Stochastic and chaotic oscillations. Kluwer, 1992, (Translation from Russian, Nauka, Moscow, 1987).
316. Nordmark A B, Non-periodic motions caused by grazing incidence in an impact oscillator. J. of Sound and Vibration, 1991, v. 145, No. 2.
317. Paidoussis M P, Li G X, Moon F C, Chaotic oscillations of the autonomous system of a constrained pipe conveying fluid. J. of Sound and Vibration, 1985, v. 135, No. 1.
318. Panovko J G, Introduction in the theory of mechanical impact. Nauka, Moscow, 1977. (in Russian).
319. Peterka F, Introduction to oscillation of mechanical systems with impacts. Academia, Prague, 1981. (in Czech).
320. Peterka F, Vacik J, Transition to chaotic motion in mechanical systems with impacts. J. of Sound and Vibration, 1992, v. 154, No. 1.
321. Pfeiffer F, Glocker Ch, Multibody dynamics with unilateral contacts. Wiley-Interscience, New York, 1996.
322. Popplewell N, Bapat C N, MacLachlan K, Stable periodic vibro-impacts of an oscillator. J. of Sound and Vibration, 1983, v. 87, No.1.
323. Sharif-Bakthiar M, Shaw S W, The dynamic response of a centrifugal pendulum vibration absorber with motion limiting stops. J. of Sound and Vibration, 1988, v. 126, No. 2.
324. Shaw S W, Holmes P J, A periodically forced impact oscillator with large dissipation. J. of Appl. Mechanics, 1983, v.90, No. 1.
325. Shorr BF, Melnikova G V Analysis of structures by direct mathematical simulation. Mashinostroenie, Moscow, 1988 (in Russian)
326. Sinopoli A, Unilaterality and dry friction: A geometric formation for two-dimensional rigid body dynamics. Nonlinear Dynanics, 1997, v. 12, No. 4.
327. Stoianovici D, Hurmuzlu Y, A critical study of the applicability of rigid body collision theory. ASME J. of Applied Mechanics, 1996, v. 63, No. 2.
328. Stronge W J, Rigid body collision with friction. Proc. Royal Society, London, 1990, A 431.
329. Stronge W J, Friction in collisions: resolution of a paradox. J. of Appl. Phys., 1991, v. 69, No. 2.
330. Thompson J M T, Ghaffari R, Chaos after period-doubling bifurcations in the response of an impact oscillator. Phys. Lett., 1982, 91A(1).
331. Thompson J M T, Ghaffari R, Chaotic dynamics of an impact oscillator. Phys. Rev. A, 1983, 27(3).
332. Vedenova E G, Manevich L I, Periodical and localised waves in vibro-impact systems with regular structures. Machine Studies, 1984, No. 4. (in Russian).
333. Veprik A M, Krupenin V L, About resonant oscillations of a system with a distributed impact element. Machine Studies, 1988, No. 6. (in Russian).
334. Veprik A M, Krupenin V L, On the solution of the system containing distributed impact elements. Machine Studies, 1985, No. 6. (in Russian).
335. Watanabe T, Forced vibrations of continuous systems with nonlinear boundary conditions. ASME J. of Mechanical Design, 1978, v. 100.
336. Wehage R A, Haug E J, Dynamic analysis of mechanical systems with intermittent motion. ASME J. of Mechanical Design, 1982, v. 104.
337. Whiston G S, Global dynamics of vibro-impacting linear oscillators. J. of Sound and Vibration, 1987, v. 118, No. 3.

338. Whiston G S, Singularities in vibro-impact dynamics. J. of Sound and Vibration, 1992, v. 152, No. 3.

339. Wood L A, Byrne K P, Analysis of a random repeated impact process. J. of Sound and Vibration, 1981, v.78, No. 3.

340. Wu S C, Yang S M, Haug E J, Dynamics and mechanical systems with Coulomb friction, stiction, impact and constraint addition-deletion II, Planar Systems. Mechanisms and Machine Theory, 1986, v. 21, No. 5.

341. Wu S C, Yang S M, Haug E J, Dynamics and mechanical systems with Coulomb friction, stiction, impact and constraint addition-deletion III, Spatial Systems. Mechanisms and Machine Theory, 1986, v. 21, No. 5.

342. Zhong Z H, Mackerle J, Contact-impact problems: A review with bibliography. Applied Mechanics Reviews, 1994, v. 47, No. 2.

343. Zhuravlev V F, Methods of analysis of the vibro-impact systems with the help of special functions. Mechanics of Solids, 1976, No. 2. (in Russian).

344. Zhuravlev V F, Motion equations of mechanical systems with unilateral constraints. Appl. Mechanics and Mathematics, 1978, No. 5. (in Russian).

345. Zhuravlev V F, Klimov D M, Applied methods in oscillation theory. Nauka, Moscow, 1988. (in Russian).

Subject Index

Absorber, impact, 243
Absorber, vibration, 79, 243, 245, 250
Amplitude pulling, 52, 157, 207, 238, 243
Autonomous system, 13, 144, 145, 151, 168, 252
Autoresonance, 166, 271, 273, 275 et seq.
Autoresonant frequency, 273, 275
Autoresonant system, 271
Axial impact, 8, 12
Backbone curve, 22, 29, 34, 35, 37, 41, 48, 51, 58, 60, 128, 130 et seq., 142, 163, 167, 170, 215 et seq., 233, 235, 236, 275, 278
Beam slapping modes, 237
Beam, vibrating, 229
Bearing, 47
Boundaries, stability, 36
Characteristic, linear static, 13
Characteristic, amplitude-frequency, 26, 27, 34, 37, 40, 47, 48, 49, 51, 58, 98, 127, 129, 131, 132, 143, 157, 162, 174, 176
Characteristic equation, 32, 44, 67, 86, 87, 89, 126, 130, 141, 145, 151, 154, 166, 167, 195, 197, 230, 231, 274
Characteristic of contact interaction, 10
Characteristics, dynamic, 76, 171
Characteristics, force, 10, 75, 76, 78, 82, 92, 102 et seq., 117 et seq., 149, 152, 167, 206, 273
Characteristics, kinematic, 1, 13
Clapping, 224
Coefficient of restitution, 14, 15, 18, 20, 23, 26, 102, 136, 156, 202, 207, 240, 242, 245, 253, 262, 269 et seq.
Collision of spheres, 11
Collision velocity, 2, 117
Compliance, dynamic, 81 et seq., 92, 193, 194 et seq., 227 et seq., 240, 273
Constraints, impact, 69, 70
Contact zone, 1, 3, 6, 8, 10, 11, 12, 199, 228
Control characteristic, 253, 262
Control function, 252 et seq.
Control point, 273, 276
Convergence, 82, 89, 93
Damping, 39, 45, 55, 59, 60, 61, 90, 116, 134 et seq., 146, 149, 156, 168, 170, 206, 213, 227, 245 et seq.
Damping, internal, 6, 7, 28, 225, 226, 231, 235, 237, 278
Decay, 143
Determinant, 32, 73, 80, 87, 155
Difference equation, 11, 31, 71, 268, 269
Dirac function, 77, 100, 111
Discontinuity surface, 70, 71, 72
Dissipation, impact, 28, 45, 46, 59, 60
Dissipation, total, 80, 82, 89
Dissipation, viscous, 45
Dissipative effects, 11, 20, 44, 45, 201
Distributed impact element, 237
Drift, 37, 125, 140, 172, 252, 261, 273

Dry friction, 15, 18, 19, 79
Elastic modulus, 1
Energy balance, 44, 59, 60, 78, 137, 168, 203 et seq., 239, 242, 276
Energy boundary of vibration, 44, 45, 46, 52, 60, 98, 127, 133, 162, 168, 172, 217, 222, 275, 278
Energy conditions for stability, 141, 152, 168, 196 et seq., 276
Equivalent linearisation, 99, 103, 194, 244, 272
Escapement, watch, 64
Excitation frequency, 28, 41, 42, 45, 48, 135, 140, 143, 157, 163, 220, 225, 227, 237
Excitation, external, 29, 45, 83, 108, 116, 118, 125, 144, 159, 160, 194, 197, 207, 240
Excitation, hard, 24, 46, 129 et seq., 157, 161, 169, 173, 207, 220, 228, 237, 264
Excitation, soft, 155, 168
Existence domains, 25 et seq., 33, 36, 40 et seq., 52, 55, 57, 59, 133, 143, 169, 170, 218, 220, 224, 236
Feedback, 271 et seq.
Filtering effect, 86, 96
First harmonic, 23, 36, 42, 43, 45, 49, 59, 96, 103, 151, 245, 274
Forced vibration, 23, 29, 36, 37, 39, 52, 55, 91, 98, 125, 132, 137, 140, 143, 136, 151, 157, 160, 174, 251, 260
Fourier series, 21, 36, 38, 48, 89, 92 et seq., 128, 132, 1327, 162, 168, 204, 220, 241, 274
Free vibration, 22, 29, 34, 37, 38, 45, 49, 52, 233, 239
Functions, generalised periodic, 89
Gain coefficient, 49, 83, 154
Generalised coordinates, 69, 80, 166, 244
Grazing, 224
Harmonic vibration, 34, 40, 42, 46 et seq., 59, 103, 133, 151, 172
Hertz equations, 8
Holonomic systems, 80
Hysterisis, 6, 10, 271
Hysterisis loops, 6
Impact impulse, 36, 89, 92, 207, 239, 242, 253, 259, 263, 269
Impact oscillator, 125, 173, 207, 238
Impact pair, 21, 75 et seq.,91, 98, 102, 125, 127, 132, 135, 139, 152, 167, 199, 206, 207, 229, 244
Impact theory, 1
Impact velocity, 140, 253, 262, 263, 264, 269, 270
Impact, constrained, 16 et seq.
Impact, oblique, 15, 16
Impact, sliding, 16, 79
Impulse-momentum equation, 2, 15 et seq., 85
Impulse response function, 11
Inertia, 14, 16, 19, 68, 136, 149, 170, 209, 230, 239
Integral equations, 10, 83, 94
Integral-differential equations, 6
Interaction, contact, 8 et seq., 757t6, 84, 94, 108
Interaction, symmetrical double sided, 75
Interference, 21, 25, 28 et seq., 35, 54, 66
Jamming, 16, 18, 65, 66
Kinetic energy, 1, 14, 80
Laplace transform, 81 et seq.
Linear oscillator, 20 et seq., 34, 37, 40, 47, 98, 125, 130, 132 et seq., 140, 198, 243
Linearisation, 66, 75, 94, 97 et seq., 125 et seq., 142, 147, 151 et seq., 167, 171, 176, 230, 243, 272, 274
Linearisation by distribution function, 99 et seq., 102
Linearisation coefficients, 98 et seq., 113, 121, 127, 128, 151 et seq., 161, 171, 176, 198, 272
Linearisation, force, 105, 121
Linearisation, harmonic, 51, 94, 98 et seq., 115, 117, 121, 125, 132,

140 et seq., 167 et seq., 194, 198, 214, 219, 230, 243, 248, 272 – 274, 277
Linearisation, motion, 105, 121
Linearisation, statistical, 100, 113, 117, 121, 123, 250
Longitudinal vibration, 3
Lower boundaries, 44, 46
Moment, central, 109
Moment problem, 256, 258
Momentum, conservation of, 4, 7, 13
Motion, stable, 141
Natural frequency, 7, 21, 38, 48, 90, 98, 127, 128, 131, 136, 150, 158, 162, 170, 199, 208, 217, 220, 238, 243, 249
Nominal point, 261, 266
Nonlinear mechanics, 29, 148, 170
Nonlinearity, hard, 206
Nonlinearity, soft, 156
Operator, dynamic compliance, 81 et seq., 193, 212 et seq., 221, 230
Optimal control function, 252 et seq.
Optimal periodic motion, 251
Optimisation, 253, 259, 261, 272
Oscillator with symmetrical stops, 132
Oscillator with viscous damping, 90
Parametric vibration, 148 et seq., 160
Pendulum, 64
Percussion machine, 53, 54, 61, 251
Periodic Green's functions, 90 et seq., 137 et seq., 202 et seq., 239 et seq.
Perturbations, small , 30, 66, 70, 72, 196, 231
Phase boundary, 19
Phase plane, 21, 38, 50, 252, 260, 262, 264
Phase point, 70
Phase trajectory, 50, 252, 262, 264, 267
Point mapping, 71, 263, 265, 267
Point mass, 64, 69, 79, 229, 237
Poisson's ratio, 2
Pulling, 48, 51 et seq., 130, 133, 156 et seq., 164, 220, 237
Random vibration, 108, 250
Regime, periodic, 39, 43, 46, 49, 51, 65, 67, 70, 151, 152, 160,253, 261, 267, 270
Regime, symmetrical, 48
Resonance curve, 29, 34 et seq., 51, 55, 58, 60, 61, 132, 142, 163, 217 et seq., 226, 236, 275
Resonance, isochronous, 35
Resonance, nonlinear, 42, 243
Resonance, parametric, 148 et seq.
Resonance, sub-periodic, 136, 202
Resonant response, 86
Response, 11, 81, 83, 86, 95, 206, 214 et seq., 240, 246, 248
Rigid body, 4
Screwthread wrench, 62
Self-sustaining system, 165, 166, 260, 263, 271
Self-sustaining vibration, 165 et seq., 173, 251, 263, 272, 275
Single degree of freedom, 64, 125, 207, 238
Sommerfeld effect, 176
Stability, 30, 32 et seq., 42, 44, 66 et seq., 92, 140 et seq.,161 et seq., 193, 196 et seq., 218, 231, 235, 248, 251, 260, 261, 263, 268, 271, 276
Stability, asymptotic, 32, 71
Stable solution, 30, 153, 196, 206, 220, 231, 236
Stationary process, 84, 110, 111 et seq.
Stereomechanics, 13, 14, 53
Stitching method, 64, 73, 136, 260
Striker, 5, 16, 53, 62, 68
Structural boundaries of vibration, 44 et seq.
Switching curve, 252, 261, 266 et seq.
Switching point, 252, 261 et seq.
Taylor series, 30, 72, 159
Transfer function, 83, 85
Transformation, 51, 67, 70, 71, 84,

85, 88, 94, 103, 111 et seq., 121, 133, 142, 148 et seq., 160, 167, 199, 201, 263, 277
Transient processes, 50, 143, 175, 261, 263
Transition, 50,51, 77, 103, 207, 238, 266
Unstable solution, 30, 142, 156, 163, 175, 223
Vibration, structural boundaries of, 44
Vibro-impact interactions in elastic systems, 212
Vibro-impact regime, 24 et seq., 35, 40 et seq., 59, 103, 105, 107,129 et seq., 156 et seq., 168, 169, 173, 201 et seq., 209, 218, 223, 225, 227, 228, 236, 245
Visco-elastic bar, 212
Viscous friction, 45, 59, 102, 108, 118, 133, 140
Waves, longitudinal, 2, 212
Wave reflection, 5
Waves, shear, 2
Waves, standing, 7, 207, 237
Waves, surface, 3

Appendix I Simulation of impact on viscoelastic limiter

1. Numerous models based on the application of mechanical elements have been developed to explain the viscoelasticity. They use successive and parallel combinations of weightless springs and dashpots to reflect the elastic and viscous behaviour of material.

Figure A1.1 demonstrates the model of axial collision of a free rigid body of mass M with the simplest viscoelastic limiter. The limiter is modeled schematically as a parallel working linear spring with elasticity coefficient c and dashpot with viscous coefficient b. Such a model describes accurately the strain and deformation occuring in a viscoelastic sample under dynamic loading. This is known as *a Kelvin-Voigt model* of viscoelastic material. The approaching velocity of a free body with mass M is denoted as v.

This model of impact has an exact analytical solution and can be conveniently used, therefore, as a test model for the development of PC-oriented simulation procedures, such as Matlab-Simulink.

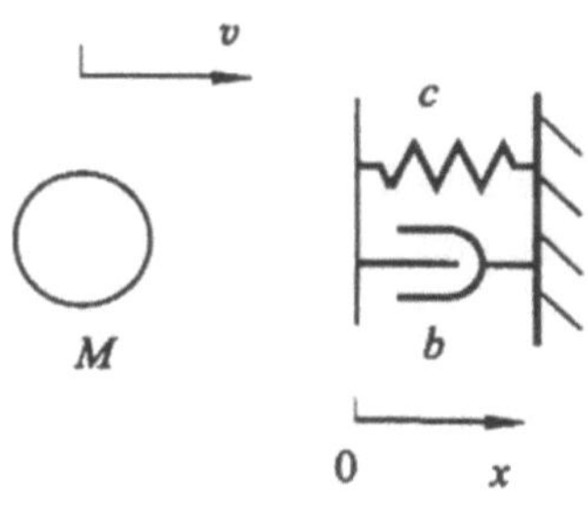

Fig. A1.1

Dynamic response of the limiter $F = cx + b\dot{x}$ governs the development of collision. The differential equation of body motion during the impact is

$$M\ddot{x} + cx + b\dot{x} = 0 \qquad t \in [0, \tau], \tag{A1.1}$$

with initial conditions: $x(0) = 0$, $\dot{x}(0) = v$, where τ is the duration of impact.

The condition of the body rebound in instant $t = \tau$ is the change of the sign of the limiter response or of the body acceleration:

$$F(\tau) = 0 \;\text{ or }\; \ddot{x}(\tau) = 0\,. \tag{A1.2}$$

The equation of impact (A1.1) has a known solution:

$$x(t) = A\exp(-ht)\sin\Omega t, \quad t \in [0, \tau], \tag{A1.3}$$

where:

$$h = \frac{b}{2M}, \quad A = \frac{v}{\Omega}, \Omega^2 = \Omega_0^2 - h^2, \quad \Omega_0^2 = \frac{c}{M}.$$

The corresponding derivatives with respect to time have been obtained as follows:

$$\dot{x}(t) = A\exp(-ht)(\Omega\cos\Omega t - h\sin\Omega t)$$

$$\ddot{x}(t) = -A\exp(-ht)\left[(\Omega^2 - h^2)\sin\Omega t + 2h\Omega\cos\Omega t\right] \tag{A1.4}$$

$$\dddot{x}(t) = -A\exp(-ht)\left[(3h\Omega^2 - h^3)\sin\Omega t + (3h^2\Omega - \Omega^3)\cos\Omega t\right]$$

The instant of impact completion can be found by solving the equation

$$\ddot{x}(\tau) = -A\exp(-h\tau)\left[(\Omega^2 - h^2)\sin\Omega\tau + 2h\Omega\cos\Omega\tau\right] = 0$$

from which

$$\tau = \frac{1}{\Omega}\arctan\frac{-2h\Omega}{\Omega^2 - h^2},$$

or, using the notation for fracture of critical damping (loss factor) $\xi = \dfrac{h}{\Omega_0}$

$$\tau = -\frac{1}{\Omega_0\sqrt{1-\xi^2}}\arctan\frac{2\xi\sqrt{1-\xi^2}}{1-2\xi^2}. \tag{A1.5}$$

The peak value of acceleration may be either at the instant collision starts (due to the immediate step-wise response $F = vb$ or at the instant when $\dddot{x}(t) = 0$. The value of the initial acceleration may be calculated as follows: $M\ddot{x}(0) = vb$ and

$$\ddot{x}(0) = \frac{vb}{M} = 2v\xi\Omega_0. \tag{A1.6}$$

The maximum value of acceleration can be achieved in the moment when $\dddot{x}(t = \tau_1) = 0$ if $0 < \tau_1 < \tau$, therefore using results of (A1.4) for the third-order derivative we have

$$(3h\Omega^2 - h^3)\sin\Omega\tau_1 + (3h^2\Omega - \Omega^3)\cos\Omega\tau_1 = 0 \text{ and } \tau_1 = \frac{1}{\Omega}\arctan\left(-\frac{3h^2\Omega - \Omega^3}{3h\Omega^2 - h^3}\right)$$

or

$$\tau_1 = -\frac{1}{\Omega_0\sqrt{1-\xi^2}}\arctan\frac{(4\xi^2 - 1)\sqrt{1-\xi^2}}{\xi(3-4\xi^2)}. \tag{A1.7}$$

By substituting the value τ_1 from (A1.7) into the expression for the acceleration from (A1.4) we obtain after rearrangement the following maximum value of acceleration

$$\ddot{x}(\tau_1) =$$

$$= \frac{v\Omega_0}{\xi(3-4\xi^2)}\exp\left[-\frac{\xi}{\sqrt{1-\xi^2}}\arctan\frac{(4\xi^2-1)\sqrt{1-\xi^2}}{\xi(3-4\xi^2)}\right]\cos\left[\arctan\frac{(4\xi^2-1)\sqrt{1-\xi^2}}{\xi(3-4\xi^2)}\right] \tag{A1.8}$$

Finally, combining (A1.6) with (A1.8), we obtain for the maximum of acceleration

$$\ddot{x}_{max} = v\Omega_0\Gamma(\xi) \tag{A1.9}$$

where

$$\Gamma(\xi) = \begin{cases} \gamma(\xi) \text{ if } \gamma(\xi) \geq 2\xi \\ 2\xi \text{ if } \gamma(\xi) < 2\xi \end{cases} \tag{A1.10}$$

with

$$\gamma(\xi) = \frac{1}{\xi(3-4\xi^2)}\exp\left[-\frac{\xi}{\sqrt{1-\xi^2}}\arctan\frac{(4\xi^2-1)\sqrt{1-\xi^2}}{\xi(3-4\xi^2)}\right]\cos\left[\arctan\frac{(4\xi^2-1)\sqrt{1-\xi^2}}{\xi(3-4\xi^2)}\right].$$

The value of $\xi = 0.5$ is the exact solution for the equation $\gamma(\xi) = 2\xi$. That means, that for the viscoelastic limiter with $\xi > 0.5$ the peak of acceleration will be at the initial instant of collision.

The peak deflection will be when $\dot{x}(t = \tau_2) = 0 \quad (\tau_2 < \tau)$, therefore with the help of expression for the first-order derivative from (A1.4) we have:

$$\Omega\cos\Omega\tau_2 - h\sin\Omega\tau_2 = 0 \quad \text{and} \quad \tau_2 = \frac{1}{\Omega}\arctan\frac{\Omega}{h}$$

or

$$\tau_2 = \frac{\arctan\dfrac{\sqrt{1-\xi^2}}{\xi}}{\Omega_0\sqrt{1-\xi^2}}. \tag{A1.11}$$

By substituting the corresponding values of the time instant τ_2 from (A1.11) into the expression for the deflection (A1.3) and after rearrangement the following peak value was obtained:

$$x(\tau_2) = \frac{v}{\Omega_0\sqrt{1-\xi^2}}\exp\left(-\frac{\xi}{\sqrt{1-\xi^2}}\arctan\frac{\sqrt{1-\xi^2}}{\xi}\right)\sin\left(\arctan\frac{\sqrt{1-\xi^2}}{\xi}\right)$$

or

$$x(\tau_2) = \frac{v}{\Omega_0}\Lambda(\xi) \tag{A1.12}$$

where

$$\Lambda(\xi) = \frac{1}{\sqrt{1-\xi^2}}\exp\left(-\frac{\xi}{\sqrt{1-\xi^2}}\arctan\frac{\sqrt{1-\xi^2}}{\xi}\right)\sin\left(\arctan\frac{\sqrt{1-\xi^2}}{\xi}\right). \tag{A1.13}$$

Figure A1.2 depicts the functions (A1.10) and (A1.13).*

By using the expression for the velocity from (A1.4) and for the instant of rebound (A1.5), we can calculate the velocity of rebound:

* These results are somewhat similar to those obtained for the maximum deflection and transmitted force for the single degree of freedom system with impulse excitation in the book "Shock and vibration concepts in engineering design," by C E Crede, Prentice-Hall, NY, 1965.

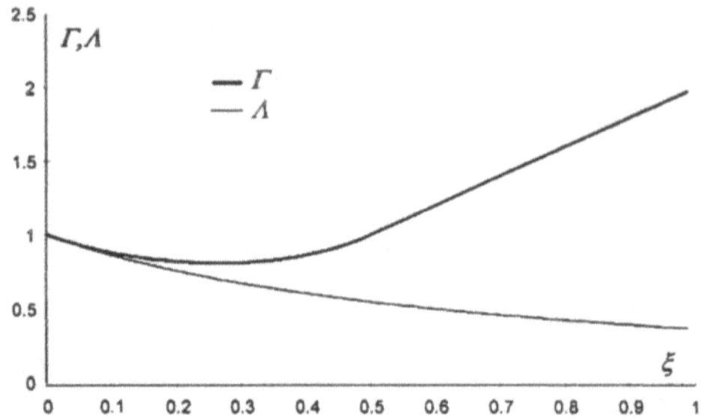

Fig. A1.2

$$\dot{x}(\tau)=v\exp\left[-\frac{\xi}{\sqrt{1-\xi^2}}\arctan\left(-\frac{2\xi\sqrt{1-\xi^2}}{1-2\xi^2}\right)\right]\sin\left[\arctan\left(-\frac{2\xi\sqrt{1-\xi^2}}{1-2\xi^2}\right)\right]$$

By definition, the coefficient of restitution will then be:

$$R=\frac{\dot{x}(\tau)}{v}=\exp\left[-\frac{\xi}{\sqrt{1-\xi^2}}\arctan\left(-\frac{2\xi\sqrt{1-\xi^2}}{1-2\xi^2}\right)\right]\sin\left[\arctan\left(-\frac{2\xi\sqrt{1-\xi^2}}{1-2\xi^2}\right)\right] \quad \text{(A1.14)}$$

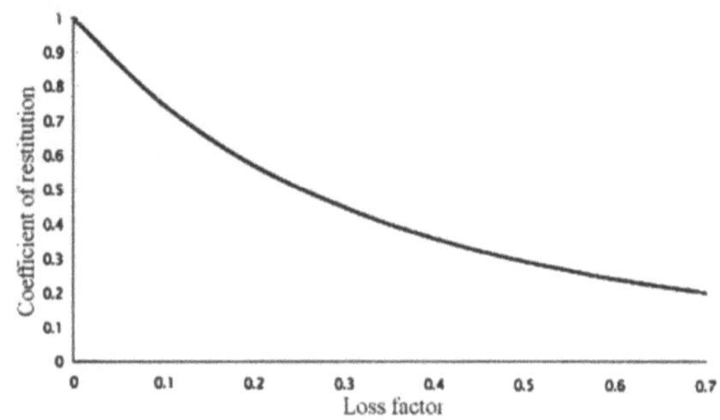

Fig. A1.3

It appears that at viscoelastic impact the coefficient of restitution is a function of the bumper loss factor ξ only.

2. For experimental validation of the model of viscoelastic impact, the experimental rig shown in Fig. A1.4 was assembled. A plastic pad (E-A-R Isodamp C-1000 plastic, 6 *mm* thick) was mounted in a vice fixture with the plain surface facing a pendulum. The weighted steel rod (∅10mm) with a rounded tip, was fixed in a bifilar suspension at a point directly above the sample vice. It was

swing up and then held in pre-start position. The height defined the approaching speed of the pendulum at impact. When released, the pendulum impacted the pad.

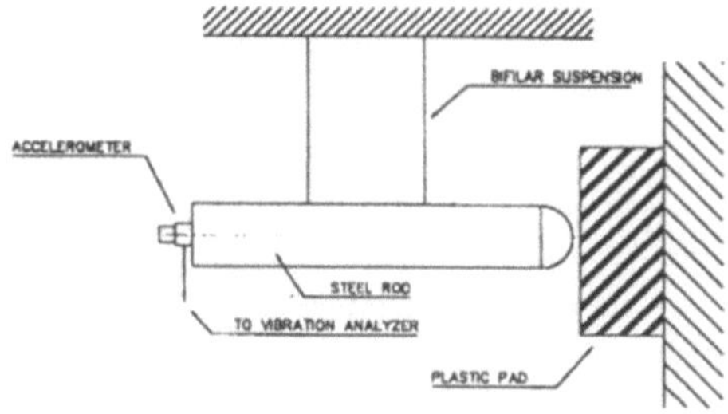

Fig. A1.4

The process of collision was registered by a miniature accelerometer (type 4393, Bruel & Kjaer) mounted on the plane edge of the rod, charge conditioner (type 5010A, Kistler, not shown) and signal analyzer (model SD-390, Scientific Atlanta, not shown). A typical example of the time history of acceleration obtained is shown in Fig. A1.5.

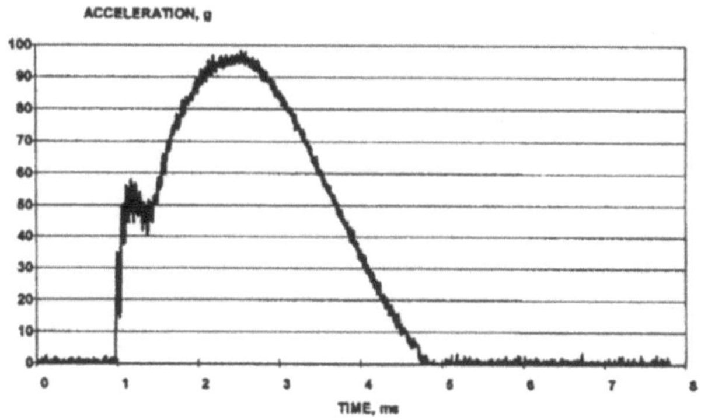

Fig. A1.5

The characteristic instants of impact, such as the start instant, instant of peak acceleration and impact duration may be easily measured by using the cursor of the signal analyser. In this manner the impact duration τ and time to peak acceleration τ_1 were measured as 3.83ms and 1.54ms respectively. Expressions (A1.5) and (A1.7) were then solved as a set of transcendental equations with unknown Ω_0 and ξ and known values τ and τ_1. Dividing (A1.5) by (A1.7) we can eliminate parameter Ω_0:

$$\frac{\tau}{\tau_1}=\frac{\arctan\dfrac{2\xi\sqrt{1-\xi^2}}{1-2\xi^2}}{\arctan\dfrac{(4\xi^2-1)\sqrt{1-\xi^2}}{\xi(3-4\xi^2)}}. \tag{A1.15}$$

Solving firstly (A1.15) for ξ and then (A1.5) for Ω_0, we obtained $\xi = 0.14$, $\Omega_0 = 755\,rad/s$.

The expression for acceleration from (A1.4) was used further with the newly determined parameters Ω_0 and ξ. The time history of acceleration was generated numerically. The initially unknown impact velocity v was identified by minimizing the integral mean square difference between measured and curve fitted shape of impact. For this particular case it was found that $v = 1.45\ m/s$. Figure A1.6 shows superimposed the almost coincident time histories of acceleration at impact obtained from the experiment ("TEST") and that obtained theoretically ("CURVE").

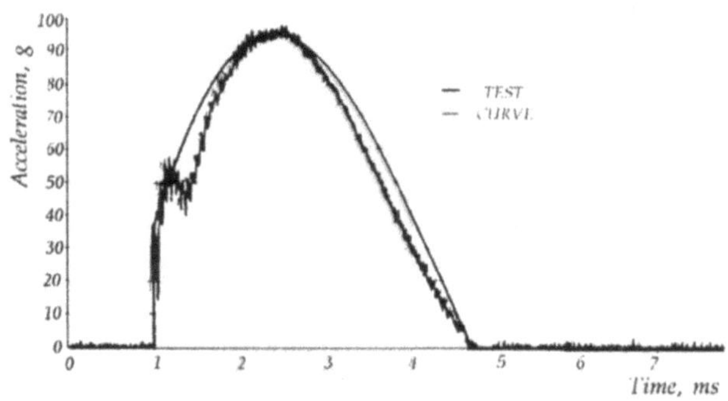

Fig. A1.6

3. The differential equation of the motion (A1.1) was valid for the finite interval of time covering the impact process. Expanding the previously obtained results for the infinite time interval the following differential equation may be written for unit mass:

$$\ddot{x}+\Phi(x,\dot{x})=0, \qquad t\in[0,\infty] \tag{A1.16}$$

with initial conditions: $x(0)=0$, $\dot{x}(0)=v$, where the force characteristic of impact interaction $\Phi(x,\dot{x})$ is as follows:

$$\Phi(x,\dot{x})=\begin{cases}\Omega_0^2 x+2\xi\Omega_0\dot{x} \text{ if } x>0 \text{ and } \Phi(x,\dot{x})>0,\\ 0 \text{ if } x>0 \text{ and } \Phi(x,\dot{x})\le 0,\\ 0 \text{ if } x<0\end{cases}$$

In accordance with (A1.17), the impact interaction begins when the impacting body crosses the zero coordinate level in the positive direction. The impact is

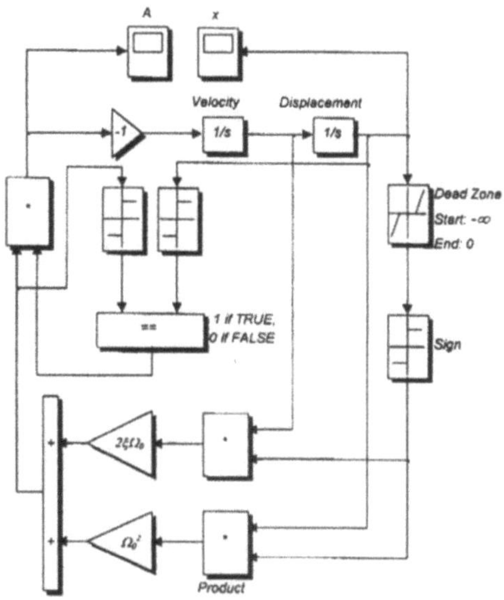

Fig. A1.7

completed when the value of the limiter response crosses its zero level in the "negative" direction.

Figure A1.7 depicts the schematics of the Simulink model representing equation (A1.16).

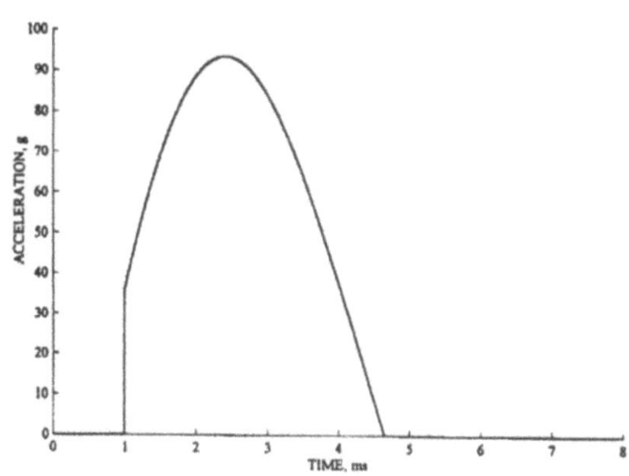

Fig. A1.8

The result of simulation with $\xi = 0.14, \Omega_0 = 755\,rad/s$ and initial conditions $x(0) = 0, v = 1.45\,m/s$ is presented in Fig. A1.8. It practically coincides with the experimentally obtained shape of impact.

Appendix II Some Applications of Autoresonant Strongly Nonlinear Vibratory Systems*

The use of resonance phenomena to develop useful mechanical work is discussed. Principles of strongly nonlinear resonant behaviour are presented, with particular reference to achievement of stable operating modes and their practical implementation in machines.
Starting from vibro-impact systems, the concept of autoresonance is developed and its implementation in a laser scanner for surfacing treatment and autoresonant ultrasonic heads for surface impact treatments and cutting are described. Development and operation of a modular robot manipulator utilizing electromechanical self excited shift and rotate modules is reported.
Finally same future directions of research and development are discussed. The potential for increased electromechanical integration within complex machines and recent research on svnchronized vibration excitation in weighted strings are briefly considered.
Keywords: Applications; Autoresonance; Homeostat; Laser surfacing; Mechatronics; Nonlinear vibration; Oscillation; Percussion machines; Q-factor; Resonance; Robotic manipulator; Surface treatment equipment; Ultrasonics; Vibro-impact.

1. Introduction

In striving towards speed and efficiency machines sometimes suffer mysterious catastrophes; most of these are the result of hardly discernible oscillatory processes. However, penetration into the amazing and complex world of oscillations also opens up new horizons for engineering. Elaboration of the technology of the future requires of scientists and engineers intensive work on the understanding and use of economic forms of motion, of which resonance proves to be one of the most efficient.

Resonance makes possible excitation and maintenance of high-energy oscillations through the use of low-power sources. It is familiar to anyone who has set a heavy swing in motion with a large amplitude by giving it light pushes. Anything that oscillates may be easily swung if the force is applied in unison with the natural oscillations. A rapid increase of the swinging amplitude is the most familiar property of resonance. It is extremely attractive to realize such highly economical motions with large amplitudes, speeds and rates of acceleration and thereby to achieve considerable force. For example, a motor car, which has stuck fast, may be freed by rocking to and fro. This explains the wide use of resonant

* From "Machine Vibration", (1992) 1:110-119

oscillations in various fields of engineering.

In traditional resonant systems, whose general arrangement is commonly represented by a load rocking on a spring under the effect of the harmonic force (Fig. 1a), all processes occur in accordance with harmonic laws (Fig. 1b) characterized by smooth alteration of the acceleration and deceleration with defined maximum and minimum of all parameters. The maximum values of the oscillatory process parameters depend on the frequency of the external periodic action in accordance with the curve known as the amplitude-frequency characteristic shown in Fig. 1c. The peak of this curve at point A represents resonance. The external periodic action coincides with the natural frequency of oscillation. The system is swung by any convenient mechanism and it readily absorbs the energy supplied from the source of excitation. This mode of operation is also optimal for the excitation source. Extremely low forces are required from the source since the system imposes virtually no load upon it, and the source functions without strain. The ratio of the natural frequency to the factor characterizing attenuation of the oscillations is termed 'the Q-factor of the oscillatory system.' The larger the Q-factor, the narrower and higher is the peak of the amplitude-frequency characteristic. For comparison, the dashed line in Fig. 1c shows the characteristic curve of a system with low Q-factor.

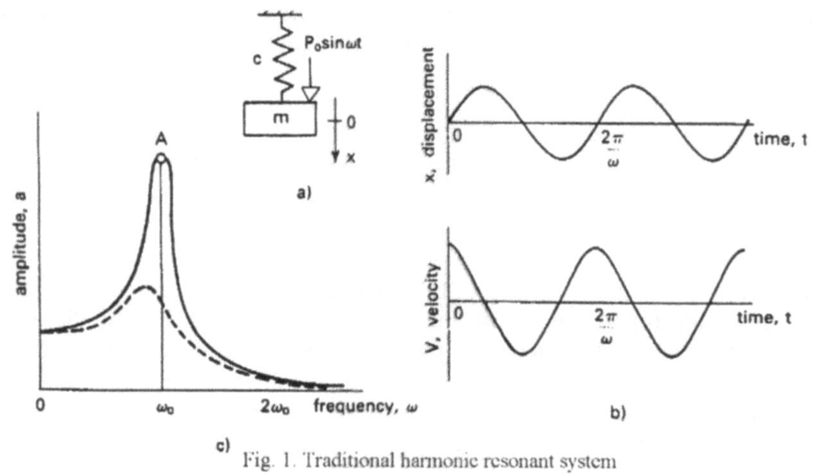

Fig. 1. Traditional harmonic resonant system

2. Vibro-impact Systems.

Is it possible to intensify the resonant oscillations and thus develop forces within the oscillation cycle which considerably exceed the conventional maximum values of harmonic curves? This may be attained through inclusion of shock into the oscillatory motion since shock is the most efficient method developed by nature for conversion of mechanical motion into force. Now we approach a new important type of mechanical oscillatory systems - the vibro-impact system. The most simple model of such a system is illustrated in Fig. 2a. Now the oscillations of the load are accompanied by periodic shocks against a fixed limiter. The smooth flow of the process is disturbed by abrupt jumps of the speed at the

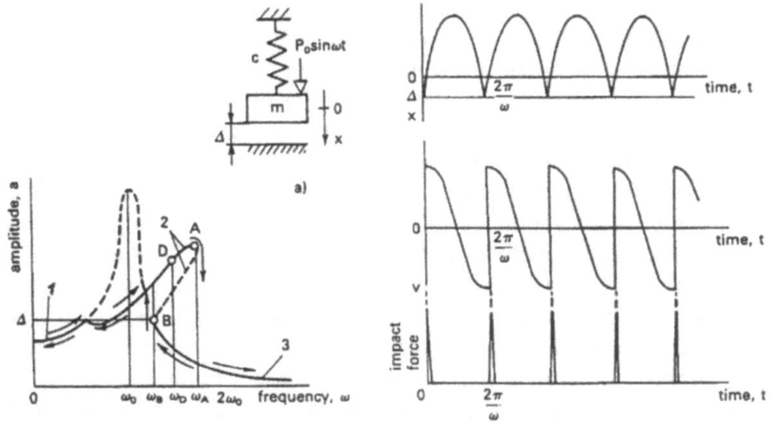

Fig. 2. Vibro-impact oscillatory system

instants of collision, with production of considerable forces of interaction (Fig. 2b). Though the time of the shock is short (in solid bodies it ranges from 10^{-3} to 10^{-5} s depending on the dimensions and configuration), the resultant forces are tremendous and capable of execution of considerable work. This work may be efficiently used in two ways: on one hand, demolition or treatment of some material (the workpiece replaces the limiter) and on the other hand, rapid reversal (change of direction of motion) of a heavy load, exemplified by the percussion tools which have been permanent companions of humanity throughout its history and remain some of the most efficient and effective work tools. From the first primitive hammers to modern forging and riveting hammers, from ancient pickaxes to up-to-date drilling installations, from wall-demolition rams to perfectly designed concrete breakers and perforators - such are the extremes of this long-established process. Both useful properties of the vibro-impact process are used to the maximum in all these machines. Large forces developed during the shock improve malleability of the hardest materials and simultaneously ensure instantaneous reversal of the striker, thereby contributing to a high rate of recurrence of the process. To ensure the maximum economy of machine operation throughout the entire cycle, the oscillation exciter (drive) should be synchronized to the bouncing load which will again be accelerated to the required shock speed in unison with its natural motion, i.e. the load should again swing in resonance but now in resonance with the frequency of the vibro-impact system.

What are the characteristics of vibro-impact resonance? For a detailed analysis of the resonance, it has been necessary to develop special methods of nonlinear mechanics, perform time-consuming calculations and make numerous experiments. Research conducted at the A .A. Blagonravov Machine Studies Institute of the Academy of Sciences of the USSR have revealed a number of amazing properties peculiar to vibro-impact resonance. We shall describe only some of these properties, commencing with analysis of the most simple model illustrated in Fig. 2a.

Bear in mind that when the limiter is installed with a certain clearance Δ relative to the spring, two kinds of oscillation are possible: without contact of the load with the limiter (motion within the limits of the clearance) and with impacts against the limiter. If such oscillations are excited by a low-power source of periodic force, e.g. by an electromagnet or other pulser, steady oscillations within the clearance limits have conventional properties of traditional oscillatory systems such as are shown in Fig. 1a. Hence, the amplitude-frequency characteristic curve (Fig. 2c) of a vibro-impact system in non-shock-motion (branches 1 and 3) coincides with the segments of the resonant curve shown in Fig. 1c.

Steady modes of motion with impacts against the limiter are no longer harmonic (Fig. 2b) therefore half of the full swing is commonly used for estimation of the oscillation amplitude. Just this magnitude is shown in Fig. 2c (branch 2). As is evident from Fig. 2c, when the excitation frequency is smoothly increased (arrows pointing to the right), the load swings to the limiter and starts striking against it, with a slight initial decrease of the full swing and with subsequent increase of the swing to the maximum values at point A corresponding to frequency ω_A close to double the frequency of free oscillation of the load suspended from the spring. If an attempt is made to further increase the oscillation frequency the vibro-impact nature of motion of the load is discontinued and harmonic oscillations within the clearance limits are set up after a short transient process. If the excitation frequency is subsequently decreased (arrows pointing to the left), this will not result in the load swinging to the previous values. The vibro-impact motion will commence by a jump at point B and resonance at point A will not be displayed.

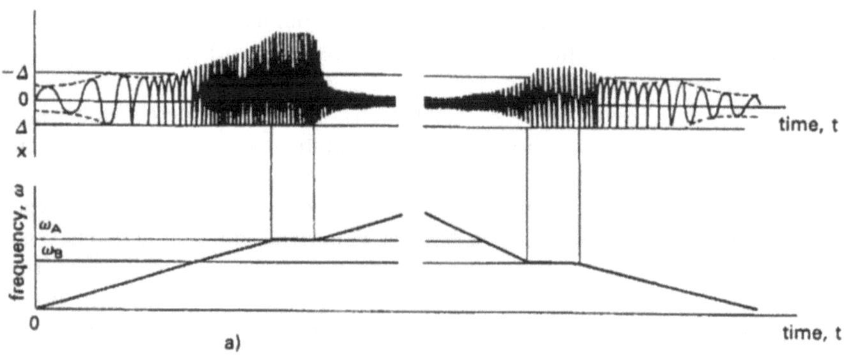

Fig. 3. Frequency pulling of oscillations

Thus, the system behaves in different ways depending on the direction of frequency change. Fig. 3 illustrates variation with time of the vibro-impact system oscillations first as the excitation frequency is slowly increased and then as it is decreased. Such behaviour is termed 'pulling of oscillations' and is peculiar to nonlinear dynamic systems whose restoring force acting on

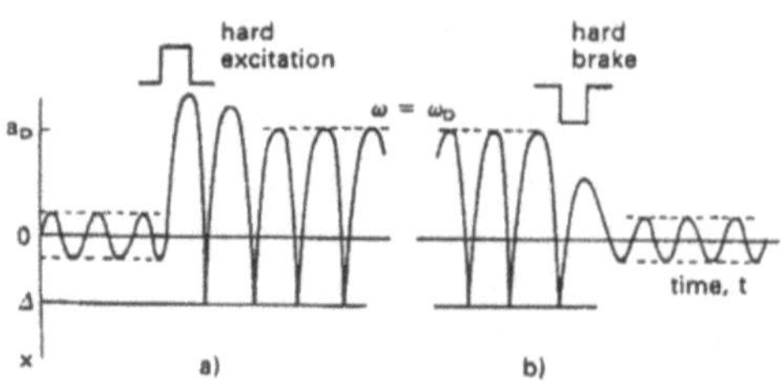

Fig. 4. Hard excitation of oscillations.

the load is not proportional to load displacement. In the given case, the non-proportionality is explained by installation of the motion limiter.

Let us now consider some other remarkable peculiarities of the vibro-impact system. It is evident from Fig. 2c, that in the frequency band corresponding to points B and A (double-valued zone of the amplitude-frequency characteristic curve) the system may realize two absolutely different types of motion for the same parameters and conditions of excitation: energetic oscillations with strong impacts (vibro-impact resonance) and weak harmonic oscillations near the equilibrium position, without contact with the limiter. Realization of one or the other behaviour depends, as we have shown, on the pre-history of the excitation process.

However, there exists another method of transition from one state to the other. Let us fix excitation frequency ω at the intermediate point ω_D (see Fig. 2c) of this frequency band at realization of the harmonic oscillations and give a strong accelerating impulse to the load in the direction of motion to ensure collision of the load with the limiter (Fig. 4a).

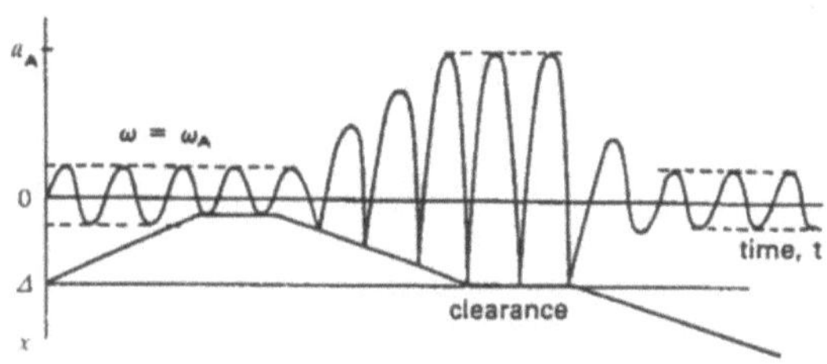

Fig. 5. Amplitude pulling of oscillations.

The vibro-impact mode of motion will be set up after a short transient process. As distinguished from smooth ('soft') commencement of the motion mode, such a method is termed 'hard' excitation of the oscillations. Having discontinued the shock oscillations of the system for an instant (e.g. by application of a deceleration impulse), we may return the system to the shockless mode of oscillation (Fig. 4b). The system apparently 'memorizes' the last action.

It is interesting 'to control' such vibro-impact mode of oscillation during displacement of the limiter (Fig. 5). Let us assume that the system experiences harmonic oscillations. When the limiter is displaced into the swinging zone, the process will be changed, and vibro-impact oscillations will originate. If the limiter is now slowly moved in the opposite direction (giving increased clearance), the vibro-impact mode of oscillation will be preserved, with an increase of oscillation amplitude. The load appears to follow the limiter, as in basketball dribbling when the player starts hitting the ball near the floor and then rises it higher and higher. Only at considerable amplitude of swing is the vibro-impact mode of oscillations disrupted and the system returns to harmonic oscillation. As distinguished from the 'frequency pulling' described earlier, this

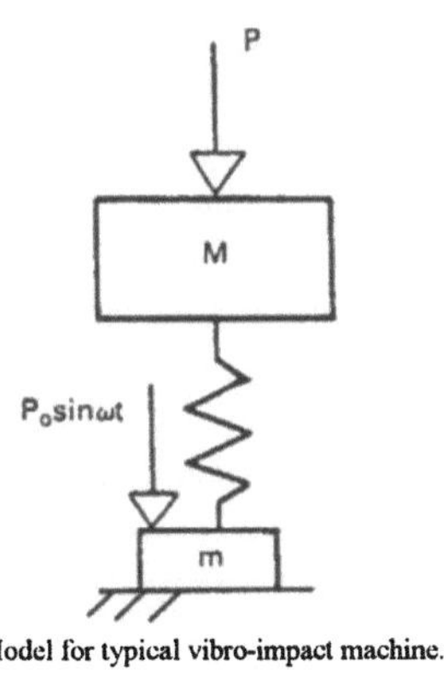

Fig. 6. Model for typical vibro-impact machine.

phenomenon is termed 'amplitude pulling.'

There are major problems associated with the transition of the vibro-impact system into the resonant mode. Point A (see Fig. 2c) corresponding to the most efficient processes lies on the boundary of existence of this mode. Accidental fluctuation will transfer the system into quite another type of oscillation from which return to the initial process is hardly practicable.

Realization of vibro-impact resonance in the system shown in Fig. 6 is even more difficult. This system approximately simulates a wide range of vibro-impact machines, in particular, hand-operated hammers, concrete breakers, tampers, riveting machines and many other appliances.

Figure 7 illustrates variation of the half-swing (amplitude) of the striker oscillations versus frequency at various feeds. With increase of the feed, the resonant frequency of the vibro-impact mode of oscillation becomes greater and the form of the amplitude-frequency characteristic curve is changed. At high rates of feed ensuring effective treatment (since the impact force impulse is proportional to the feed force), the resonant mode of oscillation again approaches the boundary of existence of the vibro-impact process (point A). Now the zone of oscillation is located to the right of the doubled frequency of the natural oscillations of the striker. Any slight decrease of the excitation frequency will result in complete discontinuation of the oscillations (see the arrows in Fig. 7).

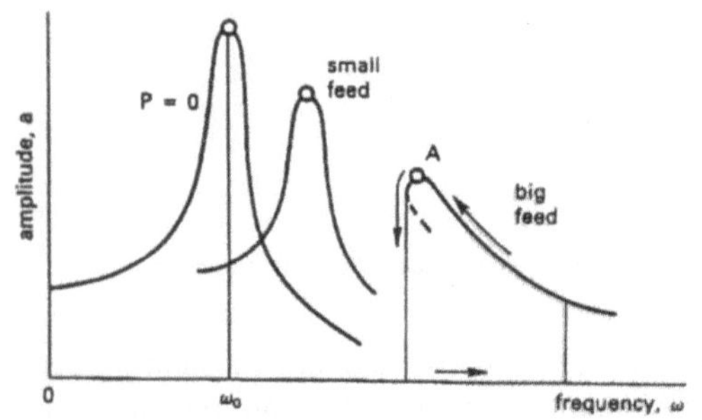

Fig.7. Effect of feed force on amplitude-frequency characteristics.

To start the system under these conditions, it is necessary to considerably increase the frequency, make transition to the vibro-impact mode of oscillation with the use of hard excitation (additional push) and, smoothly decreasing the frequency, establish resonance. This method is hardly practicable and frequently cannot be realized at all. Even automatic frequency control systems commonly used in radio engineering for tracking bell-shaped resonance (as in Fig. 1c) prove to be inefficient. This is because exploratory steps which change the excitation frequency to both sides of the effective value are continuously made in these systems and the direction of trimming is determined by the results. When the mode exists on the boundary of its realization, such a search is not permissible since it may lead to disruption and complete discontinuation of the process. Therefore, the most economic dynamic mode of operation of the system turns out to be the most inconvenient for realization.

Since reliability of operation is the main parameter of a machine, an alternative approach to development of the engineering was selected. Doing their best to avoid instability and having no clear theoretical concepts, the development engineers started empirically de-tuning the machines, withdrawing them in fact from highly efficient vibro-impact resonance to the high-stability processes associated with relatively low rates of feed and low Q-factors of oscillatory systems. As a result, efficiency is decreased, effectiveness is lost, and energy requirement is too great.

3. Autoresonance

The investigators of vibro-impact systems could not resign themselves to this situation. It was necessary to find some other method of excitation of resonant vibro-impact modes ensuring that these modes of oscillations become most stable and, if practicable, the only state of the system at all changes of its parameters and loads. And such a way was found. The researchers directed their attention to self-excited oscillation processes originating under the effect of forces depending on the motion of the system.

Let us remember how the swing is set in motion. As the swing amplitude is increased, the time between stops in the upper position becomes longer, though the speed of passage through the lower position continuously increases. Therefore, to accelerate the swing, it is necessary to change the frequency of the pushes, each time trying to push the swing in the direction of the speed of motion. Such feedback between the motion of the system and the force causing this motion is termed 'positive feedback.' In our case, speed positive feedback is meant. In practice such feedback is realized, for instance, by the use of a speed transducer (Fig. 8a) whose amplified output signal is fed to the system which is being swung through an electromagnet or other source of power. As a result, the equilibrium of such a system becomes unstable: any infinitesimal deviation gives rise to a push in the direction of the acquired speed. The system starts self-swinging in unison with its natural oscillations. Autoresonance takes place; this has been defined by Academician A A Andronov as '... the resonance under the effect of the forces depending on the motion of the system.' Speed positive feedback is the main feature of self-excited oscillation systems and autoresonance is their most characteristic behaviour.

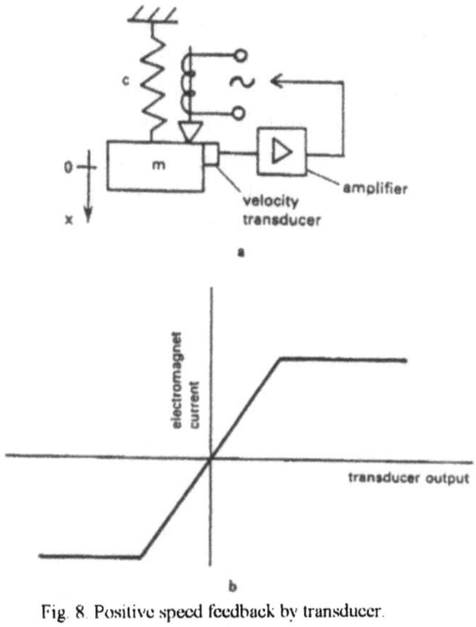

Fig. 8. Positive speed feedback by transducer.

For limitation of self-excited oscillations, it is necessary to reduce the swinging force when the system approaches the required level of oscillation. To this end, special technical means are used, for instance, 'roll-off' of the amplifier characteristic curve in the course of increase of the current fed to the electromagnet (Fig. 8b).

Man successfully uses self-excited oscillation systems. They efficiently convert source energy into the oscillatory mode of the driven member. The most common implementation is a normal timepiece - from pendulum clock to modern crystal-operated watch. It is no mere chance that the saying goes 'Reliable as a watch.' In all spheres of modern radio engineering self-excited oscillations are the foundation of diverse oscillators. Whistles, bells and many types of musical instrument are actually examples of self-excited oscillation systems. Air-operated and electric hammers are self-excited oscillation vibro-impact systems.

The progress of modern electronics and control means provides for the use of self-excited oscillation processes in machine engineering as methods of conversion of source energy into the production process. Just such a direction has been taken in the Vibration Engineering Systems Laboratory of the Machine Studies Institute whose scientists have studied nonlinear dynamic processes of machines for many years.

4. Vibro-impact Scanners

The autoresonant vibro-impact scanners that are necessary primarily for laser engineering represent one of the first developments of our Laboratory. Physical fields do not deflect a laser beam, therefore movable mirrors are used for laser beam displacement in space. Fig. 9a illustrates laser treatment of a surface using a vibro-impact scanning system. Laser beam 1 is directed into scanner mirror 2 which reflects it onto surface 3 which is being displaced at speed V. The mirror is spring-loaded relative to casing 4 (Fig. 9b). Tilting oscillations of the mirror through angle φ are generated by the use of an electro-magnet 5 and are limited by

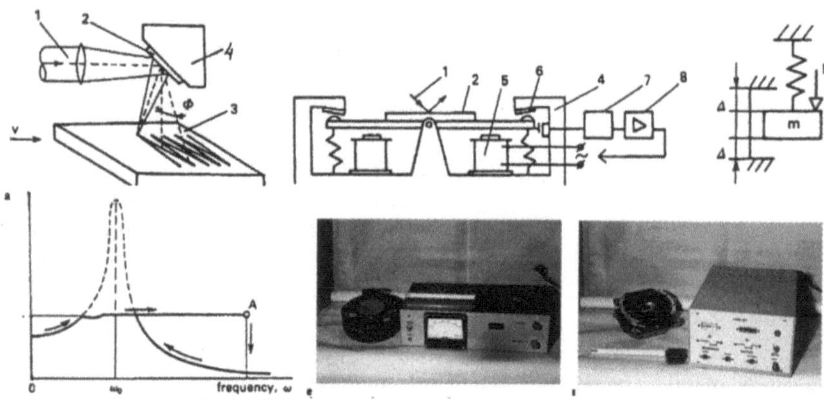

Fig. 9. Vibro-impact laser surface treatment scanner

rigid stops 6.

The mechanical model of such a system is shown in Fig. 9c. As distinguished from the system shown in Fig. 2a, now the motion of the load is limited from both sides. The amplitude-frequency characteristic curve of such a system is illustrated in Fig. 9d. The frequency 'pulling' effect described earlier becomes more pronounced in such a model. Having commenced striking against the stops, the mirror preserves the vibro-impact mode of motion with constant amplitude, equal to the amount of clearance to the limiters, up to point A, i.e. at frequencies which materially exceed the natural frequency (ω_0) of the oscillatory system arranged between the limiters (see the arrows in Fig. 9c). When this mode is realized, the motion of the mirror in time approximates the saw-tooth motion desirable for uniformity of laser treatment. Thus, high frequency, linearity and large scanning amplitudes are attained in the 'pulling' mode. The resonant nature of the process leads to a decrease in energy consumption and reduction of the dimensions of the feeding magnets.

The problem is reduced to ensuring stability of the mode, since in this case the best characteristics are also obtained at point A (Fig. 9d) on the boundary of existence of the vibro-impact mode. To this end, a positive feedback is provided (Fig. 9b): the signal from the transducer measuring mirror oscillations is converted (7), amplified (8) and fed into the electromagnet windings. The circuit is now completed and the system starts functioning in the autoresonant vibro-impact mode with characteristics corresponding to those at point A. This is the only stable state of the system.

The developed theory and methods of design of auto-resonant vibro-impact scanning systems have made it possible to preserve saw-tooth motion accurate to 1% or better for mirror oscillation amplitudes of 3° and scanning frequencies of up to 400 Hz. The diameter of the water-cooled mirror equals 80 mm. Vibro-impact scanners with such unique characteristics are patented in all industrially developed countries. Figure 9e, demonstrates scanner for gas laser, Fig. 9f for solid state laser.

Stable scanning devices for light beams provide also for making high-accuracy measurements exercising reception and transmission of information signals and image analysis. The information vibro-impact scanners designed in the Laboratory

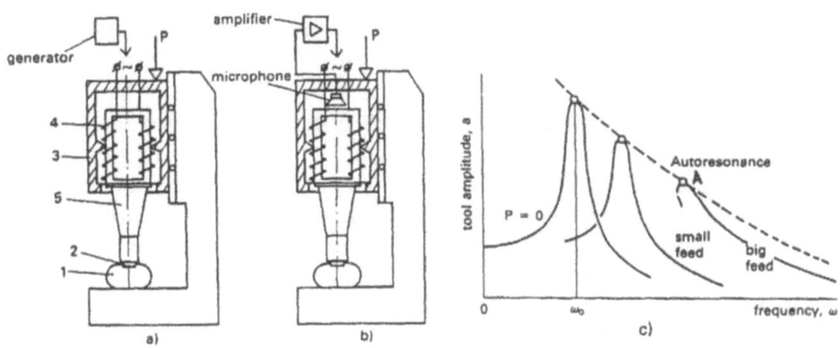

Fig. 10. Surface treatment using an ultrasonic converter

for these purposes feature high stability, small overall dimensions and mass, making it possible to build these scanners into complex electronic control instruments.

5. Autoresonant Ultrasonic Heads

As already stated in the beginning, the impact method of transmission of action from driven members of the vibration machine to the medium being treated makes it possible to increase considerably the efficiency of technological processes whose realization calls for application of great force. When this method is used, a high Q-factor of the system and hence its economy are preserved. These advantages may be realized in processes intensified by high-frequency vibration. Fig. 10 a is a diagram of an ultrasonic converter used for superposition of high-frequency force impulses on the process of treatment of article 1 with tool 2. The main component of the converter is the acoustic head consisting of casing 3, magnetic or piezostriction vibrator 4, and stepped waveguide (concentrator) 5 carrying tool 2 on its end. The waveguide is attached to the end-face of the vibrator. The acoustic head functions as follows. When a high-frequency electric impulse is fed to the input of vibrator 4, it starts making longitudinal oscillations because of the magnetic or piezostriction effect. These oscillations excite standing waves in waveguide 5. The amplitude of the standing waves at the lower end carrying the tool exceeds the amplitude of the waves in the joint between the waveguide and vibrator. As a result, the oscillations are intensified. Having pressed the head to the blank with feed force P, treatment of the article occurs by systematic separation of the tool from the surface and subsequent collision caused by the waves.

Let us remember the analysis of the mechanical model (see Fig. 6) and try to understand what happens in the course of ultrasonic treatment. In modern ultrasonic converters, the vibrators are supplied with power from powerful sources of high frequency (oscillators). The frequency is selected directly behind the boundary of the audibility threshold (somewhat above 20 kHz) to avoid unpleasant sound. The waveguide-concentrator is tuned to the same frequency. As a result, the tool starts oscillating like a load suspended from a spring. The peak-to-peak amplitude of the oscillations is low (several micrometres) since at such high frequencies the attenuation of oscillations in the material is extremely large. Nevertheless the researchers try to ensure a maximum Q-factor for the system by selection of material properties - otherwise the system will not oscillate: the solid metal of the waveguide must be made to 'breathe.'

Pressing the head to the article gives rise to the vibro-impact mode. The natural properties of the system are modified as shown in Fig. 7 and the selected frequency is found aside to the most favourable modes corresponding to vibro-impact resonance. Again progress of the technology occurred by way of ensuring stability 'at any cost.' The power of ultrasonic installations kept growing, special cooling systems were designed, and the overall dimensions of the oscillators were increased. As a result, in modern ultrasonic technological installations, power over 1 kW is required for development of feed forces of the order of 100 N. In this case, efficiency equals ~3%. Therefore, ultrasonic technology is used only when it cannot be replaced by any other means, namely for treatment of extra-hard and brittle materials (diamonds, corundum, glass, ceramics, semi-conductor materials).

Even here autoresonance provides for improvement of efficiency of ultrasonic equipment. Multi-year researches conducted by staff members of our Laboratory have resulted in development of the principles of design of autoresonant ultrasonic systems and to production of pilot installations. The principle of operation of such installations is rather simple. Instead of a master oscillator, a feedback transducer and special amplifier feeding the vibrator (Fig. 10b) are introduced into the system. Since we are dealing with ultrasound, even a microphone may be used as a transducer. As a result, positive feedback is provided, the system loses stability, and starts generating ultrasonic oscillations at the frequencies of the vibro-impact system. The conditions of treatment, feeding forces, and properties of the article material are changed. The frequency of the autoresonant vibro-impact mode is changed as well. The amplitude-frequency characteristic curve of an autoresonant system is the envelope of the peaks of the resonant curves (Fig. 10c) corresponding to various feeds. The process then becomes fully automated; it is possible to change the feed in accordance with any program - the system will automatically select the optimal mode for each feed. The energy consumption of the installation is changed as well. The efficiency may be increased to the theoretical optimum of 50%. The feeding forces have sharply increased, though the power required for this increase is negligible.

The principal features of an autoresonant ultrasonic machine consist of the dynamic connection and coordination of the oscillatory processes in the machine with the technological processes of treatment. The main efforts of the scientists have been directed towards ensuring such coordination. As a result, the machine has acquired a new property: ability to quickly and independently adapt itself to the treatment process. Now it has become practicable to exclude complex analyzing processes and their support facilities from the control system. As a result, a new type of machine has been produced - homeostat which, like a living being, responds in an optimal way to any change of the functioning conditions, while the type of behaviour - autoresonance, is not changed. This considerably extends the field of application of high-power ultrasound.

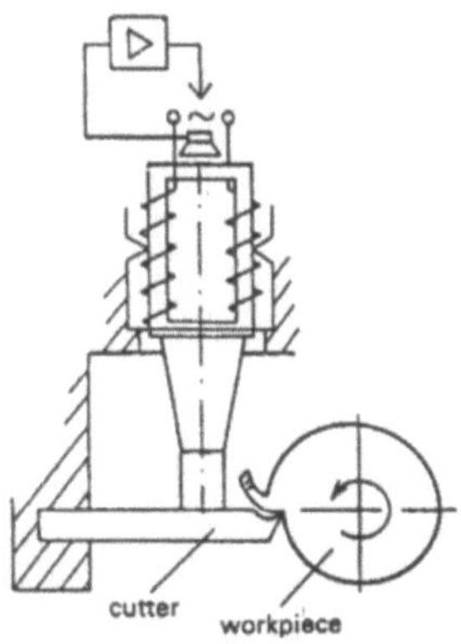

Fig. 11. Autoresonant ultrasonic cutting lathe.

If oscillations from the ultrasonic head shown in Fig. 11 are brought to the treatment zone when cutting metals, the high cutting load will be replaced by a series of micro-collisions. This will obviate the need for transmission of considerable forces via the machine drive that imposes additional loads on the transmission, supports and casing. Machines intended for plastic deformation of materials may be similarly relieved of loading. Research to boost the power of technological processes by use of vibration and ultrasonic actions is conducted in several countries. Autoresonance will bring about a technological era of intensive ultrasound.

6. Resonant Robots

Is it possible to employ the developed principles of coordination and relief of resonant systems in new types of machines? To answer this question, our Laboratory began work in a new direction associated with the production of resonant robots - manipulation systems designed for displacement of loads between assigned positions with stops made for loading and offloading or for

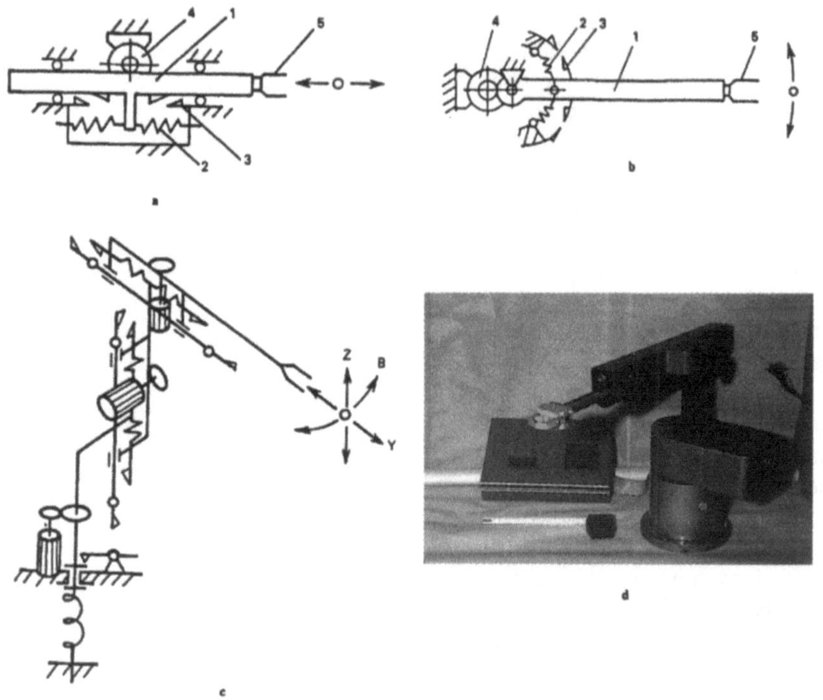

Fig. 12. Manipulation Auto-Resonance System.

performance of technological operations. To this end, the complex motion of a load in space was divided into simple operations: translational and rotational motions. Special translational motion modules (Fig. 12a) and rotational motion modules (Fig. 12b) have been designed. The robot is assembled from such modules in accordance with the 'outlet-base' principle in which the moving member of one module serves for attachment of the next module. The moving link of the last module mounts a gripper. Depending on the modules selected, the manipulator operates in the desired coordinate system. Fig. 12c, is a diagram of the MARS (Manipulation Auto-Resonance System) robot incorporating two translational modules and one rotational module.

Each module (Fig. 12) is an electromechanical self-excited oscillation system consisting of movable carriage 1 attached to the base by the use of spring 2. Latches 3 mounted on the ends of the carriage path preclude motion of the carriage in the opposite direction once it has stopped. The carriage is connected with reversible electric motor 4. Gripper 5 is installed on the end of one of the modules.

The module operates in the following way. Bring the carriage into the operating state by fixing one of the latches. To this end, select the automatic swinging mode by successive reversing of the electric motor. When a work stroke is to be made, the latch is automatically withdrawn and the carriage moves to the other extreme position under the force of the spring and electric motor, where it is again fixed by a latch. The next work stroke takes place after release of the latch and reversal of the electric motor. The program cycle of manipulator operation is realized by a controller that establishes the sequence of module operation. The autoresonant control adapts the system to change of parameters and external conditions.

The MARS robot (Fig. 12d) is four times as light and three times as movable as its industrial analogues. When the robot manipulates loads of up to 0.5 kg, the power consumption in all coordinates does not exceed 20W which is an order of magnitude less than the power consumption of fellow-robots. This consumption of power has been attained by virtue of autoresonance. All displacements represent parts of the self-excited oscillation processes interrupted by latches when the system is stopped in positions corresponding to the maximum deformation of the springs. The task of the electric motors consists only in replenishment of the energy reserves in the oscillatory systems. If the system is swung in unison, in resonance, the energy consumption is low. Other useful properties have been attained: soft start-up and smooth stopping which improves accuracy of positioning and reliability of operation. Mechanisms operating on this principle have record-breaking specific parameters such as consumption of materials, consumption of energy, as well as maximum speeds and rates of acceleration.

7. Mechatronics and Machine Engineering

Vibration engineering represents a boundless field of application of autoresonance. Control of the displacement of materials and parts, mixing, separation, grinding and many other processes may be intensified by autoresonance. The burgeoning of new technical possibilities has provided the incentive for studies in the field of nonlinear process dynamics, development of effective calculation methods, design of control facilities and manufacture of new materials.

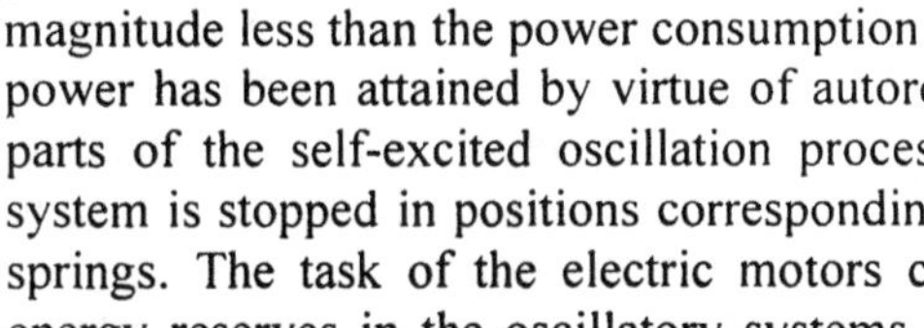

Fig. 13. Synchronised vibro-impact excitation in a string.

Such interaction is characteristic of technical progress. A peculiar feature of the present period is the ever more close juxtaposition of mechanical and electric processes in the executive and control functions. This calls for complete 'cooperation' of the above-mentioned sub-systems, their transformation into a single dynamic system. This in turn involves in-depth coordination of dynamic and algorithmic functions of the multi-link system necessary to fully utilise the merits of each subsystem. A new term 'mechatronics' has been coined for designation of this permanent unity. Developments in this direction have just commenced. There is opportunity for use of new materials, employment of superconductivity properties and creation of new methods of solution of differential equations.

A proliferation of functional qualities is attained when the control means utilise to the maximum extent the executive opportunities of the mechanical devices. This places quite new requirements on such established branches of science as mechanics. Each phase opens up new horizons for development of technology, provides the outline of new generation machines which are integrated dynamic complexes of mechanical, electric and information-calculation processes. A peculiar sort of resonance is displayed here: progress in one sphere gives a push to development in other spheres. While the design bureaux and plant shops are busy with development of next generation machines, the scientific laboratories proceed with studies of new dynamic processes and methods for their mathematical analysis, conduct experiments, and discover hitherto unknown phenomena. Such intensive work is conducted in our Laboratory.

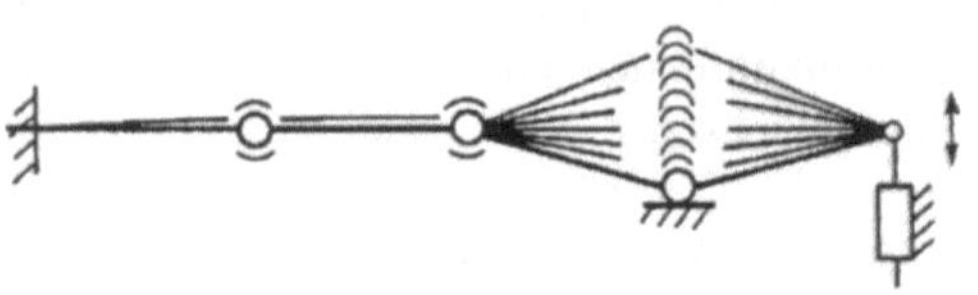

Fig. 14. Isolated vibro-impact excitation in a string.

We have recently observed experimentally new vibro-impact effects previously predicted by the theorists. The possibility of synchronization of the process of vibration excitation of a string mounting hard particles and located in the vicinity of a fixed limiter has been discovered. In this case, all the particles, as if coming to an agreement, simultaneously strike against the limiter (Fig. 13). The 'committee' of particles behaves as a single load suspended from the spring (string). The system is 'pull' in frequency and amplitude and the oscillations are excited in a hard manner (similar to the simplest system shown in Fig. 2a). Such synchronization of impacts in structural elements or in materials may lead to intensive weakening of these systems and their rapid destruction. Quite opposite behaviour of the particles has been observed under other conditions. Each particle may fall into resonant vibro-impact interaction with the limiter (Fig. 14) resulting in the same nonlinear effects whereas the remaining particles are practically immobile. The future will show what new machines and technologies will be designed on the basis of these effects.

Springer and the environment

At Springer we firmly believe that an international science publisher has a special obligation to the environment, and our corporate policies consistently reflect this conviction.

We also expect our business partners – paper mills, printers, packaging manufacturers, etc. – to commit themselves to using materials and production processes that do not harm the environment. The paper in this book is made from low- or no-chlorine pulp and is acid free, in conformance with international standards for paper permanency.

Springer

MIX
Papier aus verantwortungsvollen Quellen
Paper from responsible sources
FSC® C105338

If you have any concerns about our products,
you can contact us on
ProductSafety@springernature.com

In case Publisher is established outside the EU,
the EU authorized representative is:
Springer Nature Customer Service Center GmbH
Europaplatz 3, 69115 Heidelberg, Germany

Printed by Libri Plureos GmbH
in Hamburg, Germany